# HANDBOOK
# OF
# CONSERVATION
# AND
# SOLAR ENERGY

## Trends and Perspectives

# Books by V. Daniel Hunt

*Energy Dictionary*, 1979, Van Nostrand Reinhold, New York.

*Energy Issues in Health Care*, 1979, The Energy Institute, Washington, D.C.

*Windpower: A Handbook on Wind Energy Conversion Systems*, 1981, Van Nostrand Reinhold, New York.

*Handbook of Energy Technology: Trends and Perspectives*, 1981, Van Nostrand Reinhold, New York.

# HANDBOOK
# OF
# CONSERVATION
# AND
# SOLAR ENERGY
Trends and Perspectives

## V. Daniel Hunt
*Director*
*The Energy Institute*

**VNR** VAN NOSTRAND REINHOLD COMPANY
NEW YORK  CINCINNATI  TORONTO  LONDON  MELBOURNE

Manufactured in the United States of America

Published by Van Nostrand Reinhold Company Inc.
135 West 50th Street, New York, N.Y. 10020

Van Nostrand Reinhold Limited
1410 Birchmount Road
Scarborough, Ontario M1P 2E7, Canada

Van Nostrand Reinhold Australia Pty. Ltd.
17 Queen Street
Mitcham, Victoria 3132, Australia

Van Nostrand Reinhold Company Limited
Molly Millars Lane
Wokingham, Berkshire, England

15  14  13  12  11  10  9  8  7  6  5  4  3  2  1

**Library of Congress Cataloging in Publication Data**

Hunt, V. Daniel.
    Handbook of conservation and solar energy.

    Bibliography: p.
    Includes index.
    1. Energy conservation—Handbooks, manuals, etc.
2. Solar energy—Handbooks, manuals, etc.
I. Title.
TJ163.3.H86        333.79'16        81-11706
ISBN 0-442-20056-0              AACR2

*To my family*

# Notice

This book was prepared as an account of work sponsored by Van Nostrand Reinhold Company. Neither Van Nostrand Reinhold nor The Energy Institute nor any of their employees makes any warranty, express or implied, or assumes any legal liability or responsibility for the accuracy, completeness, or usefulness of any information, apparatus, product, or process disclosed, or represents that its use would not infringe upon privately owned rights.

The views, opinions, and conclusions in this book are those of the author and do not necessarily represent those of the United States government or the United States Department of Energy.

Public domain information and those documents abstracted, edited, or used in whole or in part, or otherwise utilized, are noted in the acknowledgments or on specific illustrations.

# Preface

The increasing scarcity and cost of energy in the United States coupled with our increasing dependence on foreign sources of petroleum threaten our economic prosperity, increase inflation at home, weaken the dollar abroad, and constrain our foreign policy options. In short, they threaten our national security. Energy conservation and the development of solar energy are important elements of our national strategy for coping with this threat both now and in the future.

Conservation, broadly conceived, is the restructuring of our market economy so that more goods and services can be produced with less energy. In essence, it is the achievement of economic efficiency in a new economic environment—one where energy is now scarce and expensive rather than abundant and cheap as it used to be. Conservation encompasses the elimination of waste, the investment in energy-efficient machines and buildings, and the changing of consumption patterns to reflect the true replacement cost of energy. Conservation, however, is not curtailment, and it should be distinguished from such short-term emergency measures as gasoline rationing. Rather, the savings from conservation come about because resources are employed more efficiently, especially in heating and cooling buildings, in powering cars and trucks, and in industrial boilers, furnaces, and electric motors, which together are responsible for most of the energy use in the United States.

The long-term potential of conservation is truly enormous. By the year 2010, it will have reduced energy use by 40 to 50 percent of what it would have been if we produced the same goods and services without conservation. Why? First, as energy prices rise, it becomes cost-effective to introduce more energy-conserving practices and measures. Second, over the longer term, as the capital stock (automobiles, furnaces, manufacturing equipment, etc.) is replaced, more energy-efficient technologies will be introduced. Third, R&D can provide new energy-conserving technologies; these enhanced technologies will be adopted as capital stock is replaced and increased. On the near term, opportunities exist for retrofitting our capital stock and making our energy-use practices more efficient.

The strong interest in solar energy has resulted from recognition of its importance as an energy resource. Solar energy includes energy from sunlight, wind, biomass, hydroelectricity, and the oceans. These resources are inexhaustible or renewable, and they could provide a significant part of our future energy requirements. For the United States, solar energy represents an energy supply that is secure against oil import disruptions, and possible slowdowns in coal production or in the use of nuclear power. Furthermore, the price of solar energy is not subject to foreign manipulation. Any environmental problems associated with using solar energy are minor compared to those associated with other energy resources.

Former President Jimmy Carter established the national goal for solar energy in his national solar message of June 20, 1979, in which he said: "We should commit ourselves to a national goal of meeting one-fifth—20 percent—of our energy needs with solar and renewable resources by the end of this century." The Domestic Policy Review analysis assumed that the United States would need 95 quads of energy by the year 2000; this suggests that meeting the president's solar goal would require the annual production of 19 quads of energy from solar energy. This would increase the energy now supplied by solar sources by a factor of almost four.

Significant barriers inhibit the achievement of the conservation and solar energy potential through the marketplace alone. Politically feasible but artificially low prices cause an underinvestment in energy conservation and solar energy. Despite phased deregulation of natural gas prices and the decontrol of oil prices, energy prices are likely to remain below their replacement value for many years to come (especially due to inadequate internalization of environmental and other social costs). Investment biases and inadequate information prevent consumers and industry from adopting cost-effective conservation and solar energy measures. Private firms are not conducting sufficient R&D because of the uncertainty of capturing its full benefits. Finally, the present United States institutional and regulatory structure hinders conservation and solar energy. An effective national strategy must involve private enterprise, individuals, and all levels of government to address these barriers.

Current legislation and ongoing funded programs will achieve less than 9.5 quads of conservation and solar energy. New initiatives are needed. The *Handbook of Conservation and Solar Energy: Trends and Perspectives* describes the current trends and perspectives for conservation and solar energy programs. However, more funding and public support are required if we are to be truly energy independent.

V. DANIEL HUNT
Fairfax Station, Virginia

# Acknowledgments

I would like to acknowledge the support of Dr. J. Michael Power, director of the Office of Policy, Planning and Evaluation of the Department of Energy. Also, special appreciation is extended to Dr. Gurmukh Gill, deputy director of the office, for his support and strategic analysis evaluations.

The material in *The Handbook of Conservation and Solar Energy: Trends and Perspectives* is based on the *Program Summary Documents for Conservation* and *Solar Energy*, which are in the public domain. This material has been edited and revised by the author. The author accepts responsibility for any errors.

A book of this magnitude is dependent upon excellent staff, and I have been fortunate. Judith A. Anderson served as the coordinator of this project as well as copy editor. Special thanks are extended to Anne Potter, Carolyn Starr, and Michelle M. Donahue for the composition and typing of the manuscript.

The majority of art, graphics, and photographs were provided through the courtesy of the Department of Energy.

# Contents

Notice      v
Preface      vii
Acknowledgments      ix

PART I / OVERVIEW OF CONSERVATION AND SOLAR ENERGY

1. The Energy Problem      3

2. Role of Conservation and Solar Energy in Resolving the Energy Problem      6
      Reduce Oil Imports      6
      Changes in Energy Supply      6
      Problems in Developing New Energy Supplies      7
      Role of Energy Conservation      8
      Capital Stock Turnover      9
      Role of Solar Energy      10
      Emergency Conservation      11

3. Government Role      12
      Rationale for Action      12
         Low Energy Prices      12
         Imperfect Information      12
         Investment Criteria      12
         Constraints on Private Research and Development      13
         Legal and Regulatory Barriers      13
      Program Goals, Study and Priorities      13
         Relation to National Energy Plan      14
         Strategy to Achieve Goals      14
         Decontrol and Deregulation of Prices      15
         Public Information      15
         Financial Incentives      15
         Research and Development      16
         Regulatory Measures      16
         Assurances of Stability      16
         Program Priorities      16

4. Program Summary      17
      Buildings Sector      17
         Conservation      17
         Solar      18
      Industrial Sector      18
         Conservation      18
         Solar      18
      Transportation Sector      19

Utilities Sector    20
    Conservation    20
    Solar    21
State and Local    21


PART II / CONSERVATION ENERGY

5. Introduction    25
    Budget Summary    25

6. Conservation Overview and Strategy    27
    Conservation Overview    27
    Conservation Potential    27
    The Federal Role in Energy Conservation    29
    Government Objectives    30
    Goverment Conservation Strategy    31
        Current Federal Strategy    31
        The Roles of State and Local Government    32

7. Buildings and Community Systems    33
    Program Objectives and Strategy    33
    Background    33
        The National Energy Outlook    33
        Energy Use in Buildings    33
    Objectives    34
        DOE Program Missions    34
        DOE Program Objectives    34
        DOE Energy Savings Goals    34
    The Federal Role    34
    Federal Program Strategy    35
    DOE Management Strategy    35
        Organization and Administration    36
        Project Implementation    36
    Major Program Thrusts    38
        Legislation Framework    38
    Programs and Milestones    39
        Program Liaison and Support (PLS)    39
        Buildings    41
        Community Systems    46
        Consumer Products    68
        Division of Federal Programs    76
    Buildings and Community Systems Funding    81
    Program Impacts    81
        Projected Energy Savings    81
        Accomplishments to Date    84

8. Industrial Programs    86
    Program Objectives and Strategy    86
        Objectives    89

The Federal Role in Industrial Energy Conservation     89
Federal Program Strategy     91
Program Management Strategy     93
Major Program Thrusts     94
Legislative Framework     94
Division of Conservation Research and Development     94
Division of Conservation Technology Deployment and Monitoring     94
Specific Program Activities     94
Conservation Research, Development, and Demonstration     94
Waste-Energy Reduction, Including Cogeneration     104
Industrial Process Efficiency     104
Conservation Technology Deployment and Monitoring     121
Financial History     127
Program Impact     128
Measures of Effectiveness to be Monitored     128
Program Validation     128
Program Options     129
Program Effectiveness     130
Activity Measures     130
Projected Energy Savings by Specific Subprogram and Fuel Type     130
Accomplishments     131
Project Selection—Analysis of Benefits and Costs     132
Environmental Impact     133
Issues     133
Activities to Resolve Environmental Issues     134
Major Program Issues     135

9. Transportation Programs     137
Program Objectives and Strategy     137
Objectives     137
Automotive Technology RD&D     139
Electric and Hybrid Vehicle RD&D     140
Transportation Systems Utilization     141
Federal Role     141
Federal Strategy     142
Program Management Strategy     142
Major Program Thrusts     143
Legislative Framework 143
Automotive Technology RD&D     143
Electric and Hybrid Vehicles (EHV)     143
Transportation Systems Utilization (TSU)     144
Programs and Milestones     144
Automotive Technology RD&D     144
Transportation Systems Utilization     151
Systems Efficiency     153
Program Funding     155
Program Impacts     155
Measures of Effectiveness     155
Projected Energy Savings     156
Analysis of Benefits and Costs     156

Environmental Impact      157
  Issues      157
Major Program Issues      157

## 10.  State and Local Programs      159
Program Objectives and Strategy      159
  Program Goal/Objectives      160
    Schools and Hospitals Grant Program      160
    Other Local Government Buildings Grant Programs      160
    State Energy Conservation Program (SECP)      160
    Energy Extension Service (EES)      161
    Weatherization Assistance Program (WAP)      162
  Federal Role      162
  Federal Program Strategy      162
  Program Coordination      162
  Program Consolidation      163
  Program Management Strategy      164
Major Program Thrusts      164
  Legislative Framework      164
  Current Program and Milestones      166
    Energy Extension Service (EES)      168
Program Funding      169
Program Impacts      169
  Measures of Effectiveness to be Monitored      169
  Projected Energy Savings      171
  Accomplishments to Date      171
  Analysis of Benefits and Costs      172
    Weatherization Assistance Program      172
    Schools and Hospital Grants Program      173
    Other Local Government Building Grant Programs      173
Environmental Impact      173
Major Program Issues      173
  Energy Management Partnership Act      173
  Regional Offices      174

## 11.  Multi-Sector Programs      175
Program Objectives and Strategy      175
  Introduction and Background      175
  Objectives      175
  Federal Role in Appropriate Technology      177
  Federal Program Strategy for Appropriate Technology      177
  Program Management Strategy      177
  Regional Solicitations      180
  Evaluation of Proposals      181
  Award of Grants      183
  Grant Review and Administration      183
  Energy Related Inventions Programs      183
  Evaluation of Inventions      184
  Type of Assistance      184
  Energy Conversion Technology      184
  Subprogram Description      184

ECT Objectives and Goals     185
Major Program Thrusts     185
Legislative Framework     185
Programs and Milestones     185
Appropriate Technology     185
Energy-Related Inventions     186
Program Funding     186
Program Impacts     187
Measures of Effectiveness to be Monitored     187
Projected Energy Savings by Specific Programs and Fuel Types     187
Accomplishments to Date     188
Analysis of Benefits and Costs     188
Environmental Impacts     188
Major Program Issues     188

12. Energy Impact Assistance     189

13. Energy Information Campaign     190


PART III / SOLAR ENERGY

14. Introduction     193
Budget Summary     193

15. Solar Program Overview and Strategy     196
The National Solar Goal     196
The National Solar Strategy     198
DOE Solar Program     200
Role of DOE in the Solar Program     200
Organization of the Solar Program     201

16. Solar Program Description by Technology
Introduction     203
Budget     204
Biomass Energy Systems     204
Introduction     204
Program Strategy     205
Program Structure and Budget     206
Program Details     207
Technology Support     207
Production Systems     209
Conversion Technology     210
Research and Exploratory Development     214
Support and Other     215
Wood Commercialization     215
Photovoltaic Energy Systems     217
Introduction     217
Program Strategy     219
Program Structure and Budget     221
Program Detail     223
Advanced Research and Development     223

Technology Development     225
Systems, Engineering, and Standards     227
Tests and Applications     229
Market Analysis     231
Market Tests and Applications     233
Wind Energy Conversion Systems     234
Introduction     234
Program Strategy     236
Program Structure and Budget     239
Program Detail     239
Research and Analysis     239
Wind Characteristics     241
Technology Development     242
Engineering Development     244
Solar Thermal Power Systems     247
Introduction     247
Program Strategy     249
Program Structure and Budget     250
Program Detail     251
Central Receiver Systems     251
Distributed Receiver Systems     253
Advanced Technology     254
Ocean Systems     256
Introduction     256
Program Strategy     257
Program Structure and Budget     258
Program Detail     259
Project Management     259
Definition Planning     259
Technology Development     260
Engineering Test and Evaluation     261
Advanced Research and Development     262
Agricultural and Industrial Process Heat     264
Introduction     264
Program Strategy     266
Industrial Sector Strategy     266
Agricultural Sector Strategy     267
Program Management     267
Program Structure and Budget     267
Program Detail     268
Systems Development     268
Market Tests and Applications     270
Market Analysis     270
Market Development and Training     271
Active Solar Heating and Cooling     272
Introduction     272
Program Strategy     275
Program Structure and Budget     277
Program Detail     277
Market Analysis     277
Systems Development     279

Market Tests and Applications    281
Market Development and Training    282
International Solar Programs    283
Passive and Hybrid Solar Heating and Cooling    283
Introduction    283
Program Strategy    286
Program Structure and Budget    289
Program Detail    289
Systems Development    290
Market Analysis    292
Market Development and Training    292
Market Tests and Applications    293
International Solar Programs    293

**17. Solar Program Description by Function    294**
Introduction    294
Technology Base    294
Introduction    294
Goals and Purpose    295
Budget    295
Program Thrust    295
Biomass Energy Systems    295
Photovoltaic Energy Systems    297
Wind Energy Conversion Systems    297
Solar Thermal Power Systems    298
Ocean Systems    299
Basic Energy Sciences    299
Environment, Health and Safety, and Social Impact    300
Introduction    300
Program Thrust    302
Biomass Energy Systems    302
Photovoltaic Energy Conversion Systems    303
Wind Energy Conversion Systems    304
Solar Thermal Power Systems    305
Ocean Systems    306
Agricultural and Industrial Process Heat    306
Active Solar Heating and Cooling    307
Passive and Hybrid Cooling 307
Social Impacts    308
Market Analysis    310
Introduction    310
Goals and Purpose    310
Budget    311
Program Detail    311
Status    311
Major Activities    312
Technology Development    313
Introduction    313
Goals and Purpose    313
Budget    314
Program Detail    314

Biomass Energy Systems    314
Photovoltaic Energy Systems    315
Wind Energy Conversion Systems    315
Solar Thermal Power Systems    316
Ocean Systems    316
Systems Development    317
Introduction    317
Goals and Purpose    317
Budget    318
Program Detail    318
Agricultural and Industrial Process Heat    318
Active Heating and Cooling    319
Passive and Hybrid Solar Heating and Cooling 319
Office of Solar Technology Systems Development    320
Market Tests and Applications    320
Introduction    320
Goals and Purpose    320
Budget    321
Program Detail    321
Status    321
Major Activities    323
Market Development and Training    323
Introduction    323
Goals and Purpose    323
Budget    324
Program Detail    324
Status and Accomplishments    324
Major Activities    325
International Solar Programs    325
Introduction    325
Goals and Purpose    327
Budget    330
Program Detail    331
Biomass Energy Systems    331
Photovoltaic Energy Systems    331
Wind Energy Conversion Systems    332
Solar Thermal Power Systems    332
Ocean Systems    333
Office of Solar Applications    333

18. Solar Program Implementation    334
Program Management    334
Management Approach    334
Participants in Federal Solar Program    334
Solar Energy Research Institute    336
Program Overview    336
Solar Energy Research Institute Permanent Facility    336
Regional Solar Energy Centers    340
Background    340
Scope of Activities    341

Status    343
   Small and Disadvantaged Business    344
   Solar Research and Development Activities    344

**19. Markets and Applications for Solar Technologies    346**
  Buildings    346
   Market Components    346
   End-Use Structure    346
   Equipment Selection Options    348
   Potential for Solar Technology in the Building Market    350
   Constraints    351
  Industry    352
   Market Components    352
   End-Use Structure    353
   Equipment Selection Options    354
   Potential for Solar Technologies    354
   Constraints    355
  Utilities    355
   Market Components    355
   End-Use Structure    356
   Equipment Selection Options    356
   Potential for Solar Technologies    359
   Constraints    359
  Transportation    359
   Market Components    359
   End-Use Demand    360
   Equipment Selection Options    360
   Market Potential    360
   Constraints    360

**20. Solar Resources    361**
  Solar Insolation    361
  Biomass Resources Potential    363
  Wind Power    363
  Ocean Energy    366
   Ocean Thermal Resources    366
   Current Velocity    366
   Ocean Wave    369
   Salinity Gradients    369

**21. Solar Energy Policy and Related Legislation    370**
  Energy Policy    371
  Existing Legislation    372
  Legislation in the 96th Congress    372

**Bibliography    373**

**Index    377**

# HANDBOOK
## OF
# CONSERVATION
## AND
# SOLAR ENERGY

Trends and Perspectives

# Part I
# Overview
# of
# Conservation
# and
# Solar Energy

# 1. The Energy Problem

Until recently, energy was relatively inexpensive in the United States. The nation had large reserves of coal, oil, and natural gas, and Americans used these resources liberally to establish a high standard of living. Low energy prices and abundant supplies induced people to build homes, cars, appliances, and factory equipment that use energy wastefully by today's standards.

In the past, energy consumption and economic growth have been closely related. Between 1960 and 1979, the United States economy grew 94 percent, from $737 billion to $1.4 trillion in constant 1972 dollars. During the same period, United States energy consumption grew 77 percent, from 44 quads to 78 quads. Before the energy crisis in late 1973, the relationship was even closer. Between 1960 and 1972, the GNP grew at an average annual rate of 3.9 percent, while energy consumption grew at an average annual rate of 4.1 percent. (See Figure 1-1.)

The American complacency over energy ended abruptly in 1973 with the OPEC oil embargo. The resultant quadrupling of imported oil prices showed us the implications of dependence on imported oil. Since then, the United States economy has been jolted by unpredictable price hikes for imported oil. These increases have been tied far more closely to OPEC politics than to the world economic situation. The economy cannot be assured of relatively smooth and steady growth as long as prices are controlled by outside political and economic forces.

Rising energy prices have been one cause of inflation in the United States. In addition, they have undermined the value of the dollar abroad, creating more inflationary pressure; they could also weaken American foreign policy. An increase in the price of oil is quickly felt throughout the economy. For example, by rendering uneconomic a portion of the country's stock of energy-inefficient capital equipment, higher energy costs have made it more difficult to reach full employment.

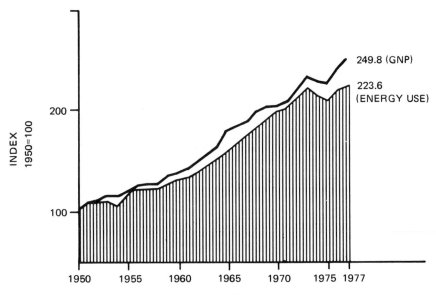

Figure 1-1. Growth of GNP and energy use.

Although the link between energy consumption and economic growth will persist, it will not be as close as it was. Energy consumption is likely to grow far more slowly than the economy. Many European countries have a high living standard yet consume much less energy than the United States. Energy prices may induce the United States economy to grow more energy-efficient.

Even though American energy consumption should grow more slowly than the GNP and well below historical growth rates, energy demand may still surpass 100 quads by 1990. In meeting this demand, the mix of energy sources will probably change. Oil and gas may provide a smaller share of the total supply. The use of coal and nuclear power may increase, while the use of other energy sources could be relatively constant even though the amount of energy they provide may rise. If the Department of Energy (DOE) and the administration aggressively pursue the development of solar energy, then by the year 2000, alternative energy sources could provide a large and rapidly rising share of United States energy. Continued administration support is needed to assure Americans that energy conservation is not forgotten.

Since 1945, petroleum and natural gas have provided more energy in the United States than has any other source. During this period,

abundant supplies from easily accessible domestic fields induced the nation to rely less on coal and more on convenient fuels. In the late 1940s, it began to augment domestic supplies of oil with imports to satisfy rising demand. By 1960, imported oil accounted for 18 percent of United States petroleum consumption, and imported gas made up 1 percent of the amount used. By 1979, these figures had risen to 48 and 6 percent, respectively.

Despite new oil and gas supplies from Alaska, offshore drilling, and synthetics, the country's dependence on imports could still grow. Some projections indicate that by 1990 imports could account for 51 percent of United States oil consumption and 17 percent of gas consumption. Other studies indicate that the world oil demand may begin to exceed supply by 1985. Recent discoveries of oil and gas, however, could push these dates into the more distant future. (See Figure 1-2.)

A worldwide exhaustion of oil and gas resources will probably never occur. As the demand for petroleum rises and the supply fails to keep pace, the price will rise. Energy users will thus switch away from petroleum, and producers will find it economic to use the currently more expensive methods of recovering oil. As the process continues, the use of petroleum will become more and more restricted. The United States therefore is

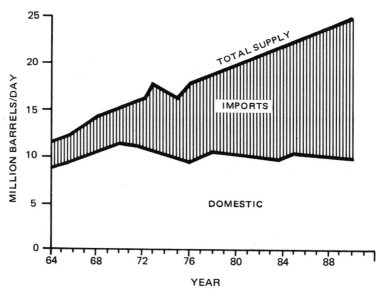

Figure 1-2. Projected United States petroleum liquids supply (projection series C).

developing alternative energy sources such as solar energy. It is also continuing to seek new ways to conserve energy. These actions will allow the economy to expand without the danger of having its energy lifeline cut.

Our supply of oil is very insecure. Over 60 percent of the world's oil supply comes from the Persian Gulf, an unstable area that cannot be relied upon for such an important commodity. The area is threatened by both internal unrest, such as in Iran and Iraq, and Soviet aggression. The situation is made even more dangerous by the geography of the Persian Gulf. About half of the world's oil exports flow through the Strait of Hormuz, which connects the Gulf of Oman and the Persian Gulf. This narrow passage could easily be closed in a war between Iran and Iraq, or other factions. The passage is so dangerous that ships transporting cargo through the strait must pay war-zone insurance rates.

The energy situation is one of the most difficult problems facing the United States. The supply of energy and its price concern almost everyone. Society is being forced to adapt very quickly to lowering the thermostat, driving smaller, more energy-efficient cars, etc., and to adopt a more frugal way of life. The transition is not easy. It involves giving up habits acquired during a lifetime, and lowering expectations for the future. We must develop solar energy, adopt conservation, and expand our other energy supply options in order to reduce our dependence on imported oil.

# 2. Role of Conservation and Solar Energy in Resolving the Energy Problem

## REDUCTION OF OIL IMPORTS

Because continued United States dependence on foreign oil is dangerous, former President Carter limited oil imports. He mandated that imports not exceed the 8.5 million barrels per day imported in 1977. Furthermore, he set a goal of reducing oil imports to about 4 million barrels per day by 1990.

Imports can be reduced in two ways: the supply of domestic energy can be increased, and the demand for energy can be reduced. The two methods complement, rather than conflict with, each other. To the extent that Americans conserve energy and use more efficient forms of energy, their need for additional supplies will be lessened.

It is often less expensive to save energy than to produce it. Thus, more efficient use of energy, coupled with expanded energy supplies when they are economically justified, will help resolve the United States energy dilemma.

Energy conservation does not mean doing without the services that energy provides. Instead, it means using energy more efficiently to provide those services. For example, homes, automobiles, and offices can use less energy while offering the same benefits. Energy use should be improved until the price of saving additional energy rises to a point where it becomes economic to produce energy through more expensive means.

## CHANGES IN ENERGY SUPPLY

The economy is gradually being transformed from one dependent primarily on fossil fuels to one dependent on renewable or inexhaustible energy sources (see Figure 2-1). The transformation involves two stages. The purpose of the first stage is to reduce dependence on conventional oil and gas resources by greatly increasing the use of coal, nuclear power, unconventional oil and gas, and synthetic fuels. This stage should see an increase in the efficiency with which energy is used.

The second stage, which can be widely implemented in the 1990s, will replace the transitional energy sources with renewable or inexhaustible energy sources derived from the sun, fusion technologies, and advanced nuclear technologies. These technologies still require considerable R&D before they are ready for large-scale deployment. They must first be proved technically sound, environmentally safe, and economically justifiable.

The transition to the first stage is already under way. Coal production rose from 613 million tons in 1970 to 776 million tons in 1979 and is expected to total 1.7 billion tons by 1995.

The Synthetic Fuels Corporation was established by Congress in 1980 to promote the development of a synthetic fuels industry in the United States. The corporation provides low-interest loans, grants, and other inducements for companies to produce these fuels. It will also create commercial, synthetic fuel production facilities of diverse types, with the aggregate capability of producing from domestic resources the equivalent of at least 500,000 barrels of crude oil per day by 1987 and of at least 2 million barrels per day by 1992.

Unconventional oil-recovery production processes will also contribute to domestic supplies. Unfortunately, even with these techniques, domestic oil production will probably not rise due to the depletion of existing fields. Barring a major discovery of domestic oil fields—an unlikely possibility—domestic oil production is expected to remain at about 10

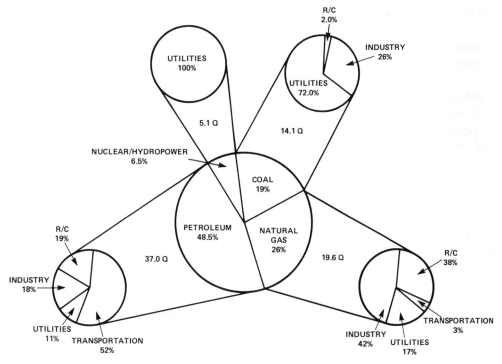

TOTAL 1977 ENERGY SUPPLY = 75.8 QUADS

Figure 2-1. Total energy supply.

million barrels per day at the end of the century.

## PROBLEMS IN DEVELOPING NEW ENERGY SUPPLIES

Difficulties with coal and synthetic fuels may prevent these fuels from achieving their potential for reducing energy imports. Significant barriers to the use of these energy sources exist. Coal, for instance, is plentiful, but it is also dirty and difficult to transport. Facilities that burn coal must invest in expensive equipment to reduce air pollution. Even so, a large increase in the use of coal will add pollution to the environment and reduce the air quality.

Two potential consequences of coal are particularly disquieting. The first is the "greenhouse effect." As coal burns, it emits carbon dioxide as well as other pollutants. Significant additional carbon dioxide in the atmosphere could warm the earth's surface enough to cause irredeemable harm. A difference of only a few degrees could melt the polar ice caps and raise the ocean level enough to flood

a majority of the earth's coastal areas. Too little is known about the effects of carbon dioxide on the environment to permit massive increases in coal burning without caution. Every effort must be taken to assure that the air quality is not further reduced.

Moreover, additional coal burning could increase the "acid rain" phenomenon. In several areas of the world, rain has fallen that was sufficiently acidic to kill the fish. Experts have linked the high acidity of the rain to the burning of coal as well as to other causes such as pollution from automobiles and factories.

Both acid rain and the greenhouse effect require additional examination before the ultimate implications of burning coal can be determined. Nevertheless, the potential for severe environmental damage could prevent coal from becoming a significantly expanded source of domestic energy.

Furthermore, the existing transportation network may not be adequate for moving greatly increased amounts of coal from the mines to the markets. The deteriorating state of American railroads and the lack of enough engines and cars could hamper operations. Coal-slurry pipelines would require large amounts of water, which is already scarce in some western areas.

Synthetic fuels may run into difficulties too. Productive capacity may not be able to expand rapidly enough to significantly reduce the United States dependence on foreign oil. The rapid buildup of a synthetic fuel industry could strain the supply of construction labor, materials, and expertise in this highly technical area.

In addition, the use of some synthetics may be limited by the design of current automobiles. For instance, in concentrations greater than 20 percent, methanol begins to corrode conventional engines. Additional R&D is needed. But if breakthroughs are not made in the near term, the use of synthetic fuels could be restrained.

Oil derived from shale faces environmental problems. Current production processes require large amounts of water and generate large amounts of waste material. The disposal of spent shale could be a major stumbling block for the oil shale industry. Suitable disposal sites may not be readily available, and public opposition may limit those even further.

Water will be an additional problem. The western areas in which shale is plentiful are already short of water for agricultural and urban uses. Water rights could thus become a major issue in shale oil development.

Moreover, the boomtown phenomenon that would result from the establishment of a full-scale shale oil industry would transform the areas. Present residents may resist such a drastic change.

## ROLE OF ENERGY CONSERVATION

Regardless of whether coal or synthetic fuels can provide a transition to renewable or inexhaustible energy sources, conservation will become a major effort in the United States. Conservation should not imply significant sacrifice, discomfort, or a reduction in the standard of living. Rather, through improvements in energy-using devices and by substituting

other factors of production for energy, the demand for energy should be reduced without a reduction in goods and services.

Improved energy efficiency does not have the glamour of the large, highly technological projects to enhance energy supplies. Efficiency is improved by highly diverse modifications in energy-using products throughout the economy. Although individual improvements may not be significant, their total could yield a great reduction in energy consumption. The energy demand shown in Figure 2-2 indicates that the automobile, space heating, and process steam account for the majority of capital stock that can be improved.

Even so, conservation should be the cornerstone of the United States near-term energy policy because it is the fastest, least expensive, and most secure source of new energy available. Improved energy efficiency will not reduce jobs and economic growth, and it could actually raise the standard of living. Finally, improved efficiency protects the environment because there is less energy extraction and pollution.

Even though considerable progress has already been made, much more could be done. Several studies have concluded that improved energy efficiency could contribute 40 to 50 percent of the nation's energy needs. Harvard's *Energy Future* project estimated that improved efficiency could reduce consumption by as much as 40 percent. The National Science Foundation estimated in *Energy in Transition 1985–2000* that an aggressive effort to increase efficiency in the use of energy could actually reduce the amount of energy used in the year 2010 to an amount less than that being used today, while maintaining healthy economic growth.

Improved energy efficiency has already begun. As Figure 1-1 showed, energy use and economic growth used to be very closely related. Between 1950 and 1973, the GNP and energy consumption grew at almost the same rate. The link between them was broken with the oil embargo in 1973 and the subsequent price rises for energy. Energy use has risen at less than half the rate of economic growth since 1973. During that time, energy con-

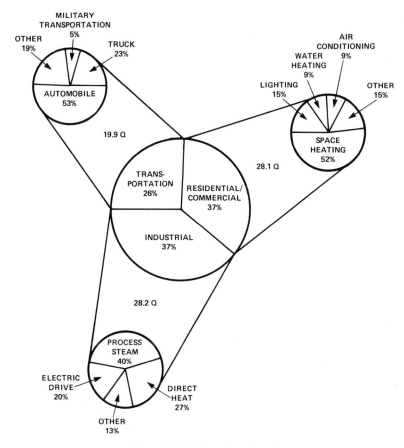

Figure 2-2. Total energy demand.

sumption has grown only 0.5 percent in the industrial sector.

## CAPITAL STOCK TURNOVER

Improved efficiency will result from replacing the existing, inefficient capital stock with new capital equipment that uses energy more productively. Gas-guzzling automobiles will be replaced with cars that go further on each gallon of gasoline. Hot-water heaters and furnaces will be replaced with models that do the same job, but with fewer Btu's. New homes will be built better, have better insulation, and derive substantial energy directly from the sun. Existing homes will be retrofitted with energy-conserving materials to prevent the loss of increasingly expensive energy. Factories will switch to more efficient processes.

Since much of the technology for accom-

plishing these tasks is already known, the job can be accomplished rather quickly. A completely new energy industry will not have to be built, nor will new transportation systems or pollution control systems.

Normal capital-stock retirement provides a ready means by which to improve energy efficiency. Capital items have known lifetimes and are replaced on a regular basis. Home heating equipment, for instance, lasts an average of 15 years; air conditioners last about 10 years; industrial boilers have an average life of 20 years; and automobiles last about 10 years. Businesses and individuals are being encouraged to replace aging capital stock before its normal lifetime is through and to replace it with capital stock that uses energy more efficiently.

In cases where capital has a very long lifetime, such as a building, owners are encouraged to modify it to provide the same service while using less energy. In this way, energy conservation can play a major role in reducing America's dependence on foreign energy.

## ROLE OF SOLAR ENERGY

Solar technologies could become vital to reducing imports. It is not certain that synthetic fuels and coal will be able to expand rapidly enough to provide the additional energy necessary to reduce oil imports to an acceptable level. Some solar technologies such as solar hot water are cost-effective and available now. Solar domestic hot water for home or industry is now cost-effective. Figure 2-3 shows a solar hot-water system retrofitted to an existing facility. Where solar energy can be used economically, it can reduce energy imports and speed the transition to renewable resources.

If other supply efforts do not live up to early expectations, solar energy may have to expand much more rapidly than is now anticipated. Rapidly rising energy costs could make solar technologies economical sooner than most experts believe. In some areas, such as solar hot-water heating, it is already competitive with electricity over the life of the system.

Solar power has considerably fewer environmental hazards than the other technologies. With solar energy there is less worry about fouling the air, destroying the land, or diverting water from other productive uses. Neither is there any need to develop exotic technologies to dispose of waste products or to anticipate catastrophic accidents.

According to the Solar Domestic Policy Review, solar energy could displace as much as 28.5 quads annually by the year 2000. However, this estimate assumes that solar energy is used in every application regardless of economic justification. Solar power could more reasonably be expected to provide 18.5 quads of energy in the year 2000. The study total presumed that oil prices would rise to $32 per barrel and that the government will take an active role in promoting solar energy. (Prices of some oil imports have already reached $37 per barrel.) Without active government support, solar energy might provide only 12.7 quads in the year 2000.

Solar technologies are different from other new supply technologies. Like conservation technologies, they are primarily decentralized and must compete in the marketplace against all other energy sources.

Like some other new supply technologies, the productive capacity in the solar industry is

Figure 2-3. Retrofitting of McDonald's to use domestic solar hot water. Arrow indicates solar hot water collector.

now quite limited. Most companies in the industry are very small and have little capital for growth. Since the companies are still fairly high-risk ventures, they do not have access to capital markets, and lenders are still wary about providing loans.

## EMERGENCY CONSERVATION

Emergency conservation may be necessary if foreign energy supplies are sharply reduced through war or embargo. In such an event, American energy consumption would have to be reduced through the control of energy-using activities. Limits might be set on the amount of gasoline that can be used or on how automobiles might be driven. The government has developed a standby rationing program, which, if implemented like other public programs, will be a failure (see Figure 2-4). Business activity might be reduced and public buildings closed.

Currently, the United States is poorly prepared for such an emergency. The Strategic Petroleum Reserve Program is inadequate, rationing of gasoline has not been sufficiently thought out, and the public is not prepared for such a radical change in energy use.

The possibility of such curtailments could be lessened if the United States initiated an intense effort now to become more efficient in energy use and to develop new supplies. To the extent that imports are reduced, any cutoff of foreign supplies will have less impact and will require less sacrifice. As a result, the "oil weapon" would be less likely to be used.

Figure 2-4. Gasoline curtailment implemented by rationing.

# 3. Government Role

## RATIONALE FOR ACTION

Government action is needed to overcome several barriers to greater energy conservation efficiency and the use of solar energy in the United States. Even though the free market alone might eventually surmount these barriers, government programs can accelerate the process without as much disruption of the economy. In addition, national security requires that the United States have a safe source of energy as quickly as possible. Thus, the government has a legitimate role in helping to smooth the transition from an oil-based economy.

The barriers to more efficient energy use and solar energy fall into five broad categories: low energy prices, imperfect information, investment criteria, constraints on private research and development, and legal and regulatory barriers.

### Low Energy Prices

As difficult as it may be to believe, energy prices in the United States are still low. For years, American consumers have been sheltered from the reality of the world energy market. However, the policy of petroleum price decontrol will serve as an incentive for energy efficiency and for the use of alternative energy sources. As energy becomes more expensive, consumers will make a greater effort to use it wisely and entrepreneurs will endeavor to increase supplies.

Even though energy prices have become completely decontrolled, the price still does not reflect energy's true cost. External costs such as environmental effects and the impact of imports on the general economy, balance of trade, and national security will still not be reflected in the price of oil. Thus, it is worth more not to import a barrel of oil than its price alone would indicate.

### Imperfect Information

Buyers and sellers must have enough information to make rational decisions. Yet, some individuals and businesses do not even know the basic concepts of energy conservation efficiency and solar energy. Even experts in those fields often have difficulty determining the relative efficiencies of competing products or the costs of alternative energy sources. Free markets cannot operate well if businesses and consumers do not make informed decisions.

### Investment Criteria

Investment decisions should be based on the discounted total cost of an item throughout its useful life. However, consumers tend to compare only the purchase prices of items. When operating costs become important, they tend to demand a relatively short payback period. Such criteria discourage investment in efficient products and in solar devices. The costs of conservation and solar energy are borne immediately, while the benefits of reduced energy consumption and lower costs accrue in the future.

Businesses also seem to demand a higher return on conservation investments than on other investments. This higher return may compensate for the risk involved in trying a new technology or the discomfort that managers feel when operating outside of their usual business area. Moreover, some very profitable investments may not be made because of a lack of capital. Inflation has hurt both individual and business income. Profits and income do not buy as much as they once did. Corporate profits are especially overstated in an inflationary environment due to inadequate depreciation and illusory earnings on inventories as well as the falling purchasing power of the dollar.

## Constraints on Private Research and Development

Private R&D in conservation and solar energy is limited, in part, by the risk that an individual company is willing to take. Even if successful, the research would take many years to become profitable. In addition, other companies may be able to reap the benefits of the research without bearing any of the costs or risks. Finally, lack of capital limits R&D even more than it limits other investments. Most capital investments bring in a relatively quick return, whereas R&D brings a return only in the distant future.

## Legal and Regulatory Barriers

Legal and regulatory problems also restrain energy efficiency and the use of solar energy. For instance, local building codes often affect conservation or prohibit solar improvements to buildings. Electricity rates that decline as the quantity bought rises act to discourage efficiency. Likewise, restrictions on the sale of electricity from cogeneration discourage this efficient means of producing electric power.

## PROGRAM GOALS

In order to respond to the preceding five barriers, the goal of the DOE Office of Conservation and Renewable Energy was to encourage more efficient use of energy and to promote the use of renewable energy sources in concert with the following important national policy objectives:

- Less reliance on imported fuels, which will reduce America's vulnerability to supply disruptions caused by world economic and political events
- A more even balance of trade, through reduced energy imports, to achieve a more stable domestic and international economy
- An accelerated transition from reliance on oil and natural gas to greater use of renewable domestic energy supplies

President Carter established the goal that the United States provide 20 percent of its energy needs through solar energy by the year 2000. It is doubtful that this goal will be met because solar funding is insufficient. The president also called for reducing the growth of energy consumption to less than 2 percent per year, a goal that has already been met and exceeded.

Achieving these two goals will reduce United States vulnerability to a cutoff of foreign oil. In the short run, the country will remain dependent on foreign oil supplies. However, as the nation becomes more efficient in its fuel use and depends more on domestic supplies, this vulnerability will diminish.

In addition to purely energy-related goals, the DOE program must meet broader social goals. They include (1) equity, (2) environmental protection, and (3) free enterprise. The first goal, equity, balances the benefits and burdens of energy policy. A policy of deregulating oil prices, for instance, will benefit some companies and individuals tremendously while imposing a heavy financial burden on others. A method must be found to spread the cost and the benefits of deregulation more evenly throughout the nation. The windfall profits tax, for example, would tax away some of the benefit that might otherwise accrue to oil producers and use the money for the benefit of the entire nation.

Protecting the environment, the second goal, could easily conflict with the goal of increasing domestic fossil and synthetic energy supplies; however, it is totally consistent with the goal of promoting the efficient use of energy and the use of solar power. A balance must be found between the nation's energy needs and the desire to preserve the environment.

During the past decade, legislation has been enacted that reflects growing public concern about the importance of maintaining a pollution-free environment. The DOE must work within both the letter and spirit of such laws in pursuing the nation's energy goals. Environmental costs should be added to fuels that degrade the environment in their extraction, processing, transportation, or use. Likewise, environmental benefits should be subtracted from the cost of conservation and solar technologies that will improve the environment by lessening the need for pollution-causing fuels.

Maintaining the free enterprise system is another important goal. The economy should not be hampered by unnecessary government interference. The government should intervene only where free markets alone would not bring about necessary changes quickly enough. Government should not act to thwart the natural workings of the economy, but should supplement them to bring about a smoother, more rapid, and more efficient transition to energy-efficient products and processes and the use of renewable resources. The free market, on its own, could go through severe disruptions while becoming energy-efficient. The government should act legitimately to mitigate the undesirable effects of resolving the energy problem and to speed its solution.

## Relation to National Energy Plan

The conservation and solar goals are designed to achieve long-term United States energy objectives. The first objective of the National Energy Plan is to reduce the nation's dependence on foreign oil and its vulnerability to supply disruptions.

The DOE Office of Conservation and Solar Energy encourages investment in energy-efficient equipment to reduce the quantity of energy consumed without reducing the United States economic growth or the welfare of its people. Such investments will enhance energy security because domestic energy will be able to provide a greater share of United States consumption. In the event of a supply disruption, the Strategic Petroleum Reserve should be able to support the economy for a limited period of time. General energy efficiency, along with effective emergency measures, will provide a greater security for the United States in the short run.

In the midterm, the National Energy Plan anticipates that the United States will shift away from oil and natural gas to new and more expensive forms of energy. As energy costs rise, energy efficiency will become an

even better investment than it is now. In addition, higher energy prices make solar energy a better investment. The DOE is encouraging the development of solar technologies now so that they will be ready to enter the market when their economics make them attractive.

Finally, the DOE is developing solar technologies that will provide inexhaustible and renewable energy to sustain healthy economic growth as other sources become depleted. Technological breakthroughs are too infrequent to be the basis of national planning. Instead, the Department of Energy is committed to an aggressive R&D program to provide a wide range of technological options.

## Strategy to Achieve Goals

The DOE strategy is a pragmatic approach to the United States energy situation. The department's programs are designed for key opportunities within each sector (i.e., residential, industrial, etc.) to promote efficient energy use and the use of solar energy. The following criteria were included in developing the DOE strategy:

- Overall national benefits must exceed overall national costs.
- Overall benefits must substantially exceed government costs.
- Specific national policy objectives must be accomplished.
- Distribution of costs and benefits must be fair.

The energy strategy relies principally on market forces. This approach helps the market system work by reducing distortions that impede the use of cost-effective energy technologies. The government should stimulate the free market because of the substantial benefits to society. Such an effort is intended to support normal market mechanisms; it is not intended to replace them.

This involves the use of enticements such as tax incentives and other forms of financial and technical assistance to the private sector. The strategy involves both broad and highly specific education and marketing programs to address market imperfections. For example, the DOE shares the cost of RD&D with the private sector to accelerate the development of innovative (high-risk, but potentially high-payoff) technologies that the private sector would not otherwise pursue on its own. This RD&D will yield advanced technologies more quickly, so that they can start to become part of the United States energy-related capital stock.

Finally, the government uses standards and regulations in cases where the private sector would otherwise not respond to cost-effective conservation opportunities.

The government cannot act alone. A successful strategy requires the combined effort of private businesses, public utilities, consumers, and state and local governments, as well as the federal government. Moreover, the DOE must act in concert with the Departments of Housing and Urban Development, Transportation, Commerce, Agriculture, and the Treasury as well as other government agencies to help resolve the nation's energy problems.

Where other federal agencies are already pursuing related R&D, the DOE must fill in any program development gaps to make it directly applicable to the energy problem. For instance, the DOE R&D program on diesel engines is relatively small because it complements a large R&D program being carried out by the Department of the Army. It would be wasteful for the DOE to duplicate the army's research.

The DOE is accelerating the movement of new conservation and solar technologies into the marketplace. However, developing improved technologies and making them available will not lead to their most rapid adoption and use. The DOE must endeavor to understand the markets to be served and to overcome the many barriers that inhibit the widespread use of a technology. It does little good in terms of energy savings to develop an improved energy technology if the technology is not accepted in the marketplace.

The DOE strategy addresses—and should gradually overcome—barriers to the use of conservation and solar technologies. While emphasis may differ depending on the char-acteristics of the sector affected, each program has elements that fall into most of the following general strategic elements.

## Decontrol and Deregulation of Prices.

Government control of energy prices has been a major cause of today's difficulties; artificially low prices have induced Americans to use too much energy and to produce too little. Part of the solution to the problem is to let energy prices rise to their replacement costs, causing producers and consumers to adjust their activities accordingly.

Changing a pattern of energy use requires a considerable amount of time, and rising energy prices will not adequately reduce the time needed to turn over the energy-using capital stock and to retrofit long-lived structures. Rapidly increasing energy prices would also result in excessive energy profits for some and would be a hardship for consumers already hit hard by inflation—especially low-income families and those living in certain areas such as the northeast and north central regions of the nation.

## Public Information. Distribution of information is a major element of any strategy that is based on freedom of individual choice. Information distribution is a low-cost method for influencing decisions that can reap immense energy savings. Information of all kinds, including advertising, publications, product labels, films, and exhibits, tells consumers and producers what they need to know to make rational decisions about energy use.

## Financial Incentives. Financial incentives, such as grants, tax credits, and loan guarantees, are a potent method for encouraging energy conservation. Because the benefits to society of many conservation and solar investments exceed the direct savings to individual investors, financial incentives have a legitimate purpose and play a critical role.

Financial assistance also lessens the impact of higher energy prices and energy development on those most adversely affected—low-income households, small and minority businesses, and rural communities.

**Research and Development.** Government-sponsored R&D provides the conservation and solar options needed to reduce energy input. Depending on the state of the technology, federal support has a variety of objectives, which include determining market requirements, technical and cost assessments, prototype development and testing, full-scale demonstrations, and associated commercialization activities. Maximum industry participation through cost sharing and broad dissemination of research results ensures that the free market is interested in the technology and provides hands-on experience in the private sector.

**Regulatory Measures.** Direct regulation of individual or business activity to improve energy efficiency or to encourage the use of solar energy should be undertaken only as a last resort. However, regulations may be necessary where the market will not provide results that are socially or economically desirable.

Setting energy-efficiency standards and regulations can force changes more rapidly than would occur through market forces. But the use of legal requirements suffers because regulators cannot be as familiar with the marketplace as individual businesses or consumers; typically, regulators have to fit everyone into simplified, average cases for rulemaking. This leads to waste and inefficiency for some people, despite the general benefits. Thus, the costs of regulatory development and enforcement as well as compliance must always be weighed against the expected benefits.

**Assurances of Stability.** In some areas, industry may not have made investments because of fear that the regulatory or legal environment will change. To the extent this is true, the intentions of the government should be conveyed to the private sector as far in advance of an anticipated shift in policy as possible. This commitment to stability implies a close relationship between government and industry.

**Program Priorities.** In pursuing its strategy, the DOE is concentrating its efforts in accordance with two criteria. The first is near-term commercial use. Priority is given to technologies that can be developed quickly and moved into the economy. This can rapidly affect energy consumption and the use of solar energy. The second is concentration on local and regional needs. No all-encompassing, single national solution to the energy problem can be expected. An approach that works well in one part of the country could fail in another part. In the past, options may have been overlooked because they were not applicable over wide areas, yet they may have provided the ideal solution to a local problem. People in various regions have different tastes, customs, traditions, and resources that can be satisfied only by a variety of conservation and solar technologies. DOE's approach gives people the freedom to choose their own ways of reducing energy.

Along with these criteria, the Office of Conservation and Renewable Energy is pursuing a long-term R&D program to provide technology products for an inexhaustible energy supply.

The key theme of all of the programs is the development of products. The DOE is taking a businesslike approach to the energy problem; items are developed to serve specific purposes and to compete in the commercial marketplace. They must be the right tools to do the job at a price that industry and consumers can afford.

A program must be accepted by citizens and industry if it is to have any positive effects. The public and industry should be involved in all stages, from the very beginning. Their participation will not only make the work easier, but will promote greater success. The more aware people are of the possibilities of conservation and solar energy, the more willing they will be to change the way they use energy.

Finally, social, environmental, and institutional problems must be identified and resolved early. It is a waste of time and resources to resolve problems late in the development process that could have been addressed earlier. Generally, the more the process has been advanced, the more difficult, expensive, and time-consuming it is to resolve problems.

# 4. Program Summary

The following brief description of the overall conservation and solar program cuts across subprogram lines. It illustrates how the efforts of the various programs and projects form a cohesive whole—and how they affect the end-use sectors and help achieve national policy goals.

## BUILDINGS SECTOR

### Conservation

Conservation programs in the buildings sector reflect the varied nature of energy consumption in public and private buildings. The buildings sector accounts for 29 quads or about 37 percent of United States energy use. By the year 2000, energy consumption in buildings could be improved up to 70 percent. The implementation of passive solar design (Figure 4-1) is a major architectural shift that conserves use of electricity.

The building conservation programs include financial incentives, information programs, standards for new buildings and appliances, and R&D. Many other federal agencies are involved. The Department of Housing and Urban Development, for instance, will operate the Solar Energy and Energy Conservation Bank, and the Department of Health and Human Services operates a substantial weatherization program.

The DOE Office of Buildings and Community Systems (BCS) is the focal point for most DOE activity, but its annual budget represents less than 10 percent of the federal government's total spending, including tax expenditures, for improving energy efficiency in buildings.

BCS has three major program thrusts, each representing a major energy end-use area:

Figure 4-1. Passive solar design concepts used to provide "natural" heating and cooling.

- *Consumer products,* which include appliances, furnaces, air conditioners, hot-water heaters, lighting equipment, heat pumps, and similar elements
- *Building systems,* which include the building envelope and the equipment in the building as it functions as an integrated energy system
- *Community systems,* which include the facilities that provide energy to a community, such as district heating and urban waste-to-energy projects, and energy-efficient community planning

BCS is also responsible for federal government energy conservation. This activity is in a fourth BCS division along with systems analysis.

## Solar

Federal efforts to encourage the use of solar energy in buildings consist of fostering accelerated research, development, and application. The federal program includes:

- Market R&D
- R&D on active, passive, and photovoltaic systems
- Engineering and systems development
- Accelerated commercialization of renewable energy systems

The primary program objective is to hasten the use of solar and renewable energy technologies in the buildings sector. This will displace oil and other depletable energy sources and speed the transition to an economy based upon domestic renewable energy sources. The objective requires a close working relationship among other offices that affect the buildings sector, including the offices responsible for developing technologies for such renewable energy sources as wood, water, and wind, as well as building conservation technologies.

Likewise, the solar buildings subprogram is responsible for developing photovoltaic technology. This effort, while primarily related to near-term buildings technology, will serve the needs of the industrial and utility sectors as well.

Efforts in the buildings sector will concentrate on technology development, testing, and demonstration, as well as market research and industry development. Efforts are geared toward creating economically competitive solar systems where technologies are not yet ready to enter the commercial market. Where technologies are nearly competitive with other energy sources, attention is focused more on market development and industry expansion.

## INDUSTRIAL SECTOR

### Conservation

The industrial sector accounts for an estimated 38 percent of the energy used in the United States. The potential for energy conservation in this sector is quite high because most industrial processes are energy-inefficient. This inefficiency is due primarily to the existing industrial plant facilities, which were built during an era of cheap energy. Due to the diversity of processes and equipment, an effective conservation effort must be tailored to the specific requirements of individual industries.

Energy use in industry has become more efficient in response to higher energy prices and a greater risk of supply interruptions. Nevertheless, significant opportunities remain for major energy savings, which can be achieved through information efforts to provide technical and economic data on conservation; financial assistance, such as tax credits, for selected conservation investments; cost-shared R&D projects to accelerate the development and demonstration of new, higher-risk energy conservation technologies; regulatory programs including the required reporting of data on energy use and the mandating of utility-rate and energy-pricing policies.

Each of these components is directed at one or more factors that inhibit effective energy saving. Since the industrial sector has responded well to higher energy prices, direct regulation does not now appear warranted.

### Solar

The solar applications for industry subprogram has two major thrusts: solar thermal energy systems and biomass energy systems. Emphasis is placed on identifying, developing, and testing solar thermal and biomass systems that have the most direct application to the industrial sector. Figure 4-2 shows a solar industrial process hot-water system produced by Acurex Corporation. Solar industrial systems can produce electricity and petroleum susbstitutes and other energy-intensive products for industry.

The solar program considers the potential application of all solar technologies to the industrial sector. Consequently, subprogram thrusts are not strictly limited to solar thermal and biomass energy systems. The approach implies a strong interaction with other solar programs to ensure an integrated effort to meet industrial solar needs. A diversified portfolio of solar technologies that have industrial applications will ensure the most rapid adoption of solar industrial processes.

## TRANSPORTATION SECTOR

The transportation sector accounts for 26 percent of the United States energy demand and consumes 53.7 percent of the nation's supply of petroleum products. Trucks and cars alone account for about 40 percent of American petroleum requirements.

During the next 10 years, major opportunities for energy savings lie in further improvements to the fuel economy of new cars and trucks as well as in the more efficient use of the existing transportation systems. This includes greater use of ride sharing and public transit, improved car maintenance, and fuel-efficient driving techniques.

In the long run, the nation's transportation system must move away from heavy dependence on petroleum. Vehicles must be capable of using alternative energy sources, such as alcohol fuels and electricity (Figure 4-3). This longer-term transition requires basic changes in the technology and structure of the transportation system.

As in other areas, the transportation subprogram involves information dissemination, financial assistance, and RD&D, which are designed to lead to the early commercialization of improved technologies. In each of these areas, close cooperation is pursued with

Figure 4-2. Solar industrial process hot-water system produced by the Acurex Corporation at the Campbell Soup Company in Sacramento, California.

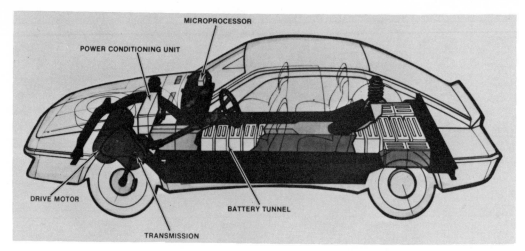

Figure 4-3. Key chassis features of experimental electric car.

the other divisions of the DOE as well as with other federal agencies. For example, the automobile fuel-economy standards are implemented by the Department of Transportation based on fuel-economy tests performed by the Environmental Protection Agency and supported by DOE's distribution of about 15 million gas mileage guides each year. A special effort by the DOE is now under way, with the Department of Transportation, to establish the federal government's long-term transportation R&D strategy. The alcohol and alternative fuel programs have been coordinated with research within the Office of Transportation programs. Gasohol (Figure 4-4), a mixture of 10% fermentation ethanol and 90% unleaded gasoline, is the only near-term synthetic fuel that can reduce our import of petroleum. It can be used in existing automobiles without modification.

## UTILITIES SECTOR

### Conservation

Electric utilities account for more than 10 percent of the nation's oil consumption and over 16 percent of its gas consumption. Because most primary energy is lost in generating electricity and generators can be powered by many fuels, electric utilities have a substantial potential both for improved efficiency and fuel switching.

Figure 4-4. Gasohol is a near-term synthetic fuel. Courtesy of Iowa Corn Production Board.

The government is encouraging utilities to use abundant domestic fuels through the Industrial and Powerplant Fuel Use Act, which forbids the use of oil or gas in new utility or industrial boilers.

The DOE is promoting conservation in the

utility sector by reducing demand by utility customers through more efficient buildings, appliances, and industrial processes. In addition, the DOE is encouraging more efficient production of electricity through cogeneration, which can almost double conversion efficiency.

## Solar

The solar power applications subprogram is developing solar technologies that can be applied to utilities. The subprogram, which consists of efforts to develop and accelerate the commercial use of energy from the wind and oceans, includes:

- RD&D of wind systems
- Market development and assessment of United States wind resources
- RD&D of ocean systems
- Development of analytical tools for comparative analysis of the impact of solar technologies on utilities and markets
- Coordination with the electric utility industry for solar thermal electric, photovoltaic, and wood-burning plants

Solar power applications seeks to (1) accelerate the development and commercialization of reliable and economically competitive wind energy systems, (2) develop the technology base and institutional links needed to encourage commercial use of ocean energy, and (3) coordinate efforts related to the utility sector with other agencies.

## STATE AND LOCAL PROGRAMS

State and local programs support national energy goals by promoting widespread energy conservation and the use of renewable resources through action by all levels of government. As part of this effort, programs have been developed in response to regional, state, and local needs to disseminate information on energy technologies that have significant energy-conserving and renewable resource potential. The effort also promotes the exchange of ideas on energy conservation needs and accomplishments among local governments, the states, and the federal government. Individual initiative is encouraged by the funding of small-scale technology projects and inventions. Finally, financial and technical assistance is provided for energy programs at the state, local, and community levels.

The Deputy Assistant Secretary for State and Local Assistance Programs is responsible for the DOE major grant programs for state and local governments, as well as for other institutions and individuals. These include institutional buildings conservation, weatherization assistance for low-income households, small-scale technology, and inventor's grants.

# Part II
# Conservation Energy

# 5. Introduction

Part II presents an overview of the energy conservation programs implemented by the Office of the Assistant Secretary for Conservation and Renewable Energy in the DOE. Together with the solar energy material in Part III, Part II covers all of the programs of the Office of the Assistant Secretary for Conservation and Solar Energy.

Consistent with the FY 1981 budget, conservation objectives, federal strategy, impacts, and issues are presented for the following programs:

- Buildings and Community Systems
- Industrial Energy Conservation
- Transportation
- State and Local
- Multisector: Appropriate Technology, Energy Inventions, and Energy Conversion Technology
- Energy Information Campaign
- Energy Impact Assistance

Table 5-1 contains a summary of the FY 1981 budget.

### Table 5-1. Energy Conservation Summary of FY 1981 Budget.

| SUBPROGRAM/PROGRAM ELEMENT | BUDGET AUTHORITY (DOLLARS IN THOUSANDS)[1] | | |
|---|---|---|---|
| | APPROPRIATION FY 1979 | APPROPRIATION FY 1980 | REQUEST FY 1981 |
| **BUILDINGS AND COMMUNITY SYSTEMS** | | | |
| Building Systems | 18,500 | 18,100 | 20,065 |
| Residential Conservation Service | 0 | 4,600 | 5,200 |
| Community Systems | 19,700 | 16,800 | 15,800 |
| Urban Waste | 8,500 | 13,000 | 10,900 |
| Small Business | 500 | 700 | 750 |
| Technology and Consumer Products | 20,350 | 29,600[2] | 22,240 |
| Appliance Standards | 4,950 | 6,000 | 7,925 |
| Federal Energy Management Program | 500 | 400[3] | 2,700 |
| Analysis and Technology Transfer | 2,800 | 5,400 | 5,900 |
| Emergency Building Temperature Restrictions Program | 0 | 3,675 | 0 |
| Program Direction | 3,533 | 5,137 | 6,120 |
| Total Buildings and Community Systems | 79,333 | 103,412 | 97,600 |
| **INDUSTRIAL ENERGY CONSERVATION (CS)** | | | |
| Waste Energy Reduction | 15,240 | 16,450 | 19,800 |
| Industrial Process Efficiency | 14,400 | 20,675 | 20,000 |
| Industrial Cogeneration | 5,000 | 11,250 | 12,000 |
| Implementation and Deployment | 3,160 | 9,800 | 4,500 |
| Program Direction | 2,193 | 2,067 | 2,600 |
| Total Industrial Energy Conservation | 39,993 | 60,242 | 58,900 |
| **TRANSPORTATION** | | | |
| Vehicle Propulsion RD&D | 47,800 | 60,500 | 55,900 |
| Electric and Hybrid Vehicle RD&D | 37,500 | 41,000 | 42,100 |
| Transportation Systems Utilization | 6,100 | 6,700 | 6,700 |
| Alternative Fuels Utilization | 5,800 | 5,300 | 5,300 |
| Program Direction | 1,949 | 2,923 | 3,000 |
| Total Transportation | 99,149 | 116,423 | 113,000 |

[1] Includes operating expenses, capital equipment and construction.
[2] Includes $9,000,000 for 40 kW fuel cell demonstration.
[3] A supplemental request for $2,300,000 has been approved by OMB and will be forthcoming.

## Table 5-1. (continued)

| SUBPROGRAM/PROGRAM ELEMENT | BUDGET AUTHORITY (DOLLARS IN THOUSANDS)[1] | | |
|---|---|---|---|
| | APPROPRIATION FY 1979 | APPROPRIATION FY 1980 | REQUEST FY 1981 |
| STATE AND LOCAL (CS) | | | |
| Schools and Hospitals Grant Program | 100,100 | 143,750 | 202,500 |
| Other Local Government Buildings Grant Program | 7,300 | 17,700 | 0 |
| Energy Management Partnership Act[4] | 0 | 0 | 151,625 |
| Energy Policy and Conservation Grant Program[4] | 47,800 | 37,800 | 0 |
| Energy Conservation and Production Grant Program[4] | 10,000 | 10,000 | 0 |
| Energy Extension Service Program[4] | 15,000 | 25,000 | 0 |
| Emergency Energy Conservation Program[4] | 0 | 0[5] | 4,072[6] |
| Weatherization Assistance Program | 198,950 | 198,950 | 198,950 |
| Program Direction | 2,980 | 7,340 | 11,437 |
| Total State and Local | 382,130 | 440,540 | 568,584 |
| MULTI-SECTOR (CS) | | | |
| Appropriate Technology | 8,000 | 12,000 | 14,100 |
| Invention Program[7] | 2,000 | 4,200 | 3,400 |
| Energy Conversion Technology | 0 | 0 | 11,000 |
| Personnel Resources | 243 | 635 | 700 |
| Total Multi-Sector | 10,243 | 16,835 | 29,200 |
| ENERGY INFORMATION CAMPAIGN | 0 | 0 | 50,000 |
| ENERGY IMPACT ASSISTANCE | 20,000 | 50,000 | 150,000 |
| TOTAL | 630,848 | 787,452 | 1,067,284 |

[4]Will be consolidated under the Energy Management Partnership Act in FY 1981.
[5]FY 1980 supplemental request is $14,072,000.
[6]State implementation grants to be included under EMPA.
[7]$2,400,000 of Invention Program, not shown here, is under Departmental Administration Budget, in FY 1981; $2,200,000 additional is available in FY 1980.

# 6. Conservation Overview and Strategy

## CONSERVATION OVERVIEW

All available evidence indicates that conservation is the most immediate and cost-effective means of dealing with current energy problems. Conservation is the elimination of waste, the investment in more energy-efficient machines and buildings, and the changing of consumption patterns, consistent with the replacement cost of energy. These actions will reduce the need for oil imports, mitigate the impact of rising oil imports, increase the energy efficiency of the United States, limit the transfer of wealth to oil-producing states and from consumers to producers of energy, and extend the lifetime of dwindling fossil-fuel supplies. Conservation will also ease environmental stress by reducing the quantity of air- and water-polluting emissions.

There is ample opportunity in the United States economy for more efficient use of energy. By applying both old and emerging new technologies, the nation can achieve an increase in productivity and general economic efficiency. However, many barriers to conservation must be overcome in order to induce significant interest in energy-conserving behavior and investment. The role of the federal government is to facilitate and promote conservation by reducing these barriers.

Federal conservation policy consists of efforts to maximize cost-effective energy-conserving investments, conduct new technology RD&D, and change end-user operating behavior, thus minimizing the need for energy curtailment and abrupt life-style shifts. While curtailment may be necessary in emergencies, the federal conservation strategy will promote new and existing cost-effective conservation technologies and personnel energy usage changes in the private and public sectors. As a result, national energy efficiency should be greatly enhanced without affecting our standard of living.

This section discusses the magnitude of conservation potential, examines the principal barriers to its implementation, and describes the overall national energy-conservation strategy. Subsequent sections describe sector-specific strategies for the end-use sectors, utilities, and cross-sectoral programs.

### Conservation Potential

Analyses of United States energy demand data over time, as well as international energy-use comparisons, support the conclusion that greater energy efficiency and reduced petroleum consumption can be achieved without adversely affecting our present or future economic well-being.

As shown in Table 6-1, macroeconomic comparisons of energy consumption in industrial countries indicate that there is room for substantial economies in the United States. Even if the figures are adjusted to allow for differences in climate, distances between economic centers, and variations in industrial activity, they will support a large savings potential. It is reinforced when comparisons are made between the efficiencies of new, available energy-consuming machines and equipment in the United States and the effi-

Table 6-1. Macroeconomic Comparisons of Energy Consumption in Industrial Countries.

| COUNTRY | ENERGY CONSUMPTION PER CAPITA (TONS OF OIL EQUIVALENT) |
|---|---|
| France | 3.5 |
| West Germany | 4.5 |
| Sweden | 6.1 |
| United States | 8.3 |

ciencies of those currently in use. On the average, the existing capital stock is about half as efficient as existing and emerging technology will allow.

Energy demand data over time, and particularly during the post-1973 period, have shown that GNP growth can greatly exceed the rate of growth in energy demand (see Table 6-2). In fact, energy use in the industrial sector has actually declined by about 1.2 percent a year during this period.

Unlike in the past, in recent years energy consumption has increased in the United States economy at a rate significantly lower than the rate of increase of the GNP. The reasons for this decline are not yet well understood. However, several major factors underlying this trend can be identified:

- *Growth and turnover of capital stock.*
  About 80 percent of United States energy is consumed in eight end-use functional demands in the three energy-using sectors (see Table 6-3).
  By the year 2000, cumulative investment in energy-related capital stock is projected to be about $9 trillion (1972 dollars) due to turnover of stock and growth in the economy. By then, the following changes will have taken place, which should improve energy efficiency: (1) About 35 percent of exist-

ing commercial buildings will have been replaced (but only 7 percent of existing one-to-four-unit residential buildings). (2) About 60 percent of existing industrial boilers will have been replaced. (3) Virtually all remaining energy-using devices, including those inside buildings, will have been replaced.

- *Availability of energy-efficient technology and consumption patterns.* The average efficiency of industrial processes, end-use equipment, and overall end-uses can be increased significantly. The amount of additional capital cost increase associated with this efficiency improvement will depend on how successful the private-sector development process is in holding down the costs of new technology.

- *Cost-effectiveness of conservation investments and altered energy consumption patterns.* The key economic considerations in any investment decision are the initial cost, the period necessary to recover investment cost (payback period), and the total economic cost over the useful life of the investment (life-cycle cost). These determinants indicate whether or how much of a technology will be purchased in the marketplace. With currently forecast world oil prices, many existing

Table 6-2.  Annual GNP and Energy Consumption Growth Rates (Percent per Year Average).

|  | 1950–1973 | 1974–1978 | 1977–1978 |
|---|---|---|---|
| GNP growth | +3.7 | +2.3 | +4.4 |
| Energy consumption growth | +3.5 | +0.9 | +1.9 |

Table 6-3.  End-Use Functional Demands by Sector (Percent of Total United States Energy Consumption).

| RESIDENTIAL-COMMERCIAL | | INDUSTRIAL | | TRANSPORTATION | |
|---|---|---|---|---|---|
| Space heating | 19% | Process steam | 15% | Cars and trucks | 18% |
| Lighting | 6 | Direct heat | 10 | Other | 8 |
| Air conditioning | 3 | Electric drive | 7 | | 26% |
| Water heating | 3 | | 32% | | |
| Other | 7 | | | | |
| | 38% | | | | |

conservation technologies, and all new technology development supported by DOE, have payback periods of 3 years or less. Should world oil prices rise above the forecast during this period, the shift toward more energy-efficient technology will be further accelerated.

Demand reduction is clearly an energy "source," and conservation investments, from a national perspective, can and should be compared with alternative fuels. Conservation investments provide long-term relief from fuel requirements (because of the long life of capital equipment and buildings), and for a wide variety of end uses compare very favorably with alternative fuel options on a cost-per-million-Btu basis. This is true for many investments (e.g., home insulation, heat pumps, industrial cogeneration) even at prices for imported oil well below the OPEC price of $37 per barrel.

## FEDERAL ROLE IN ENERGY CONSERVATION

In other than high-risk RD&D or in a national emergency situation, direct federal intervention through regulation, deregulation, and taxes on energy consumption is highly contentious; its cost is generally high in terms of protracted debate and market disruption. Government-funded RD&D and direct financial incentives are somewhat less controversial, but there are limits on the resources that can be so employed. The decentralized nature of most conservation decisions dictates substantial reliance on market mechanisms; if the market does not respond to the purported economic desirability of conservation investments (as mentioned earlier), it is because of market imperfections or inadequacies of market institutions, which can be corrected by government action.

Market response to conservation opportunities is impeded by:

- Current energy prices being lower than their replacement cost
- Uncertainty about future energy prices
- Uncertainty about investment costs
- Doubts about potential savings from

conservation investments due to (1) present energy prices and (2) lack of confidence in emerging technology
- Lack of capital
- Lack of choices because of (1) unavailability of suitable equipment and materials and (2) absence of reliable services to install and maintain them
- Uncertainty about government's intentions

The federal government's conservation policies and programs must address all of these market barriers because the speed of market response to conservation is not well known. The government's role will thus form the basis, in collaboration with state and local governments, for a coherent national conservation strategy for the next 10 years.

Advanced energy technologies will play an increasingly important role in the national conservation strategy. RD&D in the short term is essential to reduce the uncertainty of technological development and costs, and it opens important new options. In the future, more reliance will be placed on totally new technologies to further advance energy efficiency.

The current near-term (through 1990) program uses a "targets of opportunity" approach. Specific existing energy-conservation opportunities are chosen and RD&D programs designed to show operational efficacy. DOE then encourages technology transfer through demonstration and information dissemination. This strategy emphasizes near-term RD&D in order to meet immediate and obvious needs.

However, it does not systematically address the need for improvement in our fundamental knowledge, which could result in major long-term benefits. With revolutionary technological advances now in early R&D stages, substantially increased energy productivity is clearly possible, given the same energy supplies. DOE advanced-technology R&D programs emphasize study of the following areas:

- Improved combustion efficiency in several temperature ranges
- Use of what is now wasted heat through better low-temperature-differ-

ence heat exchangers and more efficient energy conversion from low-grade heat to higher grade heat (e.g., significantly improved heat pumps)

• Development of high-temperature materials to help achieve higher efficiency from processes critically dependent on thermodynamic cycles

The logic behind these approaches is straightforward and compelling. Even a 1 percent improvement overall in combustion efficiency would reduce fossil fuel use by over 300,000 barrels per day. Yet today's efficiencies are appallingly low—for example, around 50 percent for typical home oil or gas furnaces. Modest efforts are already under way to improve home furnace efficiencies by 80 to 90 percent; comparable improvements seem likely with R&D in many other areas. The potential payoff is very high, and the required R&D relatively moderate in cost.

Similarly, waste heat from utilities represents a resource exceeding 14 quads per year; waste heat from industry represents at least another 6 quads per year. Such a vast resource, although of low thermodynamic quality, warrants even larger R&D effort than it now receives.

In addition, DOE conservation emphasizes R&D and commercialization activities that can substantially and quickly improve the efficiency of most energy-intensive industrial processes. For example, the most energy-efficient processes now contemplated show that energy consumption per unit of output could be reduced 50 percent with new technology

resulting from the conservation R&D programs.

## GOVERNMENT OBJECTIVES

The overall objective of the federal government's conservation program is to encourage the adoption by the economy of relatively cost-effective conservation measures as rapidly as possible.

There have been many studies on long-term energy demand in the United States, both inside the government and elsewhere. Using CONAES,* one of the most well-known studies, and in the interest of assessing the possible results of a federal conservation strategy involving enhanced federal action for conservation and reliance on available technology through 1990, DOE projected consumption by end-use in that year. For a real GNP growth rate of 3 percent, the analysis shows the following possible energy savings in 1990 in comparison with actual consumption in 1975 and a projection of "business as usual" practices. Energy savings resulting from two scenarios are shown in Table 6-4.

The savings in the table look attainable, especially since they do not entail major technological breakthroughs. Achievement of this level of energy savings, however, entails a much sharper focus of federal action than has been the case. The United States strategy is to treat these savings as minimums to be attained and, it is hoped, exceeded, through the careful and discrete application of a range of federal actions all designed to stimulate market response. Early results from the federally

Table 6-4.  United States Energy Consumption (Quads).

|  | RESIDENTIAL-COMMERCIAL | INDUSTRIAL | TRANSPORTATION | TOTAL |
|---|---|---|---|---|
| 1975 consumption | 16.8 | 36.7 | 17.3 | 70.8 |
| 1990 no-change path[a] | 23.6 | 69.5 | 26.9 | 120 |
| 1990 possible[a] | 18.4 | 58.6 | 23.0 | 100 |
| Percent savings | 22% | 15.7% | 14.5% | 20% |
| 1990 possible[a] | 14.1 | 43.6 | 16.5 | 80 |
| Percent savings | 40.3% | 37.3% | 38.2% | 40% |

[a]*Committee on Nuclear and Alternative Energy Systems.* Scenario B, moderate efficiency improvement, scaled to 3 percent GNP growth rate, was used, interpolated to 1990, for the "possible" case. Scenario C, unchanged from present policies, was used to represent the no-change path. See Figure I-4 of CONAES study.

Source: CONAES, December 1979.

funded conservation R&D programs will provide additional savings by 1990.

## GOVERNMENT CONSERVATION STRATEGY

### Current Federal Strategy

The current government strategy represents a careful blend of all of the elements involved in a federal role (see Table 6-5). It is a pragmatic approach and is appropriate, given the multifaceted aspects of the United States energy situation. The governmental tools are tailored to the specific opportunities within each sector that must be realized to expeditiously achieve cost-effective conservation in the United States.

Criteria that have been applied in the development of the government strategy include:

- Overall national benefits must exceed overall national costs.
- Overall benefits must substantially exceed government costs (i.e., federal

Table 6-5. Summary of Conservation Policies.

| ENERGY-USING SECTOR | PRICING | INFORMATION/ DEMONSTRATION/ COMMERCIALIZATION** | R&D* | FINANCIAL INCENTIVES/ GRANTS | STANDARDS/ REGULATIONS |
|---|---|---|---|---|---|
| Residential/ Commercial | | $26.2M RCA, FEMP, FOMP, LC/NC | $57.6M | Residential Tax Credit to $300, 10% Investment Tax Credit for Commercial Buildings | Mandatory New Building Energy Performance Standards |
| | | CCEMP; ACES, ICES, EES, EMPA | Emphasis on Mid to Long Term | | Mandatory Appliance Standards |
| | | | | Weatherization Grants @ $199M Schools & Hospital Grants @ $2.5M | Mandatory Emergency Building Temperature Restrictions |
| Industry | Deregulation of Prices | $25.1M Cogeneration Systems | $33.1M | Additional 10% Investment Tax Credit for Specified Equipment | Voluntary Energy Efficiency Targets |
| | | Energy Diagnostic Centers | Emphasis on Mid to Long Term | | Mandatory Energy Use Reporting |
| | | | | | Coal Conversion of Large Furnaces and Boilers |
| Transportation | | $30.5M 400 Electric Vehicles, Fleet Tests of Gasohol; Gas Turbine Engine in Buses | $74.5M Emphasis on Mid to Long Term | Gas Guzzler Tax | Corporate Average Fuel Economy Standards: |
| | | | | | Mandatory 27.5 mpg New Car Fleet Avg. by 1985 |
| | | | | | Electric and Hybrid Vehicles to be Added |

*All budget figures are for FY 1980

RCA     – Residential Conservation Service
FEMP     – Federal Energy Management Program
FOMP     – Fuel Oil Marketing Program
LC/NC     – Low Cost — No Cost
CCEMP     – Consolidated Community Energy Management Program

**Includes only illustrative programs/projects.

EES     – Energy Extension Service
EMPA     – Energy Management and Partnership Act
ACES     – Annual Cycle Energy Systems
ICES     – Integrated Community Energy Systems

dollars must be highly leveraged by private sector investment).

- Costs of energy savings should not exceed those of alternative fuel supplies.
- Specific policy objectives must be accomplished (e.g., congressional mandates, the president's oil import goals).
- Equity considerations (fairness of distribution of costs and benefits of the policy) must be addressed.

Although government energy policy has many elements, its principal reliance is on market forces. The approach is to make the market system work by eliminating or reducing the distortions that impede the rapid and widespread adoption of cost-effective conservation technologies. This involves the use of tax incentives and other forms of financial and technical assistance to the private sector. The government strategy involves both broad and highly specific education and marketing programs to address specific market imperfections. The government engages in cost-shared RD&D with the private sector to ensure and accelerate the development of innovative (i.e., high-risk, but potentially high-payoff) conservation technologies that the private sector would not otherwise pursue on its own. Such RD&D will yield advanced technologies more quickly, so that they can become earlier a part of the United States energy-related capital stocks. Finally, the government uses the "stick" of standards and regulations in cases where the private sector would otherwise not respond sufficiently (from a national perspective) to cost-effective conservation opportunities on its own. Such measures have been deemed necessary in the residential-commercial and transportation sectors.

In sum, the government strategy uses a mixture of all of the elements just mentioned. The details are briefly outlined in Table 6-5.

The federal government cannot act alone to enhance energy efficiency in the United States. A successful program requires the combined effort of private businesses, public utilities, consumers, and state and local governments, as well as the federal government. Moreover, DOE must act in concert with the Departments of Housing and Urban Development, Transportation, Commerce, Agriculture, and the Treasury, as they interact with their constituencies.

## Role of State and Local Governments

Because conservation opportunities are frequently small-scale and dispersed, the design of successful projects typically is heavily dependent on specific local conditions such as climate, geography, fuel-use patterns, economic base, political institutions, land-use patterns and densities, and current and projected commercial and industrial development. Identification of and action on the full range of opportunities consequently depends heavily on analyses and action at state and local levels. For example, the cities of Seattle, Portland, and Davis, California, have each completed such analyses and have developed plans to reduce energy use in both public and private sectors by at least 30 percent.

The ability to conserve specific fuels and to develop specific renewable resources varies from one state to another depending on local climate, available energy resources, economic base, population distribution, and institutional structures. Furthermore, the various state legislatures and local governments have the unique authority to pass and implement energy conservation laws such as those that regulate utilities, building practices, and land use.

Not only can states develop and implement energy conservation practices, which can have an immediate and substantial effect on local energy growth rates, but they can also disseminate energy conservation information internally and serve as collection points for conservation-related information that can benefit other states, other regions, or the country as a whole. While focusing on internal energy problems peculiar to their areas, state and local governments may develop information that can be applied usefully elsewhere. Support for greater decentralized energy expertise and close communication with state and local government is thus a central feature of the overall federal-government energy conservation strategy.

# 7. Buildings and Community Systems

## PROGRAM OBJECTIVES AND STRATEGY

### National Energy Outlook

On April 29, 1977, President Carter issued a summary report on the National Energy Plan. It stated that the United States has three overriding energy objectives:

- In the short term, to reduce dependence on foreign oil and vulnerability to supply interruptions
- In the midterm, to keep United States oil imports low enough to prepare for the time when world oil production nears its limited capacity
- In the long term, to develop renewable and essentially inexhaustible sources of energy for sustained economic growth

Energy conservation, which includes the more efficient use of energy, can reduce the demand for nonrenewable energy sources in the near future. Recognizing this, the president's energy plan stated that "conservation and fuel efficiency are the cornerstones of the National Energy Plan."

The DOE is responsible for improving energy in the building and community sectors and can therefore make a significant contribution to national energy conservation goals.

In 1976, the United States imported 7 million barrels of oil per day at a total annual cost of approximately $46 billion. Over one-third of its total energy demand in 1976 was for use in residential and commercial buildings. One objective of the DOE is to help reduce American dependence on oil imports. The strategies for accomplishing these objectives include programs for conserving energy in buildings and improving community energy systems.

### Energy Use in Buildings

There are approximately 74 million residential units in the United States and 1.5 million non-residential buildings (approximately 29 billion square feet of commercial floor space). In 1978, the energy demand for these buildings amounted to the equivalent of 13.8 million barrels of oil per day. In the federal sector alone, there are approximately 400,000 buildings (2.6 billion square feet of floor space), which accounted for 350,000 barrels of oil per day equivalent (BDOE) in 1976.

Most existing buildings were constructed in an era when energy was thought to be plentiful and inexpensive. Consequently, few or no saving considerations were incorporated into their designs. As a result, an estimated 41 percent (5.7 million BDOE) of the energy used in buildings is now wasted. Thus, the DOE is working to develop methods of using energy more efficiently in several basic areas:

- In the buildings themselves
- In the appliances and products used in buildings (lighting, heating and air-conditioning systems, etc.)
- In the systems that supply energy to buildings and communities
- In the design of communities
- In the energy-consuming practices of consumers and the institutional factors (i.e., legal, economic, behavioral) affecting these practices.

In addition, DOE programs are aimed at conserving oil and gas by substituting more abundant fuels (such as coal) and renewable energy sources (urban wastes, solar, and thermal waste). In implementing these programs, DOE is encouraging the implementation of energy conservation technologies through close cooperation and coordination of programs and results with local governments, the building industry, public and private sectors, and other federal agencies.

## Objectives

DOE Program Mission. President Carter's National Energy Plan set forth a number of broad objectives for energy conservation. The overall DOE mission, in conjunction with these objectives, is to achieve short-, mid-, and long-term energy conservation and to improve energy efficiency in buildings, building components, and appliances, and community systems. In order for DOE to accomplish this successfully, it must:

- Accelerate the adoption of minimum energy performance standards for new and existing structures
- Emphasize the retrofit of existing structures to reduce energy losses and improve efficiencies of energy-using systems
- Give priority consideration to conserving energy in hospitals and schools
- Bring 90 percent of existing homes, as well as all new buildings, up to minimum energy-efficiency standards
- Develop mandatory minimum energy-efficiency standards for major appliances, such as furnaces, water heaters, refrigerators, and air conditioners
- Reduce energy use in existing federal buildings by 20 percent by 1985
- Develop integrated community energy systems that achieve high fuel-efficiency ratios
- Restrain the growth of energy demand through conservation measures and improved energy-efficient technology
- Vigorously expand the use of nonconventional sources of energy, such as urban wastes
- Encourage the development of small-scale (25 to 100 tons per day) technologies for converting urban wastes to energy
- Provide the consuming public with the information required to make rational purchase decisions
- Conserve energy in water- and waste-water-treatment facilities

DOE Program Objectives. The program is formulated around increasing energy use efficiency, developing energy substitution options (such as coal for natural gas), and providing technologies, methodologies, and processes that decrease the amount of energy required to satisfy human needs. All activities undertaken are directed at maximizing the effectiveness of energy use within a framework that is both economically and environmentally sound. Activities that accelerate and complement private sector efforts and foster the acceptance of energy-saving technology in the residential and commercial sectors are pursued as well.

DOE Energy-Savings Goals. The energy savings goals of the DOE program are to:

- Reduce the consumption of scarce fuels (oils, gas) in new buildings by 435,000 BDOE in the near term and by 2.65 million BDOE from 1975 consumption (in the midterm)
- Reduce the consumption of scarce fuels in existing buildings by 1.14 million BDOE in the near term and by 1.91 million BDOE in the midterm from 1975
- Decrease the consumption of scarce fuel resources by 5000 BDOE in the near term and by 1.5 million BDOE through the use of community systems that allow for the direct substitution of waste heat for scarce fuels (in the midterm)
- Decrease the consumption of scarce fuels an additional 130,000 BDOE in the near term and 750,000 BDOE in the midterm through the recovery of energy from municipal wastes

To attain these goals, it is necessary to determine the appropriate federal role, develop strategies to address the areas of greatest energy-savings potential, and conduct the necessary RD&D, commercialization, and implementation activities.

## Federal Role

The DOE is responsible for RD&D, commercialization, and implementation of technologies and measures that will support the government's energy initiatives. The DOE role is to complement and accelerate private-sector efforts to commercialize energy technologies at the earliest possible time through cooperative RD&D with other federal agencies. These include the Department of Housing and Urban Development, Department of Defense, National Bureau of Standards, General Services Administration, Environmental Protection Agency, and Department of Commerce.

Federal participation in energy conservation programs carried out by DOE was warranted for several reasons:

- Because over one-third of the nation's energy demand is for use in buildings of various types, energy savings in this sector will have a significant national impact.
- The building industry, which includes all elements of community development and urban waste management, is fragmented, and construction occurs in a highly segmented, incremental fashion. Therefore, it is difficult to aggregate private research funds for the development of energy-efficient technologies without federal help.
- The regulatory apparatus in the building environment is complex. It has many variations in building codes, zoning ordinances, utility regulations, taxing authorities, and management jurisdictions. The introduction of energy conservation programs is stymied by these institutional barriers, which federal efforts can help overcome.
- New energy-efficient systems and products often have higher first costs than conventional ones, thereby discouraging market acceptance. Some federal risk-sharing is needed to encourage the private sector to invest in energy-conserving products and practices that are cost-effective on a life-cycle basis.

In general, the federal role is to establish policies and priorities and conduct the necessary research, development, demonstration, and implementation (RDD&I) activities, which will accelerate the development and acceptance of energy technologies that will relieve national energy shortages.

## Federal Program Strategy

In applying the appropriate federal role to active RDD&I programs, the following strategies are being followed:

- Identification of the greatest potential energy-savings areas
- Selection of the proper federal role
- Encouragement and support of installing existing energy-efficient technologies as soon as possible
- Performance of implementation activities with supporting RDD&I activities to commercialize energy-efficient systems and technologies for the midterm and long-term
- Development and commercialization of systems that will reduce dependence on petroleum and natural gas
- Support of community processes for selecting among alternative programs and identification and development of tools for implementation
- Development and dissemination of information about new and existing systems and technologies concerning energy-efficient utilization improvement
- Promotion of the use of energy-conserving technologies and encouragement of energy-conserving practices in the facilities and operations of the federal government
- Promotion of the use of energy-conserving technologies and encouragement of energy-conserving practices by consumers (regulatory and nonregulatory emergency programs)
- Development and implementation of energy-efficient standards for new buildings and appliances
- Removal of economic and institutional barriers to the development and adoption of energy-efficient technologies and practices

## DOE Management Strategy

The objectives and strategies of DOE are carried out through a management approach that includes a well-defined organizational struc-

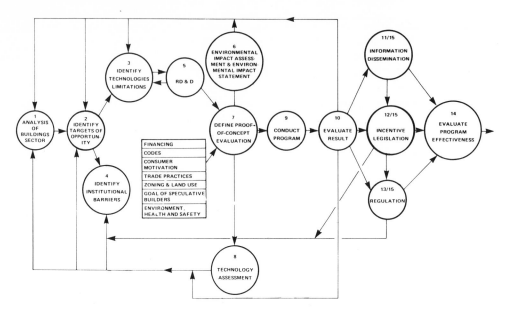

Figure 7-1. Buildings and Community Systems Program flow.

ture, decentralization of major RD&D programs, coordination with other government agencies and private sectors, and specific project implementation methods.

Organization and Administration. The DOE consists of four divisions: Federal Programs, Buildings, Community Systems, and Consumer Products. Overall management of the program is vested in the Assistant Secretary for Conservation and Renewable Energy, with program policy and planning, as well as certain key institutional activities, being administered by the DOE. The headquarters staff provides programmatic definition, direction, and policy to federal and national laboratories, and industrial contractors, in consultation with all segments of the building industry, state, local and federal agencies, other segments of DOE, and consumer groups.

As program needs arise, DOE uses federal and national laboratories, universities, private institutions, and other government agencies to assist in the achievement of its stated objectives. This is done through a decentralized management approach. The decentralization plan leaves the overall program policy and planning at headquarters, moving the project management to the field. This approach in-

cludes the use of national laboratories, alone or in conjunction with DOE operations offices, as well as other federal laboratories, to administer and manage projects.* Operations offices and national laboratories principally involved in the DOE program include Brookhaven National Laboratory for central heating and controls, and support of residential building activities; Lawrence Berkeley Laboratory for ventilation, infiltration, building envelopes, building energy analysis calculations, diagnostics, windows, and lighting; Oak Ridge National Laboratory for building insulation materials, consumer products, appliances, macroeconomics and energy analysis, and support of the Annual Cycle Energy Systems and Innovative Structures work; and Argonne National Laboratory for requisite management support for the Community Systems and Urban Waste Technology programs.

Project Implementation. DOE programs are conducted within the operational concept shown in Figure 7-1. This represents the flow of activity from program inception to comple-

---

*BCS is presently assessing the role of the national laboratories to support RD&D and commercialization activities in nonsolar areas. The final assessment may prove that a redirection of laboratory-program RD&D is necessary.

tion, along with feedback loops to ensure achievement of the goals. Technical, economic, and institutional objectives must be accomplished to ensure market acceptance of resulting processes and products. Projects may be initiated at different points in the flow diagram.

To facilitate the management of programs, various tools and techniques have been developed by which DOE can determine whether a proposed project is consistent with DOE objectives. This involves determining whether a project should be developed as a federal "showcase" demonstration effort—that is, one that accelerates private-sector conservation efforts while fostering the acceptance of energy-saving technologies.

The RD&D function of DOE depends on the rate at which resulting innovations are implemented by producers and consumers of energy technologies. The evaluation tool developed by DOE will play a part in reducing the uncertainty associated with DOE budgetary decisions that relate to market potential within Conservation and Solar Application programs. Specifically, DOE has developed a screening and scoring model to identify and prioritize appropriate conservation RD&D.

To estimate the break-even point for government involvement in RD&D, DOE calculates the rate of return to the end user and delineates commercial, technical, and institutional barriers associated with market introduction of specific energy technologies. The DOE investment criterion involves a determination of the current and projected ability of the technology to compete for scarce investment capital. This criterion centers on two parameters:

- Potential economic benefits to the public and private sector
- Risk inherent in the successful introduction of the subject technology

Conceptually, these issues are not difficult to grasp. Whenever an individual, a corporation, or the federal government makes an investment, there are perceived costs and benefits. In many instances, these costs and benefits are subject to uncertainty, which in turn can have an impact on the decision to invest or not to invest.

Federal support for private-sector RD&D can affect both of these parameters. By subsidizing the research phase, the government can reduce the overall cost of product development. Similarly, the involvement of the government can enhance the credibility and provide for greater dissemination of research results, which may broaden market exposure and increase the rate of market penetration. Undoubtedly, these factors play an important role in stimulating the private sector to seek federal support for production innovations. However, these benefits could also exist for projects that the private sector might pursue independent of federal support. Therefore, in weighing the economic costs and benefits of projects seeking federal support, it is necessary to differentiate between those projects that the private sector should undertake without federal support and those that require federal support. The objective of a government RD&D program from a commercialization perspective must therefore focus on the following:

- Improving the potential net energy and economic benefits of the project
- Reducing the commercial risk

Specifically, a federal program can influence risk elements by improving the credibility of the marketing program, stimulating interest groups to push for the adoption of the technology, creating a market within the federal establishment, and assisting in the adoption of regulations and removing barriers to commercialization. The assistance that the federal government can provide in these areas may be as important as subsidization of the research program.

In order to determine the payoffs from government intervention, DOE has assessed a portfolio of specific technologies. For each case, DOE characterizes the technology and the market, and delineates the market potential by sector and year of introduction. Technology characterization involves the determination of capital costs, operating costs, and future applications.

In essence, the evaluation system involves three stages: (1) a preliminary review to determine consistency of the project with DOE authorizing legislation; (2) a quantitative screening to estimate the rate of return to the end user before and after federal involvement, an estimate of marginal and total efficiency that involves the energy savings associated with federal involvement, and a portfolio model that ranks competing technologies that are entering common markets; and (3) a scoring model that delineates externalities associated with the project.

The DOE threshold screening model has been very useful in justifying the federal role in RD&D as well as in delineating for private-sector decisionmakers the expected rate of return and benefit ratio of emerging technologies. In some instances, this has required further consideration of financial incentives to encourage commercialization of energy technologies.

The threshold approach has three facets: (1) simulating a venture from the specific standpoint of private sector manufacturers and end users; (2) identifying and quantifying the social benefits that accompany the commercialization of an emerging technology; and (3) determining the extent of government assistance in the form of incentives required to ensure commercial development of the system. All projects that meet these criteria are then ranked using a scoring model that captures both quantifiable and nonquantifiable attributes of the selected projects.

The first stage of the screening criteria is related to the potential energy benefits that would result from the successful commercial introduction of the technology. In some instances, the purpose of the DOE involvement is to introduce the technology at an earlier date. By accelerating its introduction into the marketplace, the corresponding energy savings will be realized sooner than would otherwise be expected. The energy benefit criteria, therefore, are keyed to both the total benefits from the project and the incremental or marginal benefits attributable to the federal program.

The investment criteria focus on the role of the federal government in assisting the private sector to develop energy-conserving technologies. The criteria center on two parameters:

- Potential economic benefits to the private sector
- Risks inherent in the successful commercial introduction of subject technologies

The objectives of the DOE RD&D program are thus accomplished by:

- Improving the potential net economic benefits
- Reducing the commercial risk

## MAJOR PROGRAM THRUSTS

### Legislative Framework

General authority for the energy conservation programs of the DOE is derived from the Federal Nonnuclear Energy Research and Development Act of 1974 (P.L. 93-577) and the Energy Reorganization Act of 1974 (P.L. 93-438), which created the Energy Research and Development Administration (ERDA). The functions of both ERDA and the Federal Energy Administration (created by the FEA Act of 1974, P.L. 93-275) were subsequently transferred to DOE upon its creation on October 1, 1977, by the Department of Energy Reorganization Act (P.L. 95-91). Under these provisions, DOE has the leading federal role in conducting RDD&I activities to reduce total energy consumption and promote maximum improvement of energy efficiency in the areas of buildings and community systems.

Energy legislation has established national goals for reduced energy consumption in more specific areas of the buildings sector. These have provided direction for many of the programs being carried out by DOE. The Energy Conservation and Production Act of 1976 (Title III, P.L. 94-385) mandated the development, promulgation, and implementation by HUD of Building Energy Performance Standards (BEPS) for new residential and nonresidential buildings. Public Law 95-91 transferred the responsibility for their development and promulgation to DOE; however, the Reagan administration has terminated funding for BEPS.

Public Law 94-163, the Energy Policy and Conservation Act (EPCA), authorized the development of an energy conservation program for federal buildings and transportation vehicles. In addition, Executive Order 12003, "Relating to Energy Policy and Conservation," established specific goals for reducing energy consumption in federal buildings and required the development of a Federal Energy Management Plan (FEMP) and a federal life-cycle cost methodology. The order requires a 20 percent energy use reduction in existing federal buildings and a 45 percent reduction energy use in new buildings by 1985. The National Energy Act (NEA) also requires the development of specific energy targets for existing federal buildings. The DOE FEMP is responsible for developing guidelines to ensure federal building compliance with these energy conservation requirements.

Approved legislation also requires action on the energy efficiency of appliances. EPCA authorized the development and prescription of energy-efficiency improvement targets and test procedures for major appliances, and NEA requires mandatory minimum energy-efficiency standards for appliances.

The NEA also calls for additional programs to reduce energy consumption in residential and commercial buildings by a combination of methods, including performance standards, economic incentives, and implementation assistance. Federal activities may assist the private sector in overcoming institutional barriers by improving use efficiency in the areas of new community forms, land-use planning, on-site total energy, and integrated utility systems. DOE will provide a focus for the development of materials, processes, methods, and equipment to achieve this reduction in energy consumption. The federal government can provide leadership, by example, by making its own facilities as energy-efficient as possible.

Public Law 95-238, the Department of Energy Act of 1978—Civilian Applications, authorizes federal assistance in the form of loan guarantees for construction of demonstration facilities for the conversion of domestic coal, oil shale, biomass, industrial, or municipal wastes and other resources into energy. It also authorizes the establishment of grant, contract, price support, or cooperative agreement programs for the development of municipal waste reprocessing demonstration facilities.

## Programs and Milestones

DOE energy conservation objectives and the strategies for attaining them have been organized into a program comprised of four basic "thrust areas." The thrusts, while addressing separate component areas in the buildings and community systems sector, complement each other. The four thrust areas are buildings, community systems, consumer products, and federal programs.

The following sections address major DOE programmatic involvement.

Program Liaison and Support. The goal of program liaison and support (PLS), as a staff support function of DOE, is to conduct ongoing, two-way communications between DOE and all users of energy in residential and commercial buildings. PLS had three basic functions: (1) to act as an "information source" regarding the RD&D and legislative activities in the area of residential and commercial buildings, (2) to provide technology applications and transfer support to residential and commercial building conservation programs, and (3) to develop, test, and evaluate marketing communication activities directed toward specific target audiences.

The primary responsibilities of PLS were:

- To develop and disseminate technical and nontechnical printed material
- To conduct consumer-oriented pilot education programs
- To develop and display exhibit material
- To develop, test, and evaluate educational and training curriculum and materials
- To conduct workshops and seminars
- To develop and disseminate audiovisual presentations
- To provide technical assistance
- To develop and implement training programs

The DOE target audiences involve a diverse array of energy users, such as builders, architects, engineers, trade and service personnel, state and local decisionmaking officials, code officials, utilities, building owners and operators, educational and research institutions, real estate and financial communities, and residential consumers. Because of the diverse nature of these target audiences, their various levels of sophistication, their varying energy consumption needs, and the presence of institutional barriers that inhibit acceptance of energy savings practices, it is necessary to address them on an individual basis, by target group, in order to be effective.

A marketing communications system approach was developed in FY 1977, whereby each stage of an RD&D project is analyzed, the user audience is identified, and the barriers that inhibit user acceptance are evaluated. Based on this information, a communication plan was formulated that specifies communication packages to be developed and communication channels to be used for project and program information dissemination on a timely basis. Efforts in FY 1979–1980 continued to evaluate the DOE project and program areas and develop corresponding technology transfer plans.

Specific communications methods that are used include technical reports, conference reports, how-to user's manuals, fact sheets on state-of-the-art technologies, demonstration projects and legislative programs, exhibit material for trade shows, home-weatherization exhibit material for display at the local level, educational curriculum and material for specific educational levels, and workshops and seminars. Basic communication channels are government distribution centers such as the Technical Information Center; National Technical Information Service; the Government Printing Office; DOE's 10 regional offices; the EES; the DOE Office of Public Affairs, DOE and non-DOE-sponsored workshops, conferences, and panel discussions; and other federal agencies such as Department of Housing and Urban Development, Department of Agriculture, etc. National trade associations, trade journals, professional societies, and public- and consumer-oriented interest groups actively seek information on DOE programs for dissemination to their respective constituents.

During 1980, DOE developed and disseminated a user's manual for energy management in college facilities. Factsheets were prepared and disseminated. The factsheets provided information on the state-of-the-art technologies, including demonstration programs and legislated programs for wood-burning systems, model codes, residential conservation service, appliance efficiency labels, and consumer education programs.

Also during FY 1979–1980, the PLS participated in the major building-industry trade shows by providing exhibit material with accompanying literature on major DOE RD&D and legislative activities. It also developed home weatherization exhibit material for display at shopping centers, local energy fairs, libraries, etc., on a national basis operating out of the 10 federal regions.

The major thrusts in the 1980s will include the following:

- Develop, test, and evaluate training programs directed toward the trade industries, and educational programs that will complement legislative activities (such as the residential conservation service, installers/maintainers of HVAC equipment, etc.)
- Continue to update and identify new areas requiring attention in terms of consumer education in the form of printed material, exhibits, and consumer outreach activities
- Provide an efficient communication network to keep the building industry apprised of the state-of-the-art technologies, systems and energy management practices (primarily through trade journals and associations, professional societies, and state and local governments)

In addition, the energy accomplishments of other countries can be used to improve our own technologies. The international technology transfer activity is an international cooperative effort to facilitate practical implementation of energy conservation RD&D programs. Many foreign experiences are applicable to United States programs in the following areas: buildings and community systems, heat pumps, energy cascading, materials, heat transfer, and industrial processes. The major milestones for PLS are presented in Figure 7-2.

Buildings. The objective of the Buildings Program is to promote energy-efficient buildings by advancing the technical understanding of energy phenomena in buildings and by converting energy consciousness to everyday practice through education, incentives, codes, and standards. RD&D activities are intended to be supplementary—to do work the private sector is not doing, or cannot do. When possible, however, projects that include cost-sharing with the private sector are included.

The Buildings Program areas are organized into three major functional units: Architectural

and Engineering Systems, Buildings Codes and Standards, and Building Applications and Incentives. Each of these works closely with the appropriate groups from the building industry—technical and professional societies, building code and standard organizations, national associations, product manufacturers, other government agencies, educational institutions, and consumer groups. The major milestones for the Buildings Program are shown in Figure 7-3.

Programs in the area of Architectural and Engineering Systems provide the scientific and technical basis upon which to build various outreach activities. Buildings are considered as integrated systems of complex architectural and engineering elements. This holistic approach focuses on the interactions of systems, subsystems, and components of buildings. The various programs, though de-

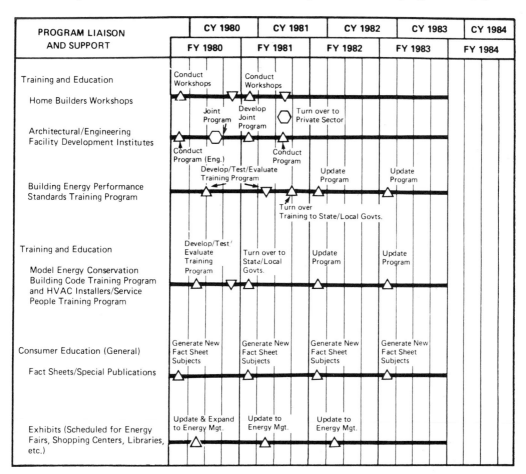

Figure 7-2. Major milestones for program liaison and support.

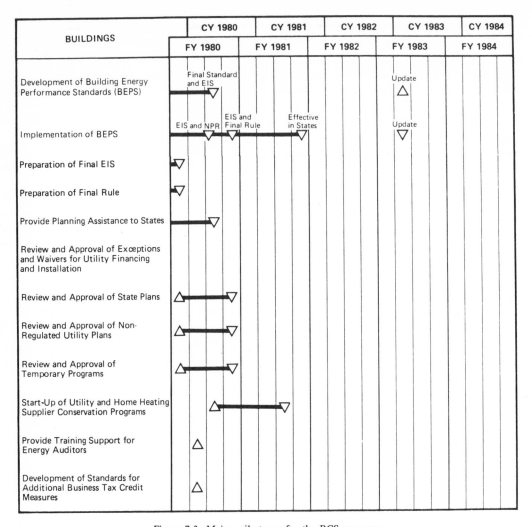

| BUILDINGS | CY 1980 | CY 1981 | CY 1982 | CY 1983 | CY 1984 |
|---|---|---|---|---|---|
| | FY 1980 | FY 1981 | FY 1982 | FY 1983 | FY 1984 |
| Development of Building Energy Performance Standards (BEPS) | Final Standard and EIS ▽ | | | Update △ | |
| Implementation of BEPS | EIS and NPR ▽ | EIS and Final Rule ▽ | Effective in States ▽ | Update ▽ | |
| Preparation of Final EIS | ▽ | | | | |
| Preparation of Final Rule | ▽ | | | | |
| Provide Planning Assistance to States | ——▽ | | | | |
| Review and Approval of Exceptions and Waivers for Utility Financing and Installation | | | | | |
| Review and Approval of State Plans | △———▽ | | | | |
| Review and Approval of Non-Regulated Utility Plans | △———▽ | | | | |
| Review and Approval of Temporary Programs | △———▽ | | | | |
| Start-Up of Utility and Home Heating Supplier Conservation Programs | △———▽ | | | | |
| Provide Training Support for Energy Auditors | △ | | | | |
| Development of Standards for Additional Business Tax Credit Measures | △ | | | | |

Figure 7-3. Major milestones for the BCS programs.

scribed as topical areas such as ventilation, infiltration, envelope systems, windows, daylighting, diagnostics, HVAC control systems, heat transfer, and energy analysis calculations, are designed to promote new understandings of these technical disciplines and their complex interrelationships.

Demonstration projects apply many of the RD&D findings in a coherent manner as well as begin the process of translating them into a viable part of the building process. In addition, in demonstration projects, technical knowledge is applied to building-type situations such as the plethora of residential and commercial building types in the United States.

Programs in the Building Codes and Stand-

ards branch translated results from the applied science and technical activities into code language and standard practices for the building community. The lack of uniform codes and standards in state and local building regulations has been a major stumbling block to the development and implementation of energy-efficient design and practices.

The major assignment to date was the development of Building Energy Performance Standards. This program was required by Title III of EPCA (P.L. 94-385), as amended by Section 304 of the DOE Organization Act (P.L. 95-91). The Reagan administration has killed the building energy performance standards program because of the additional regulatory burdens suggested.

The Residential Conservation Service seeks to reduce the energy consumption of single- to four-family residences by requiring larger electric and gas utilities* to provide information to their residential customers on energy conservation and renewable resources. Authorized by the National Energy Conservation Policy Act (NECPA) (P.L. 95-619), the Residential Conservation Service also requires covered utilities to offer energy audits, arrangements for installation and financing of measures, and the repayment of loans for purchase of measures through utility billings.

The Residential Conservation Service subprogram is to be implemented by the states, which may submit residential energy conservation plans to DOE for approval. Nonregulated utilities not included in a state plan and the Tennessee Valley Authority must submit their own plans to DOE. A state or utility may propose a temporary program, for a period of up to 3 years, which differs from the program required by the proposed rule.

The Residential Conservation Service also includes material and installation standards where necessary for safety and effectiveness; requirements for state-approved lists of suppliers, contractors, and lenders; procedures to protect consumers and provide redress; and a 3-year warranty for all products sold as a result of the program.

The residential and business energy tax credits are authorized by the Energy Tax Act, which provides tax incentives to individuals and businesses for energy-conserving equipment and devices. The energy tax credits are available through the 1985 tax year. DOE is providing assistance and advice to the Department of the Treasury and the Internal Revenue Service in the development and implementation of the tax credits.

The Residential Conservation Service subprogram and residential tax credits are projected to achieve a reduction of 0.6 quad of residential energy consumption by 1985. The business tax credit is projected to reduce industrial and commercial consumption by 0.3 quad by 1985.

*Sales of 10 billion cubic feet of gas or 750 million kilowatt-hours of electricity.

*GSA energy conservation demonstration building, Manchester, New Hampshire.* The Norris Cotton Federal Building, a new federal facility in Manchester, New Hampshire, is a "living laboratory" for testing and evaluating methods to conserve energy in office buildings. It was the first federal building designed with a high priority given to energy conservation. The building includes both recognized and innovative energy conserving technologies.

These features are being carefully evaluated and monitored for DOE by the National Bureau of Standards. Data collection and analysis began shortly after the building was occupied.

Six major areas are being evaluated:

- Energy—heating and cooling
- Solar system
- Interior lighting
- User acceptance
- Window studies
- Economic analysis

*Innovative structures program.* The purpose of the Innovative Structures Program is to investigate the economic feasibility of specific energy-conserving innovative buildings, to explore the likelihood that these new building concepts may receive public acceptance, and to determine whether significant institutional or regulatory barriers exist that might impede the adoption of those concepts that are found to be significantly energy-conserving. If appropriate, recommendations will be made and actions taken to speed up the penetration of energy-conserving designs into the marketplace.

*Energy performance standards for the design of new buildings.* In 1976, Congress passed Title III of P.L. 94-385, the Energy Conservation Standards for New Buildings Act. This law, as modified by the Department of Energy Organization Act (P.L. 95-91), directed DOE to develop energy performance standards for the design of new buildings. The Department of Housing and Urban Development was responsible for the program to implement these standards. The program was cancelled by the Reagan administration.

The standards would have taken the form of "energy budgets," which would consider buildings as complete energy-using systems. Their objective was to reduce the designed energy consumption of entire buildings without specifying individual building components. The program called for the rapid development of trial energy design standards, to be available for testing and review before 1980—the deadline for promulgation of proposed national standards.

*Model Training Program for states and localities.* The Model Training Program for states and localities is designed to reach and enlighten decisionmakers on principles of energy conservation, professionals on the technical aspects and enforcement provisions of energy conservation standards, and enforcement personnel on requirements of the energy conservation standards and code provisions that are to be applied.

*Minimum Energy Dwelling II.* The Southern California Gas Co. built and evaluated a Minimum Energy Dwelling II (MED II) to demonstrate a number of technological innovations that can satisfy the energy requirements of a home. The MED II project is an extension of the successful MED I project.

Some of the features selected in the MED II design are the insulation package, insulation doors, solar domestic hot-water heater, double-glazed windows, and architectural overhangs. An anticipated 500 homes using MED II as a prototype are being built at Mission Viejo.

The goal of this project is to reduce energy use in a typical southern California home by at least 50 percent and encourage the building industry to adopt MED's energy-saving features.

*Design of residential buildings using material thermal storage.* The project objectives are (1) to provide technical data in support of an effort to make natural thermal storage in the building structure a feasible design option that can minimize the annual cost and consumption of energy in space conditioning of residential and small commercial buildings,

and (2) to develop market-competitive residential building design concepts that use natural thermal storage in the building structure for the reduction and load-leveling of their short-term energy consumption, in order to reduce operating and demand costs.

The building(s) were built, instrumented, tested, and evaluated under existing field conditions. It is expected that the demonstration building will result in the voluntary use of new construction design guidelines by professionals and builders and will be demanded by consumers. This effort is directed at rapid implementation on a regional basis.

*Building envelopes and ventilation.* The goal of the building envelopes and ventilation program is the development of improved design practice for better energy use, safe health and comfort levels for all buildings.

The energy consumption of buildings is determined by a complicated interaction between the weather, the building envelope (the walls, windows, roof, and floor), the climate control system, the building's occupancy, and its end use. All of these elements are being addressed, both individually in laboratory studies and collectively in demonstrations.

*Energy use in office buildings.* Existing office buildings represent a major target for energy conservation measures in both operation procedures and retrofitting. In New York City, which has more office space than any other United States city, office buildings use 55 percent of Consolidated Edison's electrical energy output. A detailed examination was made of the important parameters of 900 office buildings in New York City to obtain energy use data and identify typical building types for further analysis. Case histories of various office buildings were disseminated nationwide to designers, owners, contractors, and operators to stimulate adoption of successful energy-saving building management.

*Energy use in building construction.* In order to determine energy use in the building industry for production of materials and to produce a catalog and a desktop manual of energy units required for each building type, energy consumption in the production, handling, and transportation of building materials to the construction site was traced. The energy units required for building types were prepared in catalog form for use by designers in determining optimum energy efficiency. The catalog was validated by use on selected buildings, and the information obtained will be disseminated to the building design and construction professions. A single, desktop reference is now being prepared.

*Federal laboratory collaborations for the development of energy analysis computer codes.* The objective of the project is to develop an easy-to-use, fast-running, completely documented, public-domain computer program that will ultimately be established as a national standard analytical tool for building energy-design trade-off studies. In addition, the computer program will be used for energy budget calculations for all classifications of buildings proposed in local, state, or national design standards.

*Energy conserving restaurant demonstration.* The DOE conducted an energy-conservation demonstration project at the Jolly Tiger restaurant in Colonie, New York. The restaurant's heating, ventilating, air-conditioning, and food-furnishing operations were designed to include a complete system of energy conservation and reclamation equipment. A set of monitoring instruments provided data on energy use in the restaurant.

This program yielded information that is useful in the design of energy-saving systems for restaurants. The primary objective of this project was to provide the restaurant industry with an assessment of a variety of energy-saving devices and systems. The assessment included an economic analysis as well as an energy-conservation impact analysis. It also provided field data for design decisions by engineers in any restaurant by segregating the effects of individual energy-saving hardware as well as prioritizing "packaged" energy-saving systems.

*Window analysis.* The window analysis project is developing a set of analytical and experimental tools to assess window performance. These include computer codes to calculate optical performance of multilayer thin-film coatings, heat-transfer rates of complex window systems, and energy consumption of buildings with novel window systems. Laboratory and field test equipment is also being developed, as necessary, to measure the performance of managed window systems.

Design tools are being developed to allow optimization of window characteristics and size as a function of combinations of glazing orientations, building envelope and internal characteristics, use patterns, comfort criteria, and climate. The project will determine how the most important variables influence optimal size and how to minimize energy consumption through cost-effective optimal window design. Finally, a comprehensive manual on optimal window design, and guidelines that will be of practical use to legislators, code officials, and architects, will be produced.

*Daylighting/sunlighting program.* Program planning, research, demonstration, and educational activities are in progress to accelerate widespread daylighting use. Several concepts for using reflected direct sunlight deep inside buildings are being evaluated using computer analysis, scale-model rooms, and full-size prototypes. Practical efforts to expand daylighting use will be encouraged by developing (1) daylighting availability data, (2) widely accepted simplified design methods, (3) methods to assess annual energy savings, and (4) predictability in the integration of conventional lighting-system hardware with daylighting practice.

*Office building energy management demonstration, Newark, New Jersey.* A new 26-story office building in Newark, New Jersey, was designed in accordance with ASHRAE Standard 90-75. It was built with a substantial investment in basic instrumentation for energy management. The building was the basis of a study to determine whether existing design and operational standards and premises, upon which decisions and regulations have been or will be based, are effective in producing energy conservation.

Community systems. Between resource extraction and the final end use of energy in the built environment, numerous opportunities exist for conserving energy through improved design and management of communities, improved overall system efficiency of existing utilities, and the use of energy sources within a community such as municipal and thermal wastes. Many barriers exist, however, to achieving energy conservation in communities. The mission of the community systems division is to foster the implementation of energy conservation measures and perform supportive RD&D activities in the following areas:

- Managing, planning for, and delivering energy in communities
- Using municipal and renewable resources for energy supply and conservation
- Conserving energy in community functions
- Fostering the adoption of energy conservation by the public and private sectors

In order to understand the direction and functioning of the Community Systems program, it is necessary to understand that the word *community* has a place of its own in the energy field. The essence of its meaning lies in the concept of linkage—a group tied together by some type of network. In applying the term to energy systems, *community* can be defined as a complex of buildings serving various human activities and connected by transportation, energy, and communications networks. The community's functions can be residential, commercial, industrial, or agricultural—or a combination of these. The various functions that characterize a community underscore a key consideration in any community-oriented energy policy: the requirement of energy systems capable of responding to a variety of needs and circumstances.

Communities may vary first of all in their demographic size, ranging from as little as several hundred persons to several million. The inhabitants of the community may be densely concentrated in a small geographic area, or they may be spread sparsely over a large urban or suburban area. The economy of the community may be grounded in a few key sectors, or it may be diversified and broadly based. The geographic and climatic regions are also critically important, both in terms of particular energy demands and of the alternative sources of energy that may be used to alleviate the demand for conventional fuels.

Their energy demands will also be linked to the community's level of development. Clearly, a boomtown will require a different energy system than an older area whose economic development is stable or declining. The degree of governmental development and structure is also substantially different from place to place. It is important whether the community already has the authority and the means to embark upon a serious program of energy planning, or whether this authority must be shaped from an existing division of municipal responsibilities that currently do not address coordinated energy planning.

Another factor defining community, which is critical to the implementation of the program, is the overlying political structure of the community. Once it is recognized that local government will be a major factor in achieving energy conservation at the community level, the mission statement can be expanded upon and restated.

The Carter administration developed a policy that recognized and took full advantage of the capacity of communities to contribute to a national energy program. The mission of the Community Systems Division is to assist in the mobilization of the maximum possible participation of communities in the formulation and execution of specific energy-related policies. This mission has four components:

- It will develop an effective channel of communications that will facilitate the two-way flow of information concerning national energy policy. This channel will help to develop the broadest possible consensus upon which the appropriate steps the nation as a whole, as well as the individual communities, can take to achieve a reduced consumption of nonrenewable energy supplies.

- It will assist in the creation of the necessary institutional capacity at the local government level to permit communities to plan and manage their energy futures. The division will develop planning and management methodologies and, where appropriate and feasible, assist communities in implementing them.
- It will assist in the identification, development, and application of feasible technologies in order to reduce the need for imported petroleum. These technologies will include those particularly suited to a community perspective, such as district heating, as well as those whose commercialization a community can markedly advance, e.g., municipal waste to energy.
- It will provide a focus for federal community assistance programs within DOE and other federal departments. Efforts aimed at aiding communities and individuals in such matters as economic development and employment security are already under way. The program mission is to integrate the energy perspective so that the broad national goals can be met.

Historically, energy services have been provided by single-service utilities, each providing one or more energy forms such as electricity, natural gas, oil, coal, etc. In many cases, these organizations are competing with each other for the provision of services to consumers. Individuals, land developers, and local governments, which may play a role in the development of energy-efficient communities, have little input to the planning for energy services. As a result, there is an excessive level of energy consumption and waste in existing communities. Therefore, the program strategy includes the development of mechanisms to encourage cooperation and joint involvement among development process participants in efforts to consider and implement integrated community energy-efficient options.

Four basic features of the strategy are diffused throughout activities of the community systems program: (1) consistent utilization and involvement of those community members who will initiate, approve, and implement the option; (2) recognition of the limitations of communities to institute change and to undertake high-risk activities; (3) the general need to balance community benefits and service with prudent practices; and (4) leveraging the federal resources. The capacity of communities to address energy issues will be enhanced by the interaction of their members (local governments and public organizations, private organizations and sectors) in demonstration projects and by information dissemination, technology transfer, etc. In addition, all activities address the impact on the economic and regulatory environment, both from the particular implication in the community and the overall national energy situation.

In order to show, by example, ways to equitably distribute benefits, costs, and risks among the vested interest groups of the community, a two-way flow of results, findings, analyses, etc., has been established with the financial community; federal, state, and local legislators; regulatory bodies at all levels; and representatives of these and other segments of the community. Areas developed, evaluated, and tested will include loans, price supports, loan guarantees, seed, and matching grants as appropriate.

Energy is only part of the overall operation, management, and development of the community. Energy delivery aspects on a community level are interwoven with aspects of housing, employment, transportation, environment, commerce, etc.

The effort relies heavily on teams of local community leaders and energy participants such as local governments, utility companies, chambers of commerce, financial and banking institutions, developers, equipment and system suppliers, and other related forces affecting overall community operation, management, and development.

Major milestones for the Community Systems Program are shown in Figure 7-4. Selected projects are described following the descriptions of the Community Systems Program.

The mission of the Community Service Branch is to mobilize communities to promote

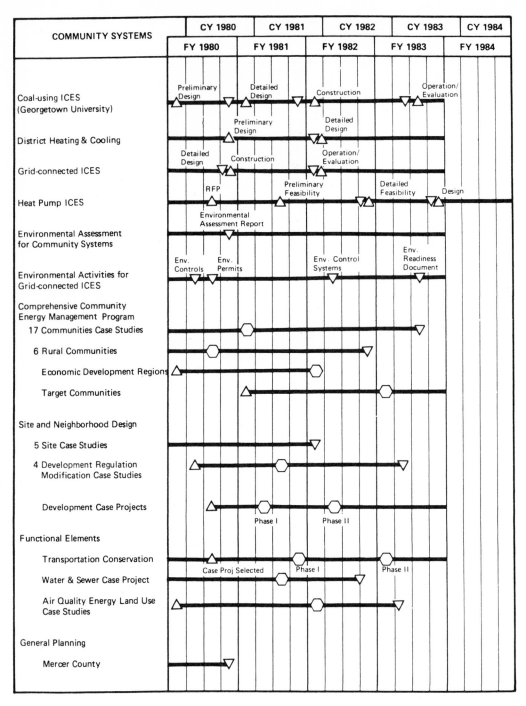

Figure 7-4. Major milestones for Community Systems Program.

and apply proven energy conservation techniques, and to use renewable resources in every sector of community life. Its goal is twofold: (1) to assist a minimum of 100 select communities over the next 3 years, to acceler-ate the introduction of energy programs, and to enhance the communities' ability to implement them; and (2) to develop energy information networks among local officials as well as among the various groups that make up

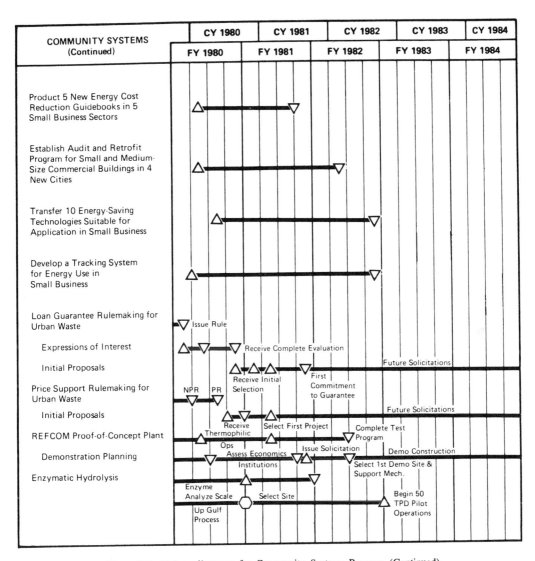

Figure 7-4. Major milestones for Community Systems Program (Continued)

communities, such as financial institutions, small businesses, industries, and utilities, and to provide a communications link between local officials and interests and federal officials responsible for carrying out energy programs directed toward assisting communities.

A strategy for the Community Systems Program has been formulated as follows:

*Provide direct assistance to communities.*

- Select the most promising energy-conservation and renewable-resource measures for inclusion in the overall community systems sales kit.

- Develop sales materials on the selected measures and on financial mechanisms directed specifically at community leaders.

- Make available to select communities trained resource teams to help them identify opportunities to implement energy measures, giving priority to those communities that are proven innovators and that have the ability to share their experiences.

- Assist targeted communities in selecting appropriate assistance packages for their specific needs by providing guidance on alternative approaches to meeting their needs.

- Help communities to resolve institutional constraints posed by federal, state, and local entities by working with federal and state agencies to develop an understanding of the overall benefits accruing to the nation from energy programs, and by advising communities on approaches to resolving local constraints.

*Identify potential financial sources and mechanisms for helping communities put energy programs in place.*

- Develop and administer cooperative arrangements with other federal agencies sponsoring community-related grant and loan programs, i.e., Department of Housing and Urban Development, Economic Development Administration, Small Business Administration, Department of Agriculture, Environmental Protection Agency, and Department of Commerce.
- Develop financial incentive programs, including drafting of the appropriate legislation, to help local communities fund energy conservation programs or use renewable resources.
- Help local jurisdictions establish links to private-sector entities such as financial institutions and utilities.

*Develop mechanisms for input by community leaders to the division's agenda for RDD&I programs and for monitoring community concerns and initiatives.*

- Select a number of representative cities in which to operate a "listening post" hotline for local officials, utilities, developers, business and industry leaders, citizens groups, etc.
- Broaden networks and organizational structures through which two-way communications can be carried on with communities.

*Carry out technology transfer, information dissemination, and educational activities.*

- Disseminate technical information generated through RD&D activities of DOE.
- For general audiences as well as for use in schools and by citizen groups, develop and distribute materials (i.e., brochures, exhibits, films) that explain energy-saving potentials from conservation and renewable resource programs.
- Develop seminars, workshops, and conferences for local officials, state regulatory commissioners, the financial community, legislators, media, etc., through which information-sharing efforts can be tailored to meet the needs of the various interests.

The principal constraints to more efficient energy management by small business are lack of information about specific actions and the economic benefits of such actions. Therefore, aims of the small-business section are:

- To ease the impact of energy-related problems on small business by developing and delivering information and providing assistance to reduce energy costs.
- To reduce gross energy consumption and its rate of growth in the small-business sector by introducing management techniques designed to increase energy-use efficiency.

The Small Business Energy Cost Reduction Program is designed to provide practical, cost-effective measures to various sectors so they can reduce their energy costs and consumption. To date, 12 sectors have been analyzed. The selection of the 12 sectors was based primarily on three criteria: (1) energy intensiveness (energy used in relation to sales or operating costs), (2) the number of small businesses in the sector and their share of the market, and (3) the lack of existing energy information and studies for the sector.

In conjunction with trade associations, guidebooks were produced and distributed for the 12 sectors:

- Laundries and dry cleaners
- Apartments
- Automobile dealers
- Commercial printers
- Auto service and repair shops
- Gasoline service stations
- General retailing
- Florists (retail and wholesale)
- Furniture manufacturing
- Wholesale distributing
- Dairies
- Bakeries (wholesale and retail)

In addition, over 15,000 businesses have attended energy cost-reduction workshops or seminars sponsored by trade associations and the Small Business Administration. (DOE provides training for the instructors.) The goal of disseminating program results to approximately 2 million businesses will be met. Eighty percent of the state energy offices and Energy Extension Service pilot states are also participating in this program. Approximately $1.5 million has been leveraged from non-DOE sources to assist in this effort.

In conjunction with the Energy Extension Service, the Small Business Section has provided on-the-job energy audit training for state Energy Extension Service personnel. With the help of trade associations, the impact of the Small Business Energy Cost Reduction Program in the first seven sectors has been evaluated.

The Energy Partnership for American Cities Program, designed to involve local city officials, business groups, and small businesses in energy management, was conducted in six cities: Columbus, Cincinnati, Miami (Dade County), Newark (in conjunction with the state energy office, Tampa, and Louisville (in conjunction with the state energy office). During FY 1979 and early FY 1980, an evaluation of the ongoing program as developed in Resource Applications was conducted, and a specific conservation and solar energy program was prepared. Under this program, a

realistic approach to disseminating energy conservation information to small businesses is being used.

During 1982 through 1984, the Small Business Energy Cost Reduction Program will be expanded to cover an average of five new commercial sectors each year, which represent a significant number of small businesses where the potential for energy savings is measurable. This program will be phased out in 1985.

The Energy Partnership for American Cities Program, in cooperation with the National Science Foundation, will pursue the objective of bringing together states, local governments, and business to reduce energy consumption in small- and medium-size commercial buildings through an audit/retrofit process. The Energy Partnership for American Cities Program, which began in 1980, will continue through 1985. By then, it will have reached a total of 30 new cities, averaging at a minimum of five new cities each year. This program will continue into 1990.

The Small Business New Technology Application Program was designed to transfer and assist in the adoption of 40 new energy conservation technologies suitable for application in small businesses. An annual average transfer of eight new energy-conserving technologies, which began in 1980, is projected to continue through 1984.

The Tracking System for Energy Use in Small Business Program will develop the necessary information data and methodologies needed by the small-business person to evaluate the building energy costs (heating, cooling, lighting, etc.) based on climate; seasonality of business; type of business; building, construction, equipment, and fuel used; and other factors.

In cooperation with the Small Business Administration, a prospective or currently operating small-business manager will be provided with "norms" indicating in percentages the energy cost for heating, cooling, lighting, etc., that is "normal" for a specific, successfully operating small business. Beginning with the feasibility study completed in 1980, the program should progress through the planning phase into full implementation in 1983. Energy norms will be published in January 1984. This program will be phased out in 1985.

State and local officials; utility, industry, and business representatives; technical and professional persons; as well as members of the financial community must be kept apprised of R&D activities in community systems technology if they are to direct the efficient use of energy and energy resources. In order to meet the needs of these diverse audiences, the Marketing/Finance Section is developing a network of communications channels through which it can reach these targeted groups with information and assistance.

In FY 1979, a 'hot line' was initiated so that local officials could receive information on energy conservation as part of the president's energy conservation initiatives. This initial service for local officials was provided in cooperation with the four organizations of local officials. In cooperation with the DOE Office of Intergovernmental Relations, a computer information system was developed that would permit local officials ready access to pertinent information developed by DOE. The questions flowing from state and local governments to the system provide the basis for Community Systems Development and other offices and divisions to identify additional information, systems, and technologies needed by local officials. Community Systems Development hopes to make this broadened system operational by FY 1981. It would be an ongoing activity.

Guidance is also being provided to local officials on developing local energy-action councils as President Carter challenged communities to do in his July 1979 speech. This activity will continue through FY 1984.

Based on the experience of the Local Official's Information Service, the hot line will be fully developed for select local leaders whose participation in community-related energy programs are crucial to their success. Directed to utility, industry, and business representatives, members of the financial community, and technical and professional persons, the hot line will also serve as a "listening post" for

identifying concerns, as well as innovative approaches being developed in communities. This service became operational in late FY 1981.

Beginning in FY 1981, the division provided to a limited number of communities (not more than 20 in the initial phase of the program) trained resource teams to help local officials and concerned citizen leaders determine priority needs and identify opportunities to implement energy conservation measures and to use renewable resources on a communitywide basis. Priority is given to communities that are proven innovators and that are accustomed to sharing their experience through information networks. This technical assistance effort will be broadened in FY 1982 to encompass 100 communities.

Technology clearinghouses will be established either at the national laboratories managing the R&D programs or through technical and professional societies beginning in FY 1981. Each clearinghouse is responsible for responding to requests on community energy conservation systems. A program of information/technology transfer directed at the technical audience such as engineers, planners, developers, consultants, plant operators, designers, and equipment manufacturers ensures better utilization of research advances and technological efforts. These groups can best be reached through technical reports, articles in professional journals, seminars, workshops, conferences, newsletters, and exhibits, etc. This section expanded its outreach efforts in this area in FY 1980 and 1981, and will continue to enhance them in FY 1982.

The technical materials were developed to explain and describe the various programs disseminated through the various federal government distribution channels such as the Technical Information Center, the National Technical Information Service, and the Energy Extension Service, as well as through programs of other federal agencies such as the Innovative Network of the National Science Foundation.

In the past, the training, information, and educational activities of the Marketing/Finance Section have been directed largely at a general, undefined audience. Although general-audience factsheets and brochures continue to be distributed, programs initiated in FY 1981 were focused on a more selective audience, namely, local leaders from community interests who are likely to play key roles in the community decisionmaking process.

A series of seminars and workshops for local officials began in FY 1980 to acquaint them with the state of the art and the potential from various technologies in community energy conservation and delivery systems. Similarly, key state legislators were invited to participate in town meetings and workshops to help develop state legislative agendas on energy matters. The initial meeting of legislators was held in cooperation with the Office of Intergovernmental Relations.

A special effort was made to familiarize city hall and statehouse reporters with the potential for energy savings and efficiencies through integrated community energy systems and the conversion of waste to energy. Since FY 1981, seminars have been held for reporters from select cities where various types of community systems have been shown to have a potential for success.

Study courses with accompanying program guides are being developed for citizen groups such as the League of Women Voters for use by its local affiliates in studying the feasibility of the various community energy-conservation techniques within their communities. This program will be in place by FY 1982. Similarly, presentations will be developed for the commissioners and staff of the state utility regulatory agencies to be presented through workshops and seminars.

The Marketing/Finance Section is responsible for coordinating review of requests from the Community Systems Program offices to the Urban Development Action Grant program of the Department of Housing and Urban Development. Criteria and guidelines have been developed for evaluating requests for grants from energy-related businesses wishing to expand their production and work force. All grants are funneled through local governments. Two specific highlights are as follows:

- The Community Energy Grants Program assists small manufacturing or service organizations that are located, or desire to relocate, in distressed areas and that require $200,000 or less for capital improvements, which would promote economic renewal and job creation.

- The Large Scale Commercialization Grants Program is carried out on a case-by-case basis, with Community Systems Development responsible for recommending to HUD technologies judged to be ready for commercialization.

The Marketing/Finance Section is promoting requests by cities for small manufacturing or service organizations located, or desiring to relocate, in distressed areas and requiring $200,000 or less for capital improvements that would promote economic renewal and job creation. However, all requests to the Department of Housing and Urban Development are considered on a case-by-case basis, with the section reviewing the proposal from the point of view of readiness for commercialization as requested by the department.

The missions of the Community Management Systems Branch are as follows:

- To develop the tools that will allow for institutional, managerial, and technical capacity in communities to deal effectively with energy problems through community energy management.
- To facilitate the incorporation of energy conservation in the design and physical development of communities by developing methods, processes, etc.
- To conduct research as required for national energy initiatives related to communities.

The strategy is to simultaneously develop and demonstrate energy management and development tools to be used by communities. The effort of the branch consists of two program elements: Community Energy Management and Planning and Development. Each element consists of several activities. For each activity, the program manager implements the full cycle of tasks required to make innovations ready for the community marketplace. The following tasks are involved in this process:

- A state-of-the-art review to identify related research and define current research requirements
- Concepts development
- Development of methodologies and analytical tools for defining planning and design approaches and assessing the energy conservation potentials of various options
- Identification of implementation barriers (organizational, financial, regulatory, managerial) and formulation of means to overcome them
- Site-specific case studies to prepare energy-efficient plans and to apply and test methodologies, analytical tools, and demonstrate the implementation of energy-conserving developments and communities
- Monitoring and evaluation of case-study projects and preparation of detailed market assessments to identify specific opportunities to implement energy-efficient options and techniques

Recent fuel shortages and fluctuations in the price of energy promoted the need for local government involvement in energy-conscious planning and management. Efforts of those who have ventured into this field have been constrained, however, by institutional barriers, inadequate reserves of trained technical personnel, a lack of adequately demonstrated methodologies and tools, and limited financial resources. In order to overcome these barriers to community energy conservation efforts, the Comprehensive Community Energy Management Program has been initiated to:

- Develop and test the role of local governments in coordinating energy management programs
- Develop and demonstrate energy management program planning and implementation
- Recommend appropriate DOE policies and federal legislative initiatives to promote energy management at the local level

The program was initiated with a grant to the city of Clarksburg, West Virginia, in 1975. Experience gained to date from this project and from the energy-planning work of other cities has been used to develop two important comprehensive community energy management methodologies. One is tailored specifically for use in small communities. The second methodology is for use in a long-term comprehensive community energy management and planning effort. The methodologies will be updated annually, reflecting the results of the communities' demonstrations.

Studies of implementation mechanisms have been initiated to provide data and develop strategies that will accelerate the acceptance and adoption of community energy systems, energy-conserving community design, and conservation programs. The following barriers must be overcome:

- Potentially higher first costs of energy-saving systems
- Possible institutional conflicts caused by the involvement of more than one utility or community in an area-wide energy conservation project
- Potential inability of a community management organization to cope with new technologies and planning processes

Major projects focus on financial instruments, regulatory arrangements, and assessments of federal programs. Of interest are financial instruments that minimize the capital costs of community energy systems. These studies investigate financing arrangements applicable to community systems and emphasize such topics as contractual arrangements, measures of investment risk, and the determination of appropriate financing instruments for different financial environments. Also being examined are ways to secure government assistance under present loan guarantee and grant programs, the use of project-financing techniques, and the mechanics of private sector–public sector joint ventures. Economic, political, and regulatory arrangements associated with the use of different cost allocation methods in community system applications are analyzed to develop guidelines for setting rates for the joint products of a community system—for example, electrical heating, cooling, and waste-disposal services. A standard prospectus framework is being developed that can be used to forecast life-cycle financial conditions for a community energy system. The legal, regulatory, and financial activities are being investigated and used to develop guidelines to assist communities in implementing energy management programs. Cataloging and assessing DOE and other federal programs affecting community energy management programs was initiated in 1980 and will be updated annually to 1983.

Beginning in October 1975, a total of 17 communities varying in size, governmental form, economic base, geographical region, climate, and energy supply were selected. These were selected to test and demonstrate the preparation of energy management programs with the following phases and tasks:

- *Organization phase:* establish advisory committees, assemble technical team, establish lines of communications and coordination, prepare project work plan.
- *Action plan preparation phase:* energy audit—inventory present and projected energy, supplies and demands, formulate energy conservation objectives, generate and evaluate energy conservation alternatives, prepare Comprehensive Community Energy Management Program (CCEMP) action plan.
- *Adoption and implementation phase:* implement the action plan. The communities completed preparation of programs in early 1981. Selected communities received funds to proceed with implementation of programs during 1981 to 1984. New demonstrations were initiated in 1980–1981 to develop and test a second generation of organization models and analytical tools. The time frame for preparing plans and initiating implementation was shortened by the experience gained, and the support of

these demonstrations will be completed by 1985. Monitoring and evaluation will proceed beyond 1985.

Six communities were selected under a joint agreement with the Department of Agriculture (in FY 1979) to develop energy management plans for rural communities. Additional communities were selected in 1980. At the conclusion of these case studies, it is anticipated that energy management will become a standard practice by all communities receiving funds from the Department of Agriculture for rural development planning.

Several Regional Economic Administration (EDA) districts were selected under a joint program with EDA. The demonstration spanned over 4 years, starting in FY 1979, and resulted in EDA requiring energy management as a standard feature of rural development districts.

The total number of CCEMP demonstrations is limited to 25 to 30 communities at any given time. A special effort was made to provide technical assistance to encourage communities to implement programs. Solicitations of support by other DOE and federal agencies to participate in the implementation were negotiated.

The Academy of Contemporary Problems, a research arm of the major public-interest groups, was selected to monitor and evaluate the demonstration projects. They will provide continuous assessment of the issues, failures, and successes of the demonstration communities. In addition, nationwide monitoring and evaluation of innovative communities in energy management has been initiated in close coordination with public interest groups and professional associations.

Based on the evaluation, a market assessment will be initiated in which the market penetration potential of systems engineering concepts and community planning approaches will be matched with communities characteristics. Models and data describing the economic competitiveness for energy supply options will be used to make forecasts.

The assessment process also identifies significant institutional barriers to, and environmental impacts of, community energy systems. Strategies and plans are being developed for the implementation of community energy systems that are compatible with current utility planning, operating, and regulatory practices in various market sectors.

Results are evaluated and information is developed on a continuous basis from the methodologies, demonstration, and monitoring and evaluation projects so that example reports can be made available to communities on how to organize for the preparation of a CCEMP, how to prepare a community energy audit, and how to evaluate and select among alternative energy programs. Special reports on successful programs of communities will be periodically prepared, and a loose-leaf manual on implementation mechanisms will be issued. Most of these reports will be updated as appropriate. The program will produce a bimonthly newsletter providing information on significant achievements. Demonstration communities are continually encouraged to participate in professional conferences and special workshops, and share information with other interested communities. All information will be disseminated by the Marketing and Finance Branch.

A community's energy consumption is determined not only by its energy supply system, but also by its location, physical structure, and functional activities. These community characteristics are affected by local growth policies, development regulations, and methods of managing community services. The objective of Planning and Development is to provide mechanisms to reduce the level and the growth rate of energy demand by incorporating conservation planning into community planning and development processes. Three types of planning on different physical and functional scales are considered: site and neighborhood; planning for functional community subsystems such as transportation and utility services; and long-range general development planning for broad patterns of land use and construction.

The objectives of the site design activity are as follows:

- To facilitate the involvement of private developers in the demonstration of technological and nontechnological energy-conserving options in new developments

- To increase the number of energy-conserving site and neighborhood developments and to provide models of successful approaches and techniques that will assist other developers and communities in similar efforts

- To increase the number of communities that prepare and adopt local regulatory procedures and measures that enhance and encourage energy conservation in the design of new developments

- To increase the number of utilities participating in the construction of decentralized systems in support of energy-conserving developments

- To evaluate the effectiveness of regulatory and design techniques for facilitating and realizing energy conservation in actual developments

The program activity focuses on projects that are necessary to overcome impediments to preparation of designs and subsequent demonstration of energy-efficient development. Prominent constraints in this area include a relative lack of existing examples of energy-efficient developments, inhibiting factors, disincentives to the incorporation of energy-conserving options in the project design and municipal review process, and significant financial risks associated with the preparation of plans and the marketing of innovative developments.

A state-of-the-art review of passive options for energy conservation in site design and a survey of recently constructed energy-conserving developments have been completed. In addition, environmental and siting regulations that affect community energy systems were reviewed, and supply technologies are being classified according to their compatibility with these regulations. The laws that control service area franchises, establish costs of services, and regulate quality of service are examples of the federal, state, and local laws and regulations that affect the construction and operation of community systems being examined. Modifications will be recommended for laws that prevent or impede the establishment of community energy systems. These studies were initiated in 1978 and were completed in 1980.

Five case-study demonstrations were initiated in 1978 in which developers prepared energy-conserving designs in parallel with conventional plans. Evaluations of these studies were completed in 1979, and three of the projects are now being considered for proceeding with plan implementation. New demonstrations based on experience gained were initiated in 1980.

Communities were selected to modify development regulation, provide incentives to developers, and remove impediments in order to design and construct energy-conserving developments. A model energy-conserving ordinance was prepared as the community effort proceeds.

Information packages were developed for design professionals, developers, public officials, and citizen organizations. These presented the findings of the five case studies in 1980. The Urban Land Institute, developers, and public-interest groups assisted in the preparation and dissemination of the packages.

Under the general development and energy master planning activity, selected cities are used as laboratories to test and demonstrate planning in the community development process. Communities of different developmental characteristics (urban, rural, new development, redevelopment) have been selected as focal points for case studies to develop and test guidelines for incorporating energy considerations into general development planning. For example, incorporation of energy conservation considerations into a redevelopment plan for an existing city is demonstrated in Atlantic City, New Jersey. In Mercer County, North Dakota, an effort was made to resolve problems caused by rapid community growth prompted by energy resource development.

A general organizational approach for managing energy-related community development and for recovering energy that would normally be wasted from nearby generating facilities is being tested by 12 local governmental entities (the county, six municipalities, and five school districts), which have formed an

Energy Development Board. The project was completed in 1980, followed by a marketing and dissemination effort in coordination with the Department of Agriculture and EDA. The Alaska State Capital Planning project has demonstrated the preparation of an energy-efficient new community plan. Energy conservation options for the new community were identified and evaluated on the basis of energy savings and sociopolitical impacts. The plans were estimated to have an economic payback period of 10 years or less with the effect of reducing energy-conservation options. Pending state legislative approval, further efforts were conducted to monitor and incorporate candidate options into the state capital design. Concepts were developed on land-use forms that are complementary to alternative energy supply and delivery systems. This was followed by an effort, which began in late 1980, to demonstrate the preparation of community energy master plans that are integrated into the general development plans of communities. This effort is tied to communities' preparation of general development plans and will be completed in 1982.

The activity consists of projects designed to develop guidelines and procedures for incorporating energy conservation considerations into the development of community functional and service subsystems. Typically, these projects are undertaken under joint sponsorship or in close coordination with relevant federal agencies, which provide technical assistance grants and loans to selected localities for subsystem development efforts. The federal agencies have been mandated to ensure that local governments consider energy-conserving options in functional elements planning. Currently, cooperative efforts are being conducted in two areas:

- *Transportation*
  1. A pilot case study was initiated with the North Central Texas Council of Governments to identify means of incorporating energy conservation considerations into the urban transportation planning process. The study began in 1978 and was completed in 1981.

2. DOE has jointly sponsored a conference with the Department of Transportation to assess the state of the art of integrating energy conservation into transportation planning practices. The conference, which was held in 1979, identified areas in which research, guidelines, and methodologies are required.

3. Based upon the results of the pilot study conducted by the North Central Texas Council of Governments and the findings of the state-of-the-art assessment conference, a request for proposals was issued by the Department of Transportation to select area-wide agencies for developing and testing energy conservation techniques related to transportation systems planning. The effort, which began in 1980, will continue for a period of 2 to 3 years.

• *Water, sewer, and air quality planning.* Two to three case studies will be initiated jointly with EPA to identify the ways in which energy conservation considerations may be included in metropolitan air and water quality management programs. The studies were initiated in 1979 and are scheduled for completion in 1982.

The mission of the Community Technology Systems Branch is to develop systems that deliver energy services to the community more efficiently, to develop processes and systems that use municipal wastes for energy and energy-intensive material recovery, and to develop systems, processes, and technologies that will conserve energy in municipal functions, such as water- and wastewater-treatment facilities.

Municipal wastes are the discards of residential and commercial activities, municipal services, and industrial nonprocess wastes that are ultimately the municipality's responsibility for disposal. These wastes are 50 percent organic material, 25 percent moisture, and 25 percent inorganic materials. Embedded in the more than 150 million tons per year that are

discarded is the energy equivalent of 3 quads: 2 quads from direct energy recovery and an additional 1 quad from recycling materials and conservation of energy in processing new products. However, in 1978, less than 1 percent of this available energy was recovered.

Implementation of energy recovery systems can be accelerated considerably through a program of RD&D of technologies that are responsive to the nation's dual needs of energy conservation and reduced disposal of urban by-products. DOE strategy is to encourage (rather than pay for) implementation through a multifaceted program designed to reduce impediments to the use of systems for energy conservation and recovery projects. The strategy reflects several policies:

• *Leverage.* The use of federal funds in support of demonstration (1) should be directed toward technologies that have wide application, or toward local conditions that are, or could be, recurrent across the country, and (2) should be the most cost-effective approach to encourage the demonstration of those projects that, if proven viable, could be widely implemented.

• *Limited role.* The federal involvement in energy recovery should neither usurp nor cause to be relaxed the reasonable responsibilities of the typical nonfederal participants in energy recovery projects. Incentives for sound organization and management of projects should be reinforced. Maximum use should be made of private institutions for the financing of desirable demonstration projects.

• *Induced delays.* The federal program should not inadvertently impede the rate of progress toward demonstration of energy recovery systems by creating uncertainty as to the magnitude and extent of federal support, or by creating a "waiting line" for limited federal funds. Furthermore, the costs of applying for technical or monetary assistance under the demonstration program, and the costs of complying with post-award requirements should not result in undue diversion of resources from efforts directed at implementing the demonstration projects.

- *Impact*. The federal demonstration program (1) should reduce the risks currently inherent in energy recovery projects, (2) should increase the capacity, and desire, of nonfederal entities to efficiently implement energy recovery projects, and (3) should establish the commercial viability of recovery systems that enhance the environment and local economic development, and that contribute the maximum amount of usable net energy from urban waste management.

The DOE approach to supporting demonstration of energy recovery is to identify and develop systems to reduce some of the most important impediments to project implementation. These include:

- Project risks
- Demanding, time-consuming, and costly project implementation processes
- Limited markets for recovered energy
- Competition from waste-disposal alternatives
- Aggregation and control of wastes
- Stringent and costly project financing procedures

The DOE program for reducing these impediments to demonstration includes methods for (1) assisting project participation in organizing, planning for, and managing their projects (Implementation Assistance); (2) providing financial and economic incentives for public and private entities to proceed with a demonstration project (Economic Incentives); and (3) supporting the capital costs of desirable demonstration projects that could not otherwise be financed (Financing Assistance). Methods used to support demonstration projects include:

- *Implementation assistance*
  1. Contracts and cooperative agreements
  2. Training and technical assistance
- *Economic incentives*
  1. Price supports, cooperative agreements
  2. Advocacy for energy and environmental regulation
  3. Tax policies
- *Financing assistance*
  1. Loan guarantees
  2. Loans
  3. Price supports

DOE is developing an effective working relationship with other federal agencies in this field. DOE and the EPA developed a memorandum of understanding in FY 1979 that uses the authorities of both agencies to support implementation. EPA will provide planning assistance through its urban policy program; DOE will cover design, construction, start-up, and financing issues. In FY 1980, DOE, EPA, and the Department of Commerce developed a joint plan for energy recovery from wastes that embodies the capabilities of the three agencies.

Technological options for converting municipal wastes into various forms of energy include mechanical, thermal, and biological processes. The mechanical process activities are directed at the development of efficient plant size reduction methods, accurate heat value measurement, and improved waste material collection, storage, densification, and transport methods to allow further use of materials through recycling or conversion to energy in various forms. Thermal process activities focus on system integration and design for improved cost-effectiveness in a variety of applications. Thermal conversion processes include combustion (both direct and co-firing) and pyrolysis. Program activities in the biological process area are directed toward research to solve design problems and optimize process parameters to maximize energy conversion efficiencies. Bioconversion processes include enzymatic hydrolysis, anaerobic digestion, and methane from landfills and other processes (humus, algae, etc.). Resource recovery and combustion techniques will be available in the near term, pyrolysis in the midterm, and biological conversion techniques in the long term.

An advanced system experimental facility for anaerobic digestion has been constructed at Pompano Beach, Florida. This is a proof-of-concept demonstration plant for the conversion of a combination of solid wastes and sewage sludge into methane-rich gas. The Pompano Beach demonstration plant processes 50 to 100 tons per day of urban wastes (on the basis of solid wastes as received). The plant uses an anaerobic digestion process in which bacteria, in the absence of air, break down organic wastes and cause the formation of methane-rich gas. The feedstock consists of the light organic fraction of solid wastes along with 5 percent sewage sludge. The plant will be operated as an experimental facility designed to test the effects of technical design parameters, establish economic parameters, and develop data on the quantities and qualities of the methane produced. Plant construction was completed and experimental operations initiated in the summer of FY 1978. Evaluation of operations ran through FY 1980. In subsequent years, construction and testing of a larger demonstration facility are planned. Commercialization of the process would have significant impact on the development of a new source of natural gas. It is estimated that a 1000-ton-per-day plant could produce enough methane to serve the gas needs of 10,000 homes, while reducing the volume of the waste processed by 70 percent.

All recovery technologies except mass burning require some degree of preprocessing to achieve an effective feedstock for conversion. These processes have generally been adapted from other industrial processes and are typically very energy-intensive and complex. DOE is sponsoring essential R&D in these unit processes to optimize their effectiveness and to develop design tools for the industry. Work was initiated in FY 1979 and will continue through FY 1982 on optimizing such processes as rotary screens for separation of components, air classification of materials, and densification.

The production of alcohol fuels from cellulosic materials in the waste stream appears promising. DOE supports enzymatic hydrolysis research, which complements other research on sugar to saccharification to alcohols. If successful, this research will support the development of pilot projects to verify technical and economic details of the process in the FY 1982 time frame. Early projections indicate economic success in the 1980s. Critical questions to be answered center on the preprocessing of the waste materials to provide a suitable feedstock and the development of reliable conversion processes tolerant of impurities.

Recovery of gas from landfills is being accelerated by a DOE-sponsored RD&D program that supports design, construction, and evaluation of 10 ongoing projects. This area of energy recovery from wastes is less capital-intensive than other systems and takes less time to institute if market conditions are satisfactory. This program was initiated in FY 1978. As more systems are operated, various technical questions will be addressed; the important first step is to gain some actual operating experience with the practice to identify research needs. Case studies were developed in FY 1980; specific research projects will continue through the FY 1982 time frame.

Every project that is implemented requires expert guidance in determining feasibility and in coordinating legal, financial, and institutional issues. DOE supports 25 communities in this important planning stage. The community project leaders are assisted by a team of technical monitors skilled in all phases of project implementation. As many common problems are addressed in the planning stage, DOE sponsors workshops that focus on specific areas of project planning, such as determining contract principles, negotiation of contracts, and technical elements of small- and large-scale systems. These workshops allow interaction among peer groups and lead to effective learning processes. Case studies of selected projects will be developed and distributed to others interested in project implementation.

In addition to the recovery of energy from wastes, the Urban Waste Technology Program supports the development of systems to accomplish municipal processes such as more energy-efficient treatment of water and wastewater. Current processes were developed in an era of cheap and plentiful fossil fuels. Therefore, little attention was given to energy efficiency in design. A program is being developed to add energy efficiency as a design element in wastewater treatment. The program will function cooperatively with EPA, which, through the Clean Water Act, will have lead responsibility for the treatment and disposal of wastes. In this area, DOE has operated a 5000-gallon-per-day (gpd) pilot plant using the ANFLOW technology for anaerobic digestion of sewage. This packed-bed, upflow reactor provides methane-rich gas and meets secondary-level effluent standards in one treatment step. Design is complete for a scale-up to 50,00 gpd at a facility that was constructed in FY 1980. This project is a joint venture with TVA; it will be evaluated 1 to 2 years following construction. Since this process appears to have promise for treating certain industrial wastes, technology transfer materials are being prepared for publication and distribution to industry.

Beginning in FY 1981, DOE plans to use loan guarantees to support construction of certain waste-to-energy demonstrations. Public Law 93-577, Section 19(y), authorizes guarantees of up to 75 percent of total project costs or 90 percent of construction costs. DOE expects to guarantee one to two projects during FY 1982. The concept for this program embodies limited-duration (approximately 5 years) support of recovered energy products at a level to be negotiated with each project sponsor, with payback of federal funds in future operating years. This program addresses the apparent negative cash-flow problem that projects suffer in the start-up period. These projects have high first cost, compared to other alternatives, even with good life-cycle economics.

The purpose of the Energy Systems Section is to create energy supply choices for communities. There are three critical reasons why the nation's energy program must have alternative community energy system options:

- Energy-conserving and self-sufficient communities mean an energy-conserving and self-sufficient nation.

- Creating energy supply options means communities can plan for and control their energy destiny.
- Tailor-made, integrated community systems ensure resource-efficient, economically competitive, reliable, and environmentally acceptable energy supplies.

By itself, the R&D of energy systems is not enough to achieve a significant national impact. These systems must be tested and applied in real community applications. Therefore, the Energy Systems Section supports communities as they pioneer in applying efficient energy system concepts using current and emerging technologies. Designs for specific sites accommodate the technical, economic, environmental, financial, and institutional factors that must be resolved in the implementation process. Lessons learned from these field activities are documented and disseminated to build the capacity of communities nationwide to plan, evaluate, design, construct, and operate community energy systems. Specific system concepts such as cogeneration district heat, with its potential for significant scarce fuel savings, are targeted for a national plan to expedite implementation. The RD&D activities of the Energy Systems Section provide the basic input to such a plan.

The systems development activities of the Energy Systems Section are divided into two categories: Central Integrated Systems, which is composed of cogeneration district heat; and Distributed Integrated Systems, which is composed of grid-connected Integrated Community Energy Systems (ICES), coal-using ICES, and heat-pump ICES.

Central Integrated Systems are suitable for existing large, dense, and contiguous urban areas. The cogeneration district heat concept conserves energy and scarce fuels by retrofitting base- or intermediate-load electric generating plants to use reject heat produced during power generation. This heat can be used to supply heating, cooling, and other thermal energy services to a ·community. Although widely used in Europe, this concept has had limited application in the United States. Market penetration studies have estimated that widespread implementation of such systems in the United States could result in 1 to 2 quads of scarce-fuel savings by the year 2005.

DOE is studying technological questions including the feasibility of various turbine-generator retrofit alternatives, distribution system routing alternatives, and the choice of various heating mediums. Distribution system insulation requirements and costs also are being examined. In addition, the impacts of cogeneration on plant electrical output and efficiency and utility system operation are being assessed.

Along with technological issues, the institutional and regulatory mechanisms for implementing cogeneration district heat systems are being analyzed. In particular, utility companies and commissions are analyzing options for expanding utility operations to include major-urban-area thermal distribution networks. District heating systems, as a part of electric utility operations, typically realize low returns on investment compared to allowed rates of return on electrical systems. Regulatory practices are being explored that allow higher rates of return on cogeneration district heat systems to enhance their attractiveness. Issues such as pricing of joint products to avoid cross-subsidization of customers, while realizing fair rates of return, are being defined. Alternative pricing schemes are now being developed in DOE studies.

Cities examining the feasibility of constructing piping networks to nearby power plants include Philadelphia, Pennsylvania; Detroit, Michigan; Green Bay, Wisconsin; Piqua and Toledo, Ohio; Newark, New Jersey; and Moorhead, Minnesota. In addition to modern and efficient technology for heat cogeneration and distribution, innovative mechanisms for financing, regulating, operating, and managing these systems are being examined by teams of community leaders, energy producers, and energy users in each city. DOE is supporting these teams to determine whether congeneration district heating is suitable for their cities. If it is, then the financing, regulatory, ownership, operation, and management arrangements that are necessary for successful implementation will have to be determined.

During FY 1980, one of these cities began program implementation. DOE will use the demonstration for widespread application of these systems in other cities in the 1985 time frame. Also in FY 1980, in cooperation with the Office of the Assistant Secretary for Energy Technology, a national plan was developed to expedite implementation of the highly fuel-efficient cogeneration district heat concept.

Distributed Integrated Systems are suitable for both existing and new developments such as high-density islands in the suburbs (shopping centers, industrial parks, high-rise residential/commercial developments, etc.). These systems are relatively small, cost less, require less time for construction, and have less noticeable impact on the physical community they serve. Therefore, the potential for accelerated near-term market penetration is greater than for a Central Integrated System. Estimates indicate that approximately 1 quad of scarce fuel can be saved by the year 2000 by applying these systems.

A typical Distributed Integrated System is the grid-connected ICES. This type of grid-connected community energy system produces both electricity and thermal (heat) energy. It is designed to meet year-round local heating and cooling demands, thereby achieving high seasonal efficiency. Maximum total efficiency is also achieved because any electricity surplus or deficit is absorbed or made up by the local utility system. Four communities are being supported by DOE to apply this system concept: the University of Minnesota is applying a coal-fired steam turbine; the downtown redevelopment area of Trenton, New Jersey, is using a gas turbine; the Health Education Authority of Louisiana designed a coal-fired steam turbine system; and Clark University is using a diesel-generator set with heat recovery.

Coal-using ICES demonstrate the development, acceptance, and implementation of integrated systems that use coal as the base fuel. The program is divided into two parts. DOE is currently supporting the first part, an initial demonstration for the Georgetown University campus. This coal-using ICES forms part of a broader university energy master plan consisting of an active energy conservation program, cooperative programs with utilities, and innovative energy systems. A prefeasibility study determined a group of 10 subsystems that were directly related to the goals of the coal-using ICES program. The second part of the program is to support small- and medium-size communities that potentially can be serviced by thermal and electrical products produced by retrofitting the energy systems and converting the power plant's primary fuel to coal. The projects promote the early implementation of community systems that use coal as a fuel, thereby conserving scarce fuels such as oil and natural gas.

Heat pump systems using free energy from the environment for heating and cooling have been applied to small, single buildings. The objective of this system development effort is to demonstrate the advantages of large, built-up heat pump systems for communities. The potential fuel-effectiveness ratio of such systems is two to three times that of conventional heating, cooling, and hot-water systems. Central heat-pump systems are particularly attractive options for small communities. Various energy sources and sinks, including the use of solar energy and reject heat, are being evaluated. In addition, the seasonal performance of central versus dispersed systems (and combinations thereof) is being analyzed, and innovative energy storage/distribution schemes are being assessed. DOE is currently funding nine contractors to develop, evaluate, and test design these concepts for selected communities. Promising concepts were selected for site application in FY 1980. The city of Scranton, Pennsylvania, is currently being funded to evaluate the potential use of mine water as a source for a central heat-pumping system.

Application of these systems in communities will continue through the early 1980s. Three communities will complete final designs of grid-connected ICES in 1980; construction is expected by 1983–1984. The coal-using system is expected to be operational in FY 1981, and the development of heat-pump centered systems is anticipated by 1985. A marketing plan using models and data describing the economic competitiveness of these energy supply options has identified 10 utility service

areas that offer high potential for large cogeneration-based district heating systems between 1981 and 1987. In addition, identification of 50 new development sites suitable for distributed integrated systems applications indicates a potential for rapid penetration of these versatile smaller systems by 1990.

The process of the R&D of energy system concepts and implementation methods is continuing. The Energy Systems Section relies on other DOE organizations for initial basic research on such subsystems and components as solar, geothermal, and fossil energy; energy storage and transmission; and energy-efficient end-use systems. This research is evaluated for application potential in the community context, and appropriate systems are selected for further development phases. Technical and economic data for successfully demonstrated and market-ready systems and components are established and maintained; and efficient methods for using that data in the analysis and design of specific systems are developed. Successful engineering approaches and options resulting from site applications projects are documented in "system-design" guides for widespread applications by engineering design professionals and energy systems planners. These activities are designed to result in a continual flow of information regarding emerging energy systems and implementing methods to key actors in communities nationally.

*Power-plant retrofit to district heating and cooling systems.* Power-plant retrofit is Phase 1 of a demonstration program to determine the technical, institutional, and economic feasibility of retrofitting base-loaded electrical utility plants for the supply of district heating and cooling to the community.

*ICES.* The ICES project involves the design, construction, and evaluation of an ICES for the Georgetown University campus. The ICES will use a fluidized bed combustor. The effort will be divided into five phases:

- Preparation of a detailed work-management plan
- Preliminary design

- Final design
- Construction
- Initial operation, testing, monitoring

*District heating for a large city.* Hot-water district heating for a large American city by means of cogeneration will be evaluated. The St. Paul-Minneapolis area has been chosen as the site area for the study. The local participants include the power company, state government, local government, and several other organizations. Oak Ridge National Laboratory, Argonne National Laboratory, and the local participants will supply technical and economic information to the contractor performing the analysis (A. B. Atomenerigi, Sweden). ORNL and ANL will also provide technical management. This effort is cofunded and managed with the DOE Office of Energy Technology.

*Heat Pump Centered Integrated Project.* The heat-pump centered ICES (HP-ICES) is a multiphase project seeking to demonstrate one or more operational heat-pump centered ICES by the end of 1983. This project will include seven phases:

- System development
- Demonstration design
- Design completion
- HP-ICES construction
- Operation and data acquisition
- HP-ICES evaluation
- Upgraded continuation

*Grid-connected ICES.* The grid-connected ICES activity is prepared for the third phase of a demonstration program leading to the construction and operation of one or more grid-connected ICES. The third phase is the preparation of detailed system designs, which include provisions for evaluation and monitoring instrumentation. The demonstration program, consisting of six phases, started in FY 1976. In FY 1977, five community teams completed a preliminary feasibility analysis on the application of the grid-connected ICES for their respective communities. The Acting Administrator for ERDA selected four of the five communities for Phase II, Detailed Feasibility Analysis and Preliminary Design.

*Pompano Beach advanced systems experimental facility.* The Pompano Beach advanced systems experimental facility (ASEF) involves the demonstration of a "proof-of-concept" scale (50 to 100 tons per day) anaerobic digestion plant using the light fraction of urban solid waste and sewage sludge to produce methane-rich gas. This effort is to prove the concept of the anaerobic digestion of a mixture of solid waste and sewage sludge, to develop design parameters, to establish economic data, and to develop data on the quality and quantity of gas produced. Urban waste is shredded and separated into light and heavy fractions. The light fraction is combined with up to 10 percent sewage sludge and is digested. The methane may be used as fuel for power generation in the locality, or cleaned for injection in natural-gas lines. Although this facility is near demonstration size, it will be operated as a true experimental facility to test the effects of parameter variations.

*Refuse conversion to methane.* The specific goals of the refuse conversion to methane (REFCOM) are to (1) establish information concerning the gas product quantities and values; (2) evaluate process and economics; (3) determine optimum design and operation parameter values for each process stage and method of operation; (4) establish a basis for comparing the process to other means of energy production and/or resource recovery from urban wastes; and (5) establish the technological and economic bases for commercial use of the process.

As produced, the product gas has potential use as a low-Btu fuel. However, if desiccated and purified, an essentially pure methane results, which may be used as a substitute for natural gas. Laboratory-scale studies performed to date by Dr. John Pfeffer of the University of Illinois, who is also technical consultant on this project, indicate that approximately 6000 cubic feet of mixed methane and carbon dioxide gas are produced per input ton of raw refuse; therefore, 3000 cubic feet of methane (a pipeline-quality gas) per input ton could be produced by this process.

This project has application to methane dessication and separation, solar heating for digesters, anaerobic digestion studies, and hydrolysis studies. Results of experimentation may be useful for studies of viscous mediums, productive use of sludges, using less energy in size reduction of urban waste, and studies of the synergy between urban waste and sewage treatment.

*Evaluation of alternative methods for combined solid/liquid waste energy recovery systems.* The objective of evaluating alternative methods is to assist in the development of innovative and workable methods for the production of energy from integrated solid/liquid waste systems and the incorporation of combined solid/liquid waste streams as replacements for fossil fuel. This program is well developed and has a high likelihood of technical success. In addition, the program will include a marketing study for recovered material and excess energy.

*Dutchess County pyrolysis program.* The objective of the Dutchess County pyrolysis program is to provide a complete, detailed engineering design for a waste-to-waste facility for Dutchess County, New York. A further objective is to obtain a final capital cost estimate, based on the engineering design, which will enable the local government to proceed with financing and construction of the facility which will recover 1.8 trillion Btu annually, or 900 barrels of oil per day.

Incoming urban waste will be sorted to remove metals and glass and will then be fed to a PUROX-type pyrolysis unit. A medium Btu fuel gas will be generated, which may be used onsite to generate electricity or may be sold to a utility in the area. The residue from the pyrolyzer will consist of granular slag that is suitable for use as a road base, etc.

*Optimum energy utilization employing waste-derived fuels in conjunction with municipal utility systems.* The purpose of the op-

timum energy utilization project is to evaluate the feasibility of using waste methane gas from a wastewater treatment plant as a supplementary fuel source in the total energy plant (supplying electricity, heat, and hot water) of a contained residential community of 24,000 people in Brooklyn, New York, with shopping and community centers, parking garages, a school, and recreational facilities.

Currently, a total energy (TE) plant supplies electricity, heating, cooling, and domestic hot water needs for the entire complex. Starrett City, Inc., the managing partner for Starrett Housing, is considering adapting the total energy plant to burn methane recovered from the nearby New York City 26th Ward Water Pollution Control Plant (WPCP). In turn, it may supply electricity to develop and test the feasibility of using currently wasted resources to generate energy for a self-contained community. DOE will also be able to analyze the potential for energy conservation and system load-balancing in a controlled setting.

*Enzymatic hydrolysis.* The objective of enzymatic hydrolysis is to convert cellulose in urban wastes to glucose with further fermentation to alcohols and other energy products. Enzymatic hydrolysis involves the conversion of cellulose to glucose, which can subsequently be converted to a food, used directly as a petroleum substitute to produce chemical intermediates, or fermented to ethanol. DOE supported efforts to develop this project to full scale at the U.S. Army Natick Laboratory. In FY 1978, various methods of size reduction were investigated for energy and fiscal economy, and the pre-pilot plant was operated to increase the cellulose feed and production concentrations.

*ANFLOW.* The purpose of ANFLOW is to develop a process to accelerate production of methane-rich gas from unconcentrated sewage sludge and possibly reduce or eliminate sludge production from conventional secondary sewage-treatment plants.

ANFLOW is a joint project between DOE,

the Norton Company, and the city of Oak Ridge, Tennessee, to install a packed-bed anaerobic digester at the city's sewage-treatment plant. On the laboratory scale, the process shows promise to drastically reduce the cost of sewage-treatment plants and the problem of waste sludge disposal. The technology can be extended to industrial wastes and, by tailoring organisms, can produce alcohols and other chemicals.

The effort has resulted in the development of a treatment system that achieves secondary-level effluent in one treatment step, and produces 700 Btu per cubic foot of gas. A 5000-gpd pilot plant has been operating for two years. Suspended solids levels are good, but biological oxygen demand is a problem. Design for scale-up to 50,000 gpd will be achieved in FY 1979. Plans are underway to install a 50,000-gpd facility at a TVA construction site in Tennessee.

*Activated carbon for sludge digestion.* The activated carbon for sludge digestion project includes conducting laboratory and small-scale pilot development of a process to enhance anaerobic digestion by the addition of activated carbon. Approximately 7500 sewage treatment plants in the United States use anaerobic digestion. Methane gas produced by the process is frequently flared but could be recovered and used as a fuel to provide power for plant operation. Following up on promising laboratory results, a pilot plant was completed in FY 1977 to determine optimum conditions for activated carbon addition to enhance gas production and recovery. During FY 1978, field demonstrations of the pilot plant were conducted. Potential future development could combine this process with a pyrolysis system producing activated carbon from the remaining sludge. Preliminary results showed best use in stressed, or "sick," digestors. It has been estimated that 20 percent of United States digestors could benefit from this technology.

*Research to improve economics and yield in the conversion to fuels and chemical feedstock.* The feasibility of using a steam size-reduction technique for preparing feedstocks for energy recovery is being investigated. Development is directed toward optimizing a new process that preheats cellulosic substances by exploding the feed cellulose into small particles. The small "opened up" particles are more accessible to enzymes. The rate of enzymatic reduction of cellulose is proportional to the surface area of cellulose. Reducing the size of particles increases the available surface area, and the rate of cellulose reduction. Available methods of size reduction are not economical.

Cellulosic material is sent into a pressure vessel, called a *steam gun*, and is treated with steam under pressure. The steam penetrates the cells and intercellular openings and upon release of pressure expands, rending the cellulose structure into small pieces.

*Small-scale systems development.* The objective of small-scale systems development is to conduct an assessment of the developmental needs and constraints for small resource-recovery systems and to identify candidate processes from the standpoints of technological, economic, and institutional constraints on the development of 25- to 50-tpd urban waste-processing systems.

This project will analyze small-scale processes (less than 100 tpd and preferably economical at 25 tpd) for urban-waste-to-energy conversion. About 45 percent of the total wastes generated in the United States originate in cities. The small-scale systems hold promise for processing waste locally, thus saving transportation energy. Combustion, pyrolysis, and bioconversion are candidate processes for urban-waste-to-energy conversion.

Follow-on projects will involve improvements in unit processes and detailed evaluation of current units.

*Commercialization of a 40-kW fuel cell.* The purpose of this project is to demonstrate the commercial feasibility of a 40-kW fuel cell module. A major program was initiated in late FY 1977 to further develop the 40-kW fuel cell (with waste heat recovery) power plant of suitable cost, reliability, and durability for use in commercial and residential total energy systems. This will lead to field demonstration of prototypes, jointly funded by DOE and the Gas Research Institute, in FY 1980.

*Energy cascading.* A market study of energy-cascading technologies, called the "Common Study," was performed. The study was a major element of the energy-cascading program of the International Energy Agency (IEA) in which DOE participated. The contract covered the following tasks: gathering data on important waste streams and potential end uses, identifying energy-cascading technologies, assessing the technical potential for cascading, assessing the implementation potential for cascading, identifying R&D projects for consideration by DOE and IEA, and drafting project annexes.

*Land-use/energy-consumption data sets.* Land-use/energy-consumption "data sets" will be important tools for energy planners and other key community-development participants. They must recommend courses of action regarding the adoption of specific technological and passive options, which should be based, in part, upon accurate assessments of their energy-related impacts.

The study is divided into two phases of activity. Phase I work identified existing data and information gaps and developed methods for obtaining missing data through field research. A methodology was formulated for developing data sets describing the energy consumption characteristics of various land-development types. During Phase II, energy consumption data associated with various types and patterns of land uses will be organized in a consistent and detailed format. Research and analysis to be performed will involve statistical analysis of data variations and trends, and estimates of the reliability of final results.

*Incorporation of energy conservation in urban transportation planning.* The objectives of this project are to identify and demonstrate options of incorporating energy conservation considerations and evaluation methodologies into ongoing urban transportation planning

and systems management. Achievement of these objectives will directly contribute to the Community Planning and Development Program objectives of helping to incorporate energy conservation objectives into ongoing community subsystems planning, development, and operation.

*Incorporation of energy conservation in comprehensive planning for rural development.* The objective of this effort is to test and demonstrate the incorporation of energy conservation considerations into comprehensive planning for rural development under assistance grants administered by the Farmers Home Administration of the Department of Agriculture. Achievement of this objective directly contributes to a major community planning and development program objective of facilitating the incorporation of energy conservation objectives into community activities in planning, developing, and operating community subsystems under federal assistance programs.

*Incorporation of energy conservation considerations in air and quality management programs.* This project will identify, through case studies, ways in which energy conservation considerations may be integrated into air and water quality management programs developed pursuant to the Federal Clean Air Act and Clean Water Act. The case studies, to be carried out by state and local agencies, will (1) identify weaknesses in existing institutional mechanisms, planning processes, and technical methods; (2) develop and apply approaches to more adequately include energy considerations; (3) identify the energy use associated with different management measures and strategies; and (4) recommend needed policies, procedures, and technical methods.

*Energy integrating master plan for Atlantic City, New Jersey.* This project will develop, analyze, and demonstrate the preparation of energy-conserving master plans for a redeveloped urban community. As a prototype, the project will contribute to the incorporation of energy conservation in redeveloped communities throughout the country. Atlantic City, because of its small size and the anticipated redevelopment growth rate, is an ideal site for a case study. The following major Phase I tasks are to be performed:

- Task 1: Preparation of energy conserving community development plan
- Task 2: Preparation of energy systems plan
- Task 3: Preparation of the energy conservation management element of the master plan
- Task 4: Preparation of the energy integrating master plan

*Rural laboratory, Mercer County, North Dakota.* The primary objective of this project is the establishment of a rural energy laboratory in Mercer County. The laboratory will permit DOE to conduct RD&D experiments in community energy design and in architectural and engineering systems through a unique institutional mechanism, the Energy Development Board. Mercer County residents will benefit through the management of rapid energy-related growth and by reductions in future energy costs resulting from the energy-conserving development plan and the demonstration projects. Tasks will include development of a growth management plan for Mercer County, and energy conservation and economic diversification plans associated with waste energy utilization.

*Organization approaches to energy-related community growth.* Phase I, which has been completed, involved the investigation of the patterns and impacts of rapid small community growth. Evaluations were made of existing European and American techniques for dealing effectively with this type of development. A general organizational methodology for successfully meeting the challenge of managing rapid small community growth was formulated and presented.

The Phase II effort will complete the development and demonstration of the general organizational methodology for managing rapid small-community growth. Specific tasks include:

- Monitoring Energy Development Board activities
- Implementing general approach
- Documenting changes in Mercer County
- Assisting Mercer County in technology transfer activities

*Comprehensive community energy management program pilot studies.* The comprehensive community energy management program (CCEMP) moved in early FY 1979 into an expanded phase of pilot-study activity, which will test the methodologies that have been produced and will extend their application to a wide range of community types. The pilot studies, which may involve as many as 17 communities, will last approximately 2 to 3 years and will cover two areas of activity:

- Development of individual community organizational arrangements for undertaking a comprehensive community energy planning effort.
- Formulation of a comprehensive community energy management action plan. This will encompass application and evaluation of methodologies and tools for preparing community energy audits, formulating energy objectives, and identifying and assessing energy conservation alternatives.

Consumer products. The objective of the consumer products division is to develop, demonstrate, and encourage the commercialization of more energy-efficient technologies in heating, cooling, and ventilating equipment, systems lighting, and appliances. Furthermore, this thrust is aimed at developing and demonstrating innovative methods of encouraging consumers to purchase and use energy-efficient products and to adopt energy-efficient practices.

The potential for energy savings in these areas is high. Over half the energy consumed in buildings is for heating and cooling. Lighting accounts for 30 percent of the energy used in commercial buildings. Major appliances (including heating and cooling equipment and hot-water heaters) account for 93 percent of residential and 70 percent of commercial building energy use. Major milestones for this thrust are shown in Figure 7-5.

Improvements and innovations being developed in the Technology and Consumer Products Branch include an efficient replacement for the screw-in incandescent lamp, integrated

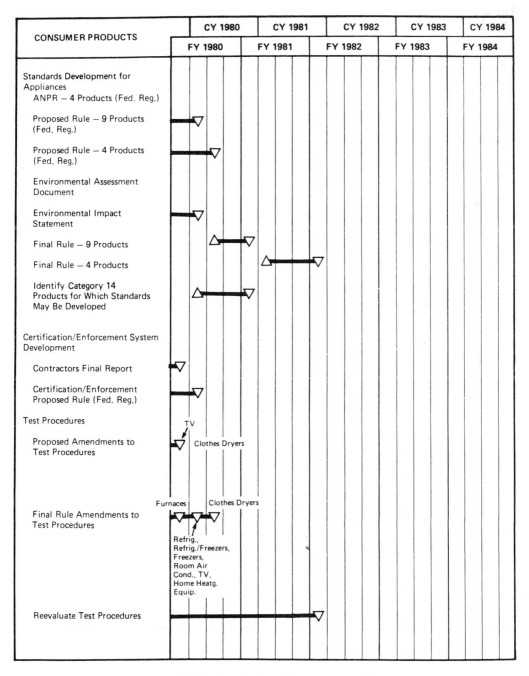

Figure 7-5. Major milestones for consumer products.

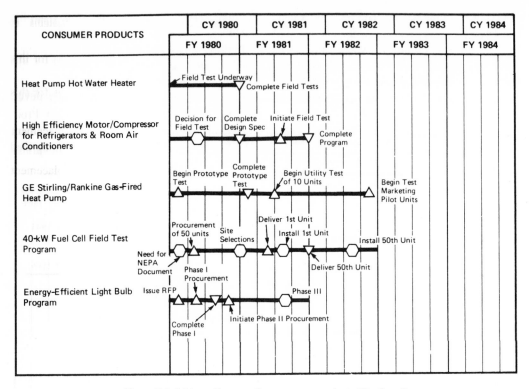

Figure 7-5. Major milestones for consumer products (Continued)

appliances with waste heat recovery from one appliance to another, and thermally activated heat pumps. Since space conditioning accounts for such a large portion of building energy demand, more efficient oil- and gas-fired equipment is being developed for use in existing and new buildings. Examples of retrofit devices and techniques being evaluated include vent and flue dampers, vent restrictors, intermittent ignition devices, high-speed flame-retention head burners, nozzle reductions, derating, and fan control. The issue of optimal sizing of new and replacement heating systems will also be examined extensively. Additional technologies will be examined and developed in the Oil Equipment Qualifications Program, which will qualify additional technical measures for incorporation into the Residential Oil Conservation Marketing Demonstration Program. It will add measures to those presently included, which are optimizing burner-firing rates, the replacement of existing burners with high-speed flame-retention head burners, and the replacement of

inefficient boilers and furnaces with more efficient units.

Since space-conditioning appliances and systems account for nearly two-thirds of the energy consumed in buildings, a large proportion of the RD&D program is directed at space conditioning. Development of furnace and boiler designs that approach the limit of efficiency improvement possible is planned to be concluded by FY 1982. Further efficiency improvements are possible by going to full condensing mode operation in furnaces and boilers, where the latent heat of combustion water is extracted usefully. It is intended that development and market introduction of full condensing mode furnaces and boilers be accomplished by FY 1985–1986. This program will involve solving significant materials, cost reduction, operational, and environmental waste-disposal problems.

The primary objective of the Thermally Activated Heat Pump Program is to demonstrate new gas- and oil-fired heat pumps that will provide high seasonal efficiency in commercial and residential facilities, particularly in the northern parts of the United States. The FY 1979 heat-pump programs include:

- A Rankine/Rankine system intended for initial commercial use, followed by eventual residential use.
- A Stirling/Rankine system that will undergo residential field testing to collect efficiency, reliability, serviceability, and cost data. Future light commercial installations are also planned.
- An absorption refrigeration cycle system and a Brayton/Rankine 7.5-ton heat pump, which will be fabricated in prototype form, field-tested, and demonstrated during FY 1980–1982.

Program activities are not limited to the technical development of more efficient products. Other efforts include procedures and methods for proper equipment selection and installation, curbing the common practice of installing oversize heating and cooling systems, and developing accurate and easy-to-use analytical tools for more efficient lighting systems design. Efforts in the appliance area are directed toward developing, evaluating, and commercializing new energy-efficient water heaters, refrigerators, room air conditioners, and integrated appliances.

A key feature of technology development programs is the joint participation of industry groups, especially manufacturers, marketers, and trade associations. Cost-sharing by private industry is an important part of these development programs, both to increase the effectiveness of the federal contributions and, more importantly, to secure practical commitment by industry to full commercialization. Many of the technologies being developed will reach the market-introduction phase within the next 5 years. It is therefore especially important to work with industry and manufacturers to provide reasonable assurance that the technical, economic, market, and institutional concepts of these technologies are sound.

The mandatory Consumer Products Efficiency Program, under the requirements of P.L. 94-163, seeks to reduce the energy consumed by major home appliances. Prior to its amendment by the National Energy Act, this was to be accomplished through the development of uniform test procedures, establishment of energy-efficiency improvement targets, and creation of a monitoring system by which progress toward target achievement could be assessed.

Energy-efficiency improvement targets and test procedures were finalized for each of 13 products covered by the legislation: refrigerators and refrigerator/freezers, freezers, dishwashers, clothes dryers, water heaters, room air conditioners, television sets, kitchen ranges and ovens, clothes washers, humidifiers and dehumidifiers, central air conditioners, and home heating equipment not including furnaces. Pursuant to provisions of the National Energy Act, the target program was deleted in favor of minimum energy-efficiency standards, which will be developed based on extensive technical and economic analysis. The individual products within each of the 13 product categories will be required to meet or exceed minimum efficiency standards to be set by DOE.

The program objective is to achieve a reduction of 0.9 quad or 5 percent of the amount of residential sector consumption in 1985. This is based on the promulgation of standards for 9 of the 13 product categories by December 1980, and standards for the remaining 4 categories by November 1981. Current program efforts associated with the establishment of minimum efficiency standards include the collection of data and the technical and economic analyses related to developing standards and the development of certification methods by which to ensure conformance with the standards. The development of new test procedures and modifications to existing text procedures represent continuing responsibilities under provisions of both P.L. 94-163 and the National Energy Act.

Program requirements in the out years, 1982–1986, include reevaluation of test procedures for all consumer-product categories by November 1981, continuous monitoring of the appliance industry to ensure compliance with DOE-established standards, implementation of a certification/enforcement program by July 1981, and reevaluation of 1981 promulgated standards by November 1986 to determine the need to amend standards due to new technology.

The objective of the market development branch is to work with the private sector to demonstrate and develop innovative methods of encouraging consumers to purchase and use energy-efficient products and to adopt energy-efficient practices. The Market Development Branch identifies barriers to energy conservation and works as a catalyst with the private sector to help initiate solutions to these problems.

The first program area is finance. DOE works with financing and real-estate organizations (1) to ensure that energy efficiency is given direct consideration in real-estate purchase and financing decisions; (2) to establish institutional policies and procedures that support conservation in buildings, especially in the federal and secondary markets' appraisal and underwriting criteria for evaluating property; and (3) to develop convenient methods of financing energy-saving improvements being tested with lenders.

A second program area tests information feedback as a potentially effective means to encourage conservation within the home. Feedback means giving homeowners immediate and understandable information about their energy use and the results of altering their energy-use patterns, e.g., changing thermostat settings. Based on positive results of earlier studies (indicating an average savings of 10 percent), all further work has integrally involved utilities and potential retailers/manufacturers in developing and demonstrating a low-cost energy cost indicator.

A third area is marketing conservation programs; for example, the Energy Cost of Ownership Demonstration (Project Payback) tests marketing strategies that will accelerate consumer acceptance of energy-conserving and energy-efficient products, despite their higher first cost or initial investment requirements. It also works with manufacturers and retailers to encourage energy-efficiency advertising in conjunction with DOE efforts. Specialized marketing programs such as the Residential Fuel Oil Conservation Marketing Program, the "low cost/no cost" approach to conservation, and the water conservation program will be implemented nationwide in 1980 and 1981. Milestones are presented in Figure 7-6.

*Vent dampers and electronic ignition devices installation standards.* Installation guidelines are to be developed for the following devices:

- Electrically actuated automatic vent dampers for gas-fired furnaces
- Mechanically actuated automatic vent dampers for gas-fired furnaces

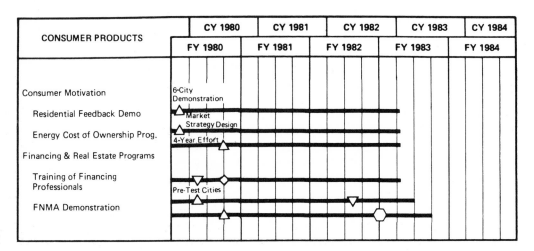

Figure 7-6. Major milestones for the Market Development Branch.

- Thermally actuated automatic vent dampers for gas-fired furnaces
- Electrically actuated automatic vent dampers for oil-fired furnaces
- Electrical ignition devices for gas-fired furnaces

A first draft has been written of an installation guideline for electrically actuated automatic vent dampers for gas-fired furnaces. This document is now being revised and supporting documentation is being written. "Standard Practice for the Installation of Replacement Oil Burners for Energy Conservation" including justification statements is being used as a format model.

*Evaluation of financing demonstration.* An evaluation of financing demonstrations has been designed to encourage the purchase of energy-efficient products in the residential sector. Support will also be required in preparing case studies to assist in the transfer of successful financing approaches to other financial institutions across the country.

The contractor will provide financing in three areas:

- Innovative Financing Programs that test methods by which convenient financing can be provided to homeowners to encourage investment in energy-conserving products
- Real Estate Training Programs that provide lending officers, appraisers, realtors, and developers with the information necessary to incorporate energy-efficiency criteria into the real-estate transaction
- Federal Financial Assistance Programs that aim to alleviate barriers to the financing of energy-efficient products. Under these programs, DOE works with federal agencies to alter their program policies and procedures to make energy efficiency a qualifying condition for financial assistance.

*Alternative metering strategies.* Economic, technical, and institutional aspects of alternative metering strategies are being analyzed.

Focal points of the analysis include individual, master, and various submetering approaches, with a view toward evaluating the effectiveness of different technical options and their relationship to human behavior.

*Real-estate training programs.* A training program, "Making the Most of Energy in Real Estate," will be delivered to real-estate professionals (appraisers, realtors, and mortgage lending officers) by appropriate trade and professional associations. The objective of the training program is to inform real-estate professionals about methods of incorporating energy-efficiency criteria into their daily transactions. The curriculum for the training program was developed and tested at the University of Georgia under contract number EC-77-C-01-8696. Twelve training programs will be held during FY 1979.

*Motivating occupants of master-metered buildings to conserve energy.* There are many buildings in the country which are master-metered for utility use. Since the cost of retrofitting such buildings with individual meters can be exorbitant, other energy conservation incentives must be developed. Motivational techniques were developed and tested in Phase I of this project. In FY 1979, these techniques have been implemented in various master-metered buildings. Data gathered regarding the effectiveness of the techniques will be analyzed, and institutional barriers to the use of the techniques will be addressed.

*Technical and economic analyses for test procedure and standards development support.* Task I, Economic Analysis and Support, involves the following activities:

- Development of life-cycle analyses and life-cycle cost equations for various products and classes
- Assessment of the economic impacts of federal test procedures and product class delineations for minimum standards
- Evaluation of the results of DOE's industry survey instrument and relation of

data obtained from survey results to Operations Research Models I and II

Efforts under Task II, Engineering Analysis and Support, include:

- Evaluation and modification of test procedures to be compatible with standards programs
- Provision of analyses and recommendations to determine the required minimum number of classes for the products
- Provision of laboratory tests as required

*Appliance efficiency standards.* Principal tasks include the following:

- Tasks I and II: Establish base case projections of appliance energy use using the ORNL residential energy-demand model, and perform preliminary assessment of consumer decisionmaking in the purchase of appliances
- Tasks III-V: Analyze weather sensitivity of heating and cooling, assess impacts of appliance efficiency standards on utility peak loads, and assist DOE in developing ranking methodology to assess alternative standards

*Analysis and review of regulatory issues related to the consumer products efficiency standards.* The following six tasks will be performed:

- Conduct an analysis of the consumer-product installation industry
- Review surveys by utilities on appliance ownership and usage
- Analyze the usefulness of the data and model developed by national laboratories
- Conduct analysis of economic issues related to consumer education program
- Review regulatory methodologies
- Evaluate and review transcripts

*Fuel oil conservation marketing demonstrations program.* The marketing program has two main components that are being adapted and implemented on a state-by-state basis:

- A consumer-education campaign to increase awareness of the benefits of the conservation devices
- An oil-dealer training program to increase industry expertise in marketing, installing, and servicing and devices

Initial pilot demonstrations are successfully underway in Massachusetts and in New York in cooperation with the state energy offices and the home-heat industry. These programs have met with considerable enthusiasm in these states, and plans are underway to expand the demonstrations statewide. DOE is now planning to extend the program to other states with a dependence on No. 2 fuel oil for residential space heating.

*Retrofit zone controls–marketing field test.* The conservation potential of retrofit zone controls for single-family homes will be investigated in three phases and then a number of homes will be monitored for at least 1 year to verify predicted savings. Three conferences (and proceedings) will be used to publish the results of each phase on a current basis.

*Screw-in circline lamp.* Phase I is the design of a solid-state ballast for a fluorescent circline lamp. The ballast will be able to dim the lamp, achieve an output equivalent to a 150-watt incandescent bulb. The circuit will be designed so that it can be hybridized, i.e., power hybrid. This will permit the system to be packaged in a small, attractive manner to please the consumer as well as obtain UL certification for portable lamp applications. The first samples are to be received by mid-1979. These units will use a discrete electronics package. Phase II will support development of an integrated circuit of the above design.

*High-efficiency refrigerator-freezer.* Arthur D. Little, Inc., with Amana Refrigeration, Inc., as a subcontractor, is developing and testing the practicality and cost-effectiveness of a refrigerator-freezer that will provide as good or better preservation service and convenience as current models but will use less than 2 kWh/day, which is substantially less than is predicted for the units resulting from DOE Efficiency Standards for 1980. Energy improvement options were identified and evaluated for their potential to save energy. A series of bench tests and computer analyses was performed, and improvements judged to be suitable were incorporated into a prototype unit that was tested for performance by the standard DOE test procedure and by other industry test procedures. A report of the R&D work is in review. A field demonstration has been initiated to test the marketability and performance of the improved units.

*High-efficiency electric water heater– Rankine Cycle.* Energy Utilization Systems (EUS) has refined a design that they had partially developed prior to this subcontracted work. This design uses state-of-the-art refrigeration components for a dedicated heat-pump system to heat water.

Performance of these heat-pump water heaters varies with supply and delivery water temperature and with ambient temperatures. Because the units do not incorporate a defrost cycle, the water is heated with a resistance heater when ambient temperatures are below 8°C. Thus, in-service performance cannot be stated simply. The tests that have been performed indicate that a coefficient of performance (COP) greater than 2 is realistic for much of the United States. EUS has prepared a report of the R&D on these units* and is proceeding with a demonstration of 100 units in a wide variety of climates and building locations. These demonstrations will be cost-shared with Mor-Flo Industries, Inc., a major water heater manufacturer, and with the 20 utilities that will install, maintain, and gather data from the demonstration units.

*High-efficiency gas-fired water heater.* Advanced Mechanical Technology, Inc., with AMTROL, Inc., as a subcontractor, is devel-

oping a residential gas-fired water heater that has the potential for achieving a service efficiency of 70 percent or more, which includes its effect on energy use for space conditioning. Development work has been started, along with evaluation of the market for such units. Following development of an efficient unit, a demonstration program will be implemented.

*Appliance test procedure.* An interagency agreement was arranged with the National Bureau of Standards (NBS) to evaluate the applicability of existing DOE test procedures to new units being developed. If the existing test procedures are found to be inadequate to indicate energy consumption in a fair and meaningful manner, NBS will develop modified or new test procedures. The only unit addressed to date is the heat-pump water heater being developed by EUS. For this type of water heater, a modified procedure will be developed and validated using a prototype unit furnished by EUS.

A new task to evaluate the separated heat-pump system, which heats water, such as the E-Tech unit, is being negotiated. Further areas that may need evaluation are the applicability of the existing procedures to the units being developed by the refrigerator-freezer and gas-fired water heater projects.

*Demonstration of Stirling/Rankine gas-fired heat pump.* Demonstration of the heat pump is cosponsored by the Gas Research Institute, General Electric, and DOE, with the work being performed by General Electric. This project uses a free-piston Stirling engine coupled with an inertial type Rankine-cycle compressor. This project should result in the marketing of the gas heat pump during the early 1980s. The market for this heat pump is initially intended to be the residential community and ultimately light commercial applications.

*Demonstration of a heat pump with a high seasonal performance factor.* This project will demonstrate a new gas-fired heat pump for commercial use, which, when taken over the entire seasonal load in the north central

---

*Research and Development of a Heat Pump Water Heater*, prepared by ORNL by Energy Utilization Systems, Inc., Vols. 1 and 2, ORNL/Sub-7321/1 and ORNL/Sub-7321/2, August 1978.

United States, will provide an exceptionally high coefficient of performance. The system will use a steam-driven turbine driving a fully modulating Rankine-cycle heat pump system. Feasibility of cost, manufacturing, and reliability will be demonstrated.

*Aerial measurement of heat loss.* This work involves the development, testing, and evaluation of methods of determining true surface temperatures from airborne infrared thermographic data of buildings. Such methods would permit rapid, large-scale, low-cost quantitative analysis of building heat loss.

Infrared radiation emitted from a building depends both on surface temperature and material. Thermal emittance of common building materials and coatings will be measured, and the degree of accuracy of measurement necessary for accurate temperature assessment will be established.

*Occupancy-based control systems.* The potential energy impact and savings associated with the use of Occupancy Detector Based Control Systems in four representative climatic regions of the country and various residential, business, and service building types will be studied. Demonstrations will be conducted in four retail stores in the Boston area on the control strategy of outside air allowed into the buildings based on the number of occupants. Relevant building codes and industry, health, safety, and comfort standards will be surveyed. An analytical model will be developed. Constraints presented by institutional barriers will be analyzed.

*Manhattan Plaza evaluation.* The effectiveness of an energy management system installed in an 800-unit multifamily apartment building in New York City will be evaluated. The operation of the unit will be compared to an identical 800-unit building situated side by side that does not have the system installed.

System operation has been monitored from August 1978 to the present date. Substantial savings have been demonstrated. The building owners have exercised their option to purchase a second system for the unequipped tower, which was installed in December 1979

and was partially funded by New York State Energy Research & Development Agency.

*Annual Cycle Energy System implementation program.* The annual cycle energy system (ACES) provides space heating and cooling and domestic water heating for residences, offices, and commercial buildings by means of a heat pump with thermal storage and, where required, solar assistance. The source of the heat for the heat pump is an insulated water-ice tank. At the beginning of the heating season, the tank contains all liquid. As the heating season progresses, the water is converted to ice. At the end of the heating season, most of the water has been frozen, providing stored ice sufficient for air conditioning for the summer. Future work will see the varous demonstrations through to completion, will further the development of second and third generation ACES equipment, and will emphasize those aspects of the program that will best promote the commercialization of the concept.

*ACES–office buildings.* The ACES office buildings project is the first application of the ACES concept in large office buildings. The facility will be used to educate the 17 member states and identify the best potential demonstration of the ACES concept. The state of Maryland submitted the best proposal for the ACES office building. The ACES concept is being used for an Elkton, Maryland, office building. To date, the ACES demonstration program indicates that larger, commercial-size ACES installations have the potential to be effective.

*Division of Federal Programs.* The Division of Federal Programs is responsible for encouraging the earliest possible acceptance of new and existing means for improving the efficiency of energy use in buildings and community systems. This division provides the transition from RD&D to market acceptance and commercialization for new and existing buildings and community systems technologies. Toward this end, it is responsible for providing overall economic and legislative analysis, program evaluation and measurement, and technology utilization and applications support to DOE. The division is also responsible for performing technology assessments and environmental assessments; developing overall marketing and commercialization strategies for DOE; developing new program initiatives for DOE; and ensuring efficient energy utilization by the federal government. Specific milestones for this effort are given in Figure 7-7.

The primary objectives of the Systems Analysis Branch are to assist DOE in determining programs that are to be undertaken and to assist ongoing programs in effecting the earliest possible acceptance of new and existing means for improving the efficiency of energy use in buildings and community systems. To meet this objective, the following functions must be performed on a continuing basis:

- Provide a decisionmaking structure for policy actions of DOE
- Provide a vehicle for strategy development
- Provide project evaluation and a framework for resource allocation
- Provide economic, regulatory, legislative, and systems analysis support for the entire DOE office, including the Residential Conservation Service and

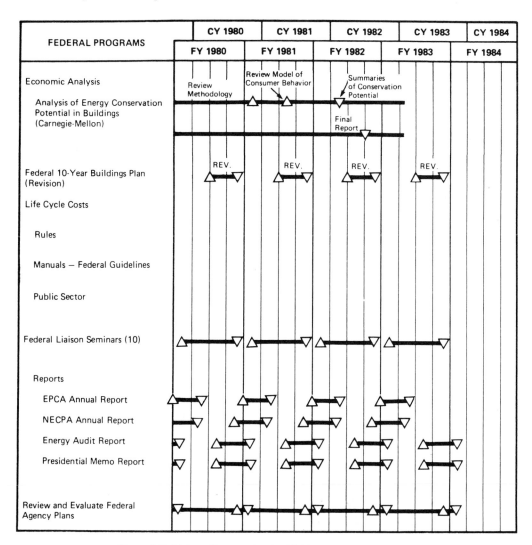

Figure 7-7. Major milestones for the Division of Federal Programs.

the Emergency Building Temperature Restrictions

- Provide inter- and intra-agency coordination of information and analyses requirements

The Systems Analysis Branch is responsible for development and implementation of all economic, regulatory, legislative systems, decision, and resource allocation analyses for the entire DOE. To meet these objectives, mathematical energy demand models are used extensively. Application of these models varies by type of tasks being accomplished. Using demand models and evaluation techniques, program priorities and trade-off decisions are made. The funding of projects that have the highest value and have the best promise of adoption by the private sector are selected. Another type of application of demand models is in the use of developed systems/economic analyses and input/output modeling for the best estimation of primary and secondary impacts of regulatory actions. Quantitative estimates of such items as regional energy savings—by fuel type and structure, employment, GNP, and environmental effects—are used in the regulatory process to establish levels or extent of regulation and to ascertain whether the same results can be achieved by other approaches, e.g., incentives.

The usefulness of these models in the management decision process is tempered with the understanding that the data used are of varying quality and that the model structures have no provisions for subjective considerations. Consequently, data improvement, model refinements, and expansions are ongoing. Examples of ongoing work for FY 1981 and beyond are the development of commercial buildings data having more statistical confidence, and the planned development of structural models for urban waste and the community systems sectors. Market response data will be developed from planned (e.g., appliance standards) and ongoing surveys plus National Energy Act-authorized programs (Residential Conservation Service, etc.). In FY 1980–1985 strategy, policy and legislative initiatives will be identified in areas such as

rebates, leasing arrangements, and removal of institutional barriers. Another area of effort will be regulatory analysis and standards promulgation.

Planning success can be measured by the progress made toward the DOE goals. To facilitate the evaluation responsibilities of this branch and to measure its own level of success, measures of national conservation progress have been defined and will be developed in FY 1980 and beyond. Furthermore, a retrospective analysis procedure will be developed and used in assessing program/project success during and after its commercialization.

FEMP is aimed at increasing energy efficiency in all federal operations including federal buildings. The federal government, a substantial energy consumer, accounts for about 2.5 percent of the total United States energy use. The buildings and facilities areas (approximately 400,000 buildings with a total area of 2.6 billion square feet) consume approximately 45 percent of all energy used by the federal government. DOE has the lead role in coordinating activities to reduce energy consumption in government facilities and operations. The program is conducted in close cooperation with other agencies, particularly the General Services Administration and the Department of Defense.

Executive Order 12003 requires DOE development of a 10-year plan for energy conservation in federal buildings and the development of an overall energy conservation plan by all agencies. The order sets goals of 20 percent energy-use reduction in existing buildings and 45 percent in new buildings by 1985. Guidelines and life-cycle costing methodologies are being developed for issuance to all executive agencies for use in agency plan preparation. These plans are to be submitted to DOE for approval within 6 months after issuance of the guidelines.

In order to demonstrate federal leadership in energy conservation, this branch will pursue the following strategies over the next 5 years:

- Provide remaining reports on the FEMP as required by legislation and executive order

- Realign and strengthen the federal government program to meet all mandated requirements for energy conservation and efficiency within the federal sector
- Develop the methodology and procedures based on demonstrated federal experiences for public sector use for (1) conservation of total energy use, (2) reduction in use of critical fuels, and (3) utilization of renewable energy sources.
- Develop demonstrated and efficiency energy-conservation and efficiency methods for public sector use in the near term (1979–1985).
- Use and demonstrate near-term backstop technologies for improved federal energy efficiencies in the near term and subsequent public sector use in the midterm (1985–2000).
- Develop and implement highly visible and cost-effective demonstrations of renewable resources in government buildings, facilities and general operations
- Develop the methodology to use replacement cost pricing in the FEMP
- Develop a more effective transportation program for the federal government and subsequent public sector
- Develop and implement a continuing technical evaluation process to identify new emerging technologies and existing energy-saving products and services for potential federal and public-sector use
- Develop and disseminate guidelines and criteria for reducing the use of critical fuels to include (1) switching from oil to natural gas and (2) use of coal in the new boiler installations.
- Develop an ongoing program for public education regarding proven federal government initiatives to (1) conserve energy use through energy efficiencies in buildings and general operations and (2) reduce consumption of critical fuels and use of noncritical renewable fuel sources

The major thrust of the Emergency Programs Branch is the Emergency Building Temperature Restrictions (EBTR) Plan (devel-

oped by President Carter and approved by Congress on May 10, 1979).

Nearly one-quarter of the energy consumed in the United States is used for space heating and cooling and production of hot water. Under the EBTR regulations, thermostats in most nonresidential buildings must be set no lower than 80°F for cooling, no higher than 65°F for heating, and no higher than 105°F for hot water. All nonresidential buildings in the country, including federal buildings, are covered by the plan with exemptions for hospitals, other health-care facilities, lodging facilities, elementary schools, nursery schools, and day-care centers.

Exemptions are also provided for buildings or portions of buildings containing special equipment, manufacturing, industrial, and commercial processes, or materials that need specific environmental conditions, or where health codes require higher hot-water temperatures.

It is estimated that approximately 5 million buildings are covered by the plan. Of these, about 75 percent are equipped with relatively simple heating and cooling systems, similar to those found in the average home. For the operator of such buildings, compliance with provisions of the regulations requires setting thermostats to 65°F when heating, or 80°F when cooling. To accommodate the various types of complex space-conditioning systems where simple thermostat settings are not possible, the program provides additional alternative methods to meet the 65°F and 80°F temperature requirements. EBTR also provides for individual state exemptions from the federal plan if the state can show that it has a comparable state plan in effect.

The Energy Policy and Conservation Act provides for civil penalties of up to $5000 for each violation of the regulations and criminal penalties of up to $19,000 for each violation. The regulations authorize DOE to delegate, to state governors who request it, the authority to implement, administer, monitor, and enforce the proposed rule.

Successful implementation of the program has resulted from active cooperation of the state and local government institutions. The power of state and local governments to educate and persuade their citizens has been tantamount to acquainting the public with the regulation requirements, its emergency conservation goals, and the necessity for voluntary compliance. Trained state and local health, fire, and building inspectors familiar with local conditions and local buildings are playing a crucial role in assisting in the program's enforcement.

*Consumer Education Program.* The Consumer Education Program focuses solely on the needs of the residential user. Informational material and specific subject-related outreach activities are developed primarily utilizing existing consumer networks. This program will focus on several areas of interest for the residential consumer. These include Appliance Label Consumer Education, Home Energy Management, and Home Weatherization and Efficiency Technologies. Approaches that will be used include regional workshops (Appliance Label Program), television/radio public-service announcements, how-to and informational factsheets (all of which are regularly updated), exhibit material for group presentations and library displays, media articles (newspapers and popular consumer journals), slide/tape presentations and traditional educational curriculum for secondary/postsecondary and college programs. Existing communication networks that will be used will include consumer and public-interest groups, educators and educational institutions, the media, DOE outreach networks, other federal organizations (such as Departments of Agriculture and of Housing and Urban Development), state and local governments, etc.

*Technical training and education.* To ensure that state-of-the-art technologies and applications and legislative programs are accurately and expediently "transferred" to the practitioners in the building sector, several technical training programs are planned in FY 1980. The approach is to design, test, and evaluate training programs including curriculum material for model energy-conservation building code, Building Energy Performance Standards, home building industry, training/licens-

ing for HVAC installers and service people, architectural faculty development, and engineering faculty development.

*FEMP projects.* Argonne National Laboratory provides technical support to the FEMP program in the following major areas:

- Development of mandatory building standards for federal buildings
- Development of building performance targets for federal buildings
- Analysis and evaluation of new technology possibilities for potential FEMP applications

The effort related to mandatory buildings standards is being initiated in FY 1980 to meet the legislative requirements specifying that DOE develop mandatory standards for lighting efficiency, thermal efficiency, buildings insulation, hours of operation, thermostat controls, and retrofit planning. The objectives of this critical effort at Argonne will concentrate on completion of the retrofit planning, lighting, and thermal efficiency standards.

The activity in FY 1980 to develop the energy performance targets for the various types of federal buildings is a new initiative to meet the investing legislative requirements. The activity will consist of an analysis of federal-agency energy audit data and the determination of typical energy consumption per square foot for the various building types. The targets will be based upon typical retrofit potentials for each building type and the typical consumption rates.

An analysis of new technology developments will be conducted in FY 1980 to evaluate the cost-effectiveness of testing and demonstrating new technology developments that may have significant new applications as FEMP initiatives.

There is a continuing project to prepare general guidelines for energy conservation by federal agencies in general operations. During FY 1979, a coniderable effort was made to produce the first documented guidelines. In FY 1980, the activity will continue the development of guidelines by adding the following areas not addressed in FY 1979:

- Maintenance guidelines
- Fuels utilization guidelines
- Contingency planning guidelines
- Guidelines for operations targets in selected agencies

The Life Cycle Costing (LCC) methods and procedures activity consists of additional developmental efforts to augment the completed LCC rulemaking accomplished in FY 1979. A life-cycle cost pamphlet will be prepared in FY 1980 for public-sector use. In addition, LCC seminars will be conducted for federal agency participation and education.

## BUILDINGS AND COMMUNITY SYSTEMS FUNDING

The DOE funding profile, shown in Table 7-1, reflects historical, current and projected levels through FY 1981.

## PROGRAM IMPACTS

### Projected Energy Savings

Many of the energy-conserving technologies that are being supported by DOE would be developed with difficulty, if at all, if federal support was unavailable. In some instances, the purpose of federal involvement is to introduce the technology at an earlier date than would occur in the marketplace without federal support. By accelerating introduction of a technology into the marketplace, the corresponding energy savings will be realized sooner than otherwise possible. The criteria for evaluating energy conservation RD&D opportunities and selecting projects for DOE support include an estimation and evaluation of the energy benefits or savings to be realized from federal support. The energy-benefit criteria are therefore keyed to both the total benefits from the project and the incremental or marginal benefits attributable to the federal program.

Table 7-1. Budget Profile for Buildings and Community Systems.

| | BUDGET AUTHORITY (DOLLARS IN THOUSANDS) | | |
|---|---|---|---|
| PROGRAM AREA | APPROPRIATION FY 1979 | APPROPRIATION FY 1980 | REQUEST FY 1981 |
| BUILDINGS | | | |
| Building Systems | 17,600 | 17,350 | 19,565 |
| Residential Conservation Service | 0[1] | 4,600 | 5,000 |
| COMMUNITY SYSTEMS | | | |
| Community Systems | 19,400 | 16,550 | 15,550 |
| Urban Waste | 8,500[2] | 13,000 | 10,100 |
| Small Business | 500 | 700 | 750 |
| CONSUMER PRODUCTS | | | |
| Technology and Consumer Products | 20,350 | 29,600 | 22,040 |
| Appliance Standards | 4,950 | 6,000 | 7,925 |
| FEDERAL PROGRAMS | | | |
| Analysis and Technology Transfer | 2,800 | 5,400 | 5,900 |
| Federal Energy Management Program | 500 | 400[4] | 2,700 |
| Emergency Building Temperature Restriction Program | 0[3] | 3,675 | 0 |
| PROGRAM DIRECTION | 3,533 | 5,137 | 6,120 |
| SUBTOTAL OPERATING EXPENSES | 78,133 | 102,412 | 95,650 |
| CAPITAL EQUIPMENT | 1,200 | 1,000 | 1,950 |
| TOTAL BCS | 79,333 | 103,412 | 97,600 |

1. Funded from internally reprogrammed DOE funds in the amount of $1,000,000.
2. Does not include $5,500,000 of Program Development and Demonstration funds.
3. Funded from internally reprogrammed DOE funds in the amount of $4,300,000.
4. A supplemental request for $2,300,000 has been approved by OMB and will be forthcoming.

Oak Ridge National Laboratory has developed residential and commercial energy-use models for DOE in evaluating energy impacts from 1985 to 2000. These computer models have generated baseline energy use projections for the residential and commercial buildings sectors. The models simulate energy use at the national level for different fuels, end uses, and building types. The models are sensitive to the major demographic, economic, and technological determinants of household and commercial fuel use.

The energy demand models have generated another set of energy use projections based on two additional factors: the conservation provisions in the National Energy Plan and the RD&D conservation activities being conducted by DOE. Table 7-2 presents the energy-use projections for 1985, 1990, and 2000 for both the baseline case and the conservation case. The resulting estimated energy savings between baseline 1985 and 2000 under the NEP and RD&D scenarios (expressed as quads/year) are:

| Buildings Sector | 1985 | 2000 |
|---|---|---|
| Residential | 2.4 | 4.3 |
| Commercial | 1.1 | 4.8 |
| Total | 3.5 | 9.1 |

Energy savings resulting from the development and implementation of urban waste-to-energy systems are projected to be 0.27 quad by 1985. The energy impact from replacing scarce fuels with wastes is projected to be 2.72 quads by the year 2000, plus an additional 0.1 quad savings from wastewater activities by 1990.

The projected energy savings resulting from the technology development programs supported by DOE will be influenced by several factors. Potential energy savings will increase with the infusion of more federal funds to support technology development efforts and to accelerate the pace of ongoing programs. Greater support of commercialization and technology transfer activities will also promote the wider adoption of energy saving

## Table 7-2. Buildings Energy Use Projections.[1]

| BUILDING AND ENERGY TYPES | QUADS/YEAR | | |
|---|---|---|---|
| | 1985 | 1990 | 2000 |
| BASELINE—RESIDENTIAL | | | |
| Electricity[2] | 11.3 | 13.2 | 16.9 |
| Gas | 5.7 | 5.7 | 5.3 |
| Oil | 0.1 | 2.2 | 2.4 |
| TOTAL[3] | 19.7 | 21.6 | 24.9 |
| BASELINE—COMMERCIAL | | | |
| Electricity[2] | 17.5 | 9.5 | 15.2 |
| Gas | 2.2 | 2.6 | 3.5 |
| Oil | 2.3 | 2.8 | 4.4 |
| TOTAL[3] | 12.1 | 14.9 | 23.1 |
| WITH NEP AND RD&D—RESIDENTIAL | | | |
| Electricity[2] | 10.3 | 11.9 | 14.4 |
| Gas | 4.7 | 4.5 | 4.1 |
| Oil | 1.7 | 1.7 | 1.7 |
| TOTAL[3] | 17.3 | 18.6 | 20.6 |
| WITH NEP AND RD&D—COMMERCIAL | | | |
| Electricity[2] | 7.0 | 8.6 | 13.3 |
| Gas | 1.9 | 2.0 | 2.3 |
| Oil | 1.9 | 2.1 | 2.7 |
| TOTAL[3] | 11.0 | 12.7 | 18.3 |

[1] Does not include Urban Waste Program
[2] Electricity is in primary energy, 11,500Btu/kWh
[3] Totals include use of "other" fuels (liquefield gases, coal) not included elsewhere

technologies at the earliest possible time, resulting in greater energy savings nationwide.

Additional projected energy savings within DOE are summarized as follows. Commercial introduction of new building-design concepts will reduce home energy consumption by as much as 50 percent in a single-family unit. The Minimum Energy Dwelling (MED), for example, is using roughly half of the energy required for a conventional home. The MED prototypes will lead to commercialization of hundreds of similar new homes in southern California. Market introduction of the MED began in 1980.

Programs aimed at the federal buildings sector should also produce substantial net energy savings. As mentioned previously, the federal government is a substantial energy consumer, accounting for approximately 2.5 percent of total United States energy use. Almost half of the energy used by the federal government is consumed in approximately 400,000 buildings with a total area of 2.6 billion square feet. Implementing the requirements of Executive Order 12003, the Energy Policy and Conservation Act (EPCA), and the National Energy Act (NEA) will result in energy savings of 128,000 BDOE by 1985. By the year 2000, the savings will have increased to an estimated 184,000 BDOE. Current programs for federal buildings include constructing more energy-efficient buildings, retrofitting existing buildings, and implementing various energy management measures and techniques.

Potential direct energy savings associated with each functional area of the planning and design process are not readily apparent. One of the objectives of the program is to identify energy consumption and potential savings for each activity. In addition, each activity area constitutes part of a framework for implementing systems applications to achieve communitywide energy savings. The program thus establishes an institutional framework for adoption and acceleration of the implementation of integrated energy-efficient options and energy conservation technologies.

Energy conservation options being incorporated in project planning and design of new site and neighborhood developments will result in energy savings through reduced energy demands from the subcommunity level. Between 1980 and 1990, more than 120,000 separate developments varying in area from 10 acres to over 2000 acres could be affected directly by ongoing and planned programs. Several energy-efficiency community development regulations and energy-conserving designs will be developed for adoption and use by communities and developers. The program could result in adoption of energy conservation features by about 10 percent of the projected number of new developments in 1985 and by 40 percent of the new developments in 2000. The widespread application and transferability of the energy conservation options being disseminated will realize energy savings of 100,000 BDOE by the year 2000. Overall, the program is estimated to have the potential of saving 5 quads by the year 2000.

Energy savings from implementation of improved municipal energy systems are available through conservation from increased efficiencies and through direct energy recovery from use of municipal wastes as energy resources. The total potential is difficult to estimate and is dependent on many site-specific factors, but it is estimated to be about 0.4 quad by 1985 and about 5 quads by 2000.

Energy recovery from wastes can contribute 2 quads from direct energy recovery and an additional quad from materials recovery leading to conservation in material processes. District heating and integrated energy systems are expected to provide up to 2 quads by the year 2000.

Energy savings in the technology and consumer products RD&D area will accrue from development and commercialization of a broad range of technologies. For example, it is estimated that the concerted near-term conservation effort underway in this program area can result in a 30 percent reduction in the electrical energy consumed by lighting. This related to better than 1 quad or 293 billion kWh of energy saved per year.

The Annual Cycle Energy System (ACES) currently being demonstrated appears promising as a new design concept that will be incorporated in many new homes beginning in 1981. This represents an acceleration of 10 years. ACES implementation efforts in both the residential and commercial sectors will result in an estimated savings of 10,000 BDOE by 1985 and a significant 200,000 BDOE by 2000.

Advanced thermally-activated pump developments will improve the efficiency of these units by reducing the amount of energy required for operation. Market penetration will include both the commercial and residential sector as well as the new and replacement installations. Installed gas furnaces generally operate at seasonal efficiencies of 40 to 60 percent. The gas heat-pump options will at least double this efficiency (the average installed seasonal efficiency of a Stirling engine heat pump being developed is 1.4 in heating and 1.0 in cooling). The market penetration potential, which is roughly 1.5 million units for 1985, should reach 2 million by 1990. Of this total, the estimated number of gas-fired heat pumps in operation will be 55,000 by 1985 and 240,000 in 2000. Estimated energy savings resulting from commercialization of thermally-activated heat pumps are 10,000 BDOE by 1985 and 200,000 BDOE by 2000.

In the consumer marketing and financing area, the Energy Cost of Ownership (ECO) Program, aimed at promoting consumer adoption of this concept when making purchasing decisions, is currently being demonstrated in six pilot cities. When expanded on a nationwide basis, the ECO market potential can be expected to increase. As energy prices increase (thereby increasing energy-related costs of ownership and maintenance), further savings will be derived from the ECO Program. Energy savings resulting from the program are estimated to be 100,000 BDOE by 1985 and 600,000 BDOE by 2000. Other targeted marketing programs relating to low-cost/no-cost conservation actions should translate into savings of 1 million BOE per day by 1985.

There are currently about 12 million oil-fired space-heating systems in use in the United States. Seasonal furnace efficiency averages 55 percent; however, there is the potential for raising average efficiency to 70

percent with the use of retrofit fuel-saving devices that are now commercially available. By using appropriate fuel-saving devices, annual savings could be as much as six barrels of oil per residence. By the year 2000, the cumulative nationwide fuel-oil savings could total 300,000 barrels of oil per day. DOE program activities to stimulate widespread use of these devices will result in an estimated savings of 200,000 BDOE by 2000.

The financing programs encourage home buyers to invest in energy-efficient measures, including insulation, storm windows, and automatic setback thermostats. These programs are aimed at single-family homes, which comprise the largest share of buildings and have an annual turnover (sale) rate of approximately 12 percent. An estimated 20 percent of the energy used in a home could be saved by implementing such measures. An estimated 50,000 BDOE is expected to be saved by 1985 through energy-efficient measures financed in this manner. By 2000, the potential savings are estimated to be 200,000 BDOE.

By encouraging the use of energy-saving technologies and management techniques, the Small Business Energy Conservation Program will result in reduced business-sector energy consumption by an estimated 100,000 BDOE by 1985 and 300,000 BDOE by 2000. These figures are based on the assumption that approximately 70 percent of the sector will participate in the program, and energy use is reduced by 20 percent. This energy impact will be the result of both an accelerated and wider use of energy saving methods, many of which are being developed in other DOE programs.

## Accomplishments

The key accomplishments of the DOE to date are summarized below:

- Developed, implemented, and promulgated proposed BEPS for new buildings as mandated by P.L. 94-383.
- Twenty-five states have implemented training programs for building-code of-

ficials. The training curriculum was developed through the National Conference on State Building Codes and Standards on the "Model Code for Energy Conservation in Buildings."

- Test procedures proposed for heat pumps and administrative procedures leading to final rulemaking are under way.
- Seventeen communities completed the first year of effort to prepare a comprehensive community energy management program.
- Grid-connected ICES for Health Educational Authority of Louisiana (HEAL) in New Orleans, University of Minnesota, Clark University and Trenton are in final design.
- The 5000-gallon-per day ANFLOW sewage-treatment pilot plan in Oak Ridge, Tennessee, has completed 2 years of operation. Promises 75 percent energy savings in small-scale treatment facilities, producing methane-rich gas and meeting secondary level effluent standards.
- A 50-to-100-ton-per-day facility for digesting mixed urban waste and sewage sludge to produce methane was completed and operated for 1 year at mesophilic (95°F) baseline conditions. The plant is expected to produce 400 million cubic feet of methane per year.
- Twenty cities were selected to begin developmental work to lead to the demonstration of a broad range of urban waste technologies. The cities chosen represent a broad range of sizes and stages of completion of project planning.
- An additional prototype of Stirling/Rankine heat pump for residential application was generated in the laboratory to collect data on the unit's efficiency, reliability, serviceability, and cost of operation. The 120-ton Rankine heat-pump demonstration system was fabricated and operated in the laboratory. Testing was initiated on prototype units of the absorption cycle system and the Brayton/Rankine 7.5-ton heat pump.

- A program to characterize the performance of all commercially available residential oil-fired heating equipment and refit options was completed. Approximately 200 units were tested. Prototype equipment with high-efficiency potential was also identified.
- Completed prototype design and evaluation of a high-efficiency refrigerator-freezer that is 33 percent more efficient than existing models. A 50-unit consumer-oriented demonstration is underway.
- Initiated an oil equipment marketing pilot program in Springfield, Massachusetts, targeted at 50,000 oil-heated homes involving 25 dealers in direct marketing; initiated a program in Nassau and Suffolk counties of New York involving 50 dealers targeting 175,000 homes, using direct-mail marketing techniques. Completed the planning for statewide implementation to follow the pilot demonstration.
- "Project Payback," an innovative test program conducted jointly with retailers and manufacturers to encourage consumers to consider the economic advantages of buying more energy-efficient products, was run in six cities.
- A demonstration involving four American and two Canadian energy-usage display meters has been completed. The meters were installed in hundreds of homes for test purposes. Use of these meters has demonstrated a 15 percent electricity savings in the test homes.
- Proposed Rule for Residential Conservation Service (Title II, NECPA) was published.
- The federal government, in its internal operations, saved 44 MBDOE since the base year of 1975 relative to the Federal Energy Management Program.
- Advance Notice of Proposed Rulemaking relating to establishment of standards for nine product categories was published.

# 8. Industrial Programs

## PROGRAM OBJECTIVES AND STRATEGY

The industrial sector consumes an estimated 36 percent of all the energy used in the United States and most industrial processes are extremely inefficient in their use of energy. In some direct-heating applications, efficiencies are as low as 10 to 15 percent. Some of the more efficient processes, such as steel-making, are only about 30 percent efficient. The inefficiency is due primarily to the industrial complex being developed during several decades of abundant and low-cost energy. Figure 8-1 on the following page shows the increase in industrial energy consumption from 1947 to 1975, as well as the increased dependence on petroleum and natural gas consumption. Figure 8-2 shows the energy consumption by industrial sector.

Although it is not possible thermodynamically to achieve 100 percent efficiency in industrial processes, it has been estimated that 30 to 50 percent of industrial energy could be saved with universal application of existing, emerging, and advanced technologies. Such an achievement would represent 10 to 20 percent of the total United States energy consumption.

Widespread reconstruction of existing plants to use today's best available technology is clearly not feasible for economic and financial reasons. The existing capital stock in industry is estimated to have a present value of $750 billion. Selective retrofitting of the most promising current technologies is practical, however, and, in the longer term with increasing energy prices, industry is likely to develop and adopt more energy-efficient tech-

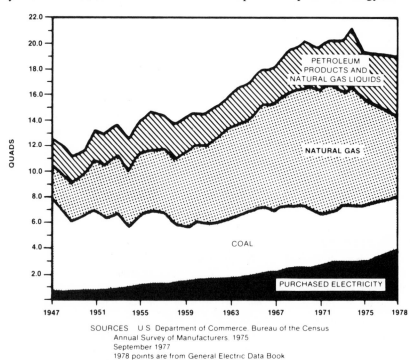

Figure 8-1. Industrial energy consumption for fuel uses.

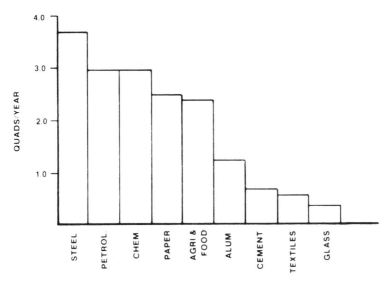

Figure 8-2. Industrial energy consumption by industrial sector.

nologies. The major issue is whether the rate of improvement will be adequate to meet national energy goals. Industry will traditionally wait for energy price increases before developing and adopting more energy-efficient processes. Furthermore, energy costs represent a small part of the product costs, even in the most energy-intensive industries. It would appear that the private-sector development and adoption of processes with increased energy efficiency would generally be in series with energy price increases, therefore incurring a delay in realizing the savings. Accelerated energy conservation is constrained by economic and technological uncertainties. Questions of corporate strategy and capital allocation to R&D and actual cost-reduction investments are linked to economic factors such as fuel price regulations, taxes, and tax credits. In addition, the technical and economic feasibility of many important conservation technologies, including some existing technologies, has yet to be proven in industrial operating environments.

A federal program targeted at mitigating these economic, technological, and institutional uncertainties and risks can have major impacts on industrial options for using selected existing but inadequately used technologies; for energy-efficiency improvements; and investment decisionmaking for energy conservation. Consequently, the DOE Indus-

trial Energy Conservation Program focuses on the following:

- Existing but underutilized technologies for which a federal action can be identified to stimulate implementation
- New technologies from RD&D to provide advanced technology with proven economic and technological feasibility in operating environments
- Incentives, such as tax credits, to provide economic advantages for industrial actions in the national interest
- Other legislated actions to establish requirements and motivation for action by industry
- A market-oriented commercialization effort to ensure accelerated transfer of technology focused on specific related industrial end users and maximum implementation of these technologies

A major issue in federal policy and program planning is estimating the effect of federal action over what industry is likely to achieve on its own. To assess the value of a federal program, the major constraints to accelerating the rate of industrial energy conservation have been identified and evaluated, leading to the identification of alternative federal actions in industrial energy conservation.

Industry has conserved and will continue to

conserve energy on its own, and several factors indicate that industry will move independently to save energy: a historical trend over the past 20 years averaging about 1 percent savings annually; an acceleration of this trend since 1972 resulting from sharply increased energy prices and initiation of voluntary energy management programs; and future incentives for conservation in the form of higher energy costs and relatively greater economic advantages for investments in energy conservation.

However, maintaining these trends in conservation by private industry alone may be limited. According to industry reports, the recent accelerated improvements have been achieved through housekeeping and small investment retrofits for which the potential is now somewhat depleted. Further results depend principally on larger capital investments, and the trend toward significant improvement in energy efficiency will require overcoming more challenging technical, economic, and institutional barriers.

The extent to which federal programs can accelerate industrial energy conservation in the future depends on the effectiveness of the overall program in mitigating major barriers in the energy-intensive industries.

There are over 310,000 manufacturing concerns in the United States and each has unique characteristics—consumption, technology base, financial capability, degree of innovativeness, and investment decisionmaking. The Industrial Energy Conservation Program focuses on processes applicable to all industry (a horizontal thrust) and on processes of the most energy-intensive industries (a vertical thrust) to increase the energy efficiency and to substitute more abundant fuels for scarce natural gas and oil.

A 13.5 percent reduction in industrial energy per unit of manufacturing production is possible in the 1972–1980 period, based on technological feasibility and economic practicability (voluntary industrial energy efficiency improvement targets, part of the Energy Policy and Conservation Act, Title III, Part D). The program seeks, therefore, to remove the technological and economic barriers to achieving even greater savings. More specifically, the estimated impact of the program will be 1.5 quads in 1985 and 5.5 quads in 2000 for RD&D activities alone. The specific savings that will result from the new effort to stimulate the application of the existing but underutilized technologies and those resulting from EPCA and the NEA have not been specifically determined but are expected to be significant.

Achievement of the voluntary energy conservation goals established in response to EPCA would result in a savings of over 2.8 quads annually in 1980, as shown in Table 8-1, but it is not possible to determine what portion, if any, is due to federal involvement.

Table 8-1.    Savings Resulting from the Energy Policy and Conservation Act.

| INDUSTRY | FINAL NET TARGET (% REDUCTION IN ENERGY/ OUTPUT) | ENERGY SAVED IN TARGET YEAR—1980 QUADS/YEAR |
|---|---|---|
| Chemicals and Allied Products | 14 | 0.67 |
| Primary Metal Industries | 9 | 0.48 |
| Petroleum and Coal Products | 12 | 0.46 |
| Stone, Clay, and Glass Products | 16 | 0.28 |
| Paper and Allied Products | 20 | 0.32 |
| Food and Kindred Products | 12 | 0.14 |
| Fabricated Metal Products | 24 | 0.14 |
| Transportation Equipment | 16 | 0.11 |
| Machinery, Except Electrical | 15 | 0.11 |
| Textile Mill Products | 22 | 0.13 |
| | | 2.84 |

## Objectives

The federal objectives for energy conservation in the industrial sector are as follows:

- Achieve maximum penetration of *existing* and *new* energy conservation technologies in as short a period as possible
- Substitute, where possible, abundant fuels for scarce fuels
- Minimize the energy loss embodied in waste streams of all types (discarded products, materials, and energies)

## Federal Role in Industrial Energy Conservation

The federal role relative to carrying out the legislated mandates is obvious. For instance, the EPCA (P.L. 94–163) and the NEA are being fully supported, as other legislated activities will be during the 5-year period.

The government role in documenting and disseminating information pertaining to industrial conservation is similar to past efforts of the Department of Commerce or the Department of Agriculture. The key difference, however, between those efforts and the commercialization effort of the Industrial Energy Conservation Program is the very selective nature of the industrial technologies chosen for development. The technologies that are energy-conservative but underutilized by the private sector will be identified and analyzed to determine why the lack of market penetration exists and to ascertain whether the technology needs federal actions to stimulate its increased use. The federal role relative to existing technology is, therefore, primarily one of analysis and dissemination of pertinent information to the specific end-use industries. Some of these underutilized technologies will require proof-of-concept demonstrations to show the merits, whereas tax credits or other incentives might be the answer in other instances.

The federal role in industrial energy conservation, particularly RD&D, is not clearly understood. Many think of large industries with billions of dollars and large research staffs and do not immediately understand why there should be any federal funding of RD&D for saving energy in industry. Many immediately conclude that industry will do the necessary RD&D on its own and that federal monies can be more effectively distributed elsewhere.

Industry will, of course, achieve significant energy savings on its own; historically, it has averaged a little over 1 percent per year improvement in annual energy savings. It is important, however, to examine more closely the industrial capital and RD&D investment decision processes to see what industry will *not* do without federal stimulus.

Major industrial capital investment decisions are strongly influenced by factors beyond those of simple profit maximization, although rate of return is the single most important element in a capital investment decision. A company will not risk shutdown or loss of market position for small gains in expected profit. Investments that favor growth or retention of market are normally preferred over those with lower operating costs, even if both offer the same opportunity to generate profits.

Advanced energy conservation technologies are usually considered high-risk projects by industry since they are unproven in industrial environments. Process changes directed toward energy savings usually affect other process parameters as well. Every change in an industrial process relates to changes in risk of slowdown or failure, which may far outweigh the potential energy and cost savings to be gained.

Capital investment budgets of the industrial sector are allocated by widely varying priorities, which are generally grouped as "mandatory" and "discretionary." Energy conservation investments can be in either category, although they are most often placed in the discretionary group unless they relate to continued energy supply or survival. The cost of energy constitutes a relatively small part of product costs in the energy-intensive commodity industries, as shown in Table 8-2; energy conservation investments are generally considered only after investments are complete for product or market development, Occupational Safety and Health Administration and Environmental Protection Agency require-

Table 8-2.   Relative Energy Costs for All
Manufacturers, 1975.

| COST CATEGORY | COST ($ BILLIONS) | FRACTION OF VALUE OF SHIPMENTS |
|---|---|---|
| Purchased Energy | 23.19 | 2.28% |
| Wages and Salaries | 209.96 | 20.63% |
| Materials | 558.52 | 54.87% |
| Other Cost and Profit | 226.18 | 22.22% |
| Value of Shipments | $1,017.85 | 100.00% |

Source: Bureau of the Census, U.S. Department of Commerce, Annual Survey
of Manufacturers, 1975.

ments, and capacity improvements. Energy
conservation investments, as confirmed in a
1977 survey by Energy and Environmental
Analysis, Inc., of capital budgeting practices,
are treated in much the same manner as other
investments. However, they are often placed
in a category requiring a much higher return
on investment because of the associated risks.
Industrial decisions are made with management judgment applied after some form of
quantitative analysis.

Of the energy conservation options being
considered, industry will more likely pursue
those involving low-to-moderate technical and
economic risk and high return on investment,
and those relating to continued energy supply.

A survey of corporate R&D spending of
683 American companies provides some significant insights as to which industries are
dominant in overall R&D.* Table 8-3 displays

*Business Week, July 2, 1979.

some of these data. Two steel companies, for
instance, account for nearly 70 percent of the
total R&D expended in the steel industry.
Similarly, three companies conduct 43 percent
of the R&D for the widely fragmented textile
industry.

It would appear, therefore, that the results
of energy conservation R&D conducted by
industry on its own would likely be held proprietary by a few dominant companies,
whereas federally cost-shared RD&D results
would be available to all industries. Government involvement thus enables equitable and
more widespread dissemination of new energy
conservation technology.

Targeting RD&D efforts by the federal government requires a closer analysis of the purpose of the private-sector R&D expenditures,
more specifically, identifying the industries
that are investing strongly in energy conservation on their own. A recent analysis of this
type revealed that the petroleum refining and

Table 8-3.   Industrial Sector R&D Expenditures.

| INDUSTRY | TOTAL R&D ($ MILLIONS) | TOTAL R&D (% OF PROFIT) | % SHARE OF R&D BY DOMINANT CO'S | DOMINANT COMPANIES |
|---|---|---|---|---|
| Chemical | 1705.6 | 41.3 | 59.0 | Dow, Dupont, Carbide, Monsanto, Cyanamid |
| Food Processing | 393.9 | 15.3 | 29.7 | CPC Int'l., General Foods, General Mills, CPC Int'l. |
| Metals & Mining | 164.3 | 14.3 | 34.4 | Alcoa |
| Aluminum Only | 96.7 | 16.8 | 58.5 | Alcoa |
| Paper | 162.3 | 12.9 | 64.6 | Weyerhaeuser, Kimberly-Clark, Scott |
| Steel | 132.7 | 19.9 | 67.5 | U.S.S., Bethlehem |
| Textiles | 37.3 | 14.0 | 43.4 | Burlington, Albany Int'l., Fieldcrest |

chemical industries are directing significant R&D funding to energy conservation, and the aluminum industry allocates a significant portion of R&D investment to energy-efficiency improvements. These facts dictate that a greater degree of care be given to the development of a federal role in involvement with these industries and that additional analysis be required to avoid redundancy of effort. This does not necessarily mean, however, that there should be no federal role.

Federally cost-shared RD&D will increase the rate of private-sector R&D expenditures and will significantly accelerate the introduction of new higher risk, higher potential programs with energy savings earlier in time and at significantly less federal cost than many of the supply options. The federal participation with key industries ensures that the RD&D is performed by the most competent talent available in the nation's leading research-oriented corporations and ensures wide dissemination of the RD&D results. The federal leadership enables development of cooperative inter-industry projects such as the energy-integrated industrial park, which may not be pursued by industry alone. In addition, the federal involvement will help industry understand and more readily adapt to required regulations.

In summary, there is a role for the federal government in industrial energy conservation through the techniques of RD&D, economic incentives, and industrial reporting programs. The emphasis would be on identifying existing but underutilized technologies, developing new energy-saving technologies that are not redundant to the efforts of industry alone, and stimulating the early implementation of such results.

## Federal Program Strategy

The basic strategy is a program of cost-shared RD&D of selected energy-conservation technologies directed at processes that apply to a wide spectrum of industries and processes that are specific to the most energy-intensive industries. Together with a strong emphasis on engineering development and full-scale demonstration in industrial environments, significant program effort is placed on the

identification and transfer of existing but underutilized technologies, processes, and techniques to achieve energy conservation in the industrial sector. At the levels of basic and applied research, the program is closely coordinated with the Office of Energy Research. Activities are selected on the basis of high energy-saving potential, acceleration of implementation, nonredundancy with efforts of private industry, the degree to which benefits accrue to fragmented industry without research funds, and the degree and appropriateness of cost-sharing. Candidate projects are selected based on extensive analysis of risk, cost, and benefit.

Selectivity and focus are key elements of the strategy. For the earliest maximum impact, the projects must be characterized by high energy savings, nearer term to realize the savings, nonredundancy to industry efforts, potentially acceptable to industry after the completion of federal actions, and environmentally acceptable and operationally safe. The existing but underutilized technologies and practices will be screened to include only those that have high conservation potential and for which a reasonable federal role can be established (incentives, assistance, demonstration, etc.). The body of a new conservation technology will be derived from technologies developed by the Office of Energy Research or other federal research organizations that are considered ready for commercialization or for demonstration in the industrial sector and from the Industrial Energy Conservation RD&D program.

The projects for Industrial Energy Conservation RD&D are primarily engineering developments of proven concepts and not basic research. These projects are usually initiated as proposals from the private sector and academia. They are carefully screened and prioritized by a rigorous evaluation of cost, energy savings, clear establishment of a federal role, competitive market penetration analysis, and environmental impact. Only those not being pursued by the private sector on its own are considered, and cost-shared contracts are initiated on the most promising activities. Every attempt is made to obtain cost-shared relationships with representative end-user companies for the demonstrations; with equipment manufacturers having the capacity to supply the market once the project is successful; and with representative trade associations. Having such a cost-shared program provides a vested interest of the elements of the private sector, which will ultimately implement the technology. Having industrial "knowns" actually performing and contributing to the project has the effect of accelerating the market penetration once it is successful.

The RD&D is directed at two industrial targets: (1) the energy-intensive generic technologies having wide application across the industrial spectrum and (2) the energy-intensive processes of the most energy-intensive industries, of which a few constitute a major portion of all industrial consumption. By this approach—horizontal and vertical thrust—the relatively low per-unit energy-savings ideas, with very large numbers of applications, and the relatively high per-unit energy-savings ideas, with a small number of applications, are both captured and selected.

The degree of federal effort required to effect suitable impact varies, depending on the technology and range of applications. The industry-specific technologies, for example, are expected to require fewer demonstrations than the generic technologies that have a diverse number of substantially different applications. Some efforts, such as waste heat recovery, are expected to require second- and third-generation technologies that build upon the results of preceding development to cover all applications.

The critical task of moving existing and new technologies expeditiously into the marketplace and into the processes of industry is equally important. The objectives of industrial energy conservation cannot be achieved unless the private sector itself puts the results to work. The Office of Industrial Programs recognizes that its activities and initiatives must impact the decisionmakers in the private sector to be successful. Energy can only be saved where it is used.

The process of getting the technologies implemented by the private sector—called technology transfer, commercialization, market-

ing, outreach, etc.—is complex for the industrial sector. Unlike other sectors, such as transportation or in residential/commercial, there is no broad, readily understood market. The industrial market is highly diverse because each industry has different requirements, capital conditions, asset turnover rates, and differing degrees of innovativeness. Therefore, it is not effective to disseminate the particulars of a given technology to industry in general since most of those reached by such methods will not be concerned with that specific technology.

Each market must be analyzed to assess its particular needs, timing, and other characteristics. A federal action that is effective with one industry is not necessarily effective for another industry with different characteristics. Close integration with appropriate trade associations provides a valuable avenue for planning the proper federal actions for specific technologies within specific industries. The planning of commercialization starts with the beginning of the project, and the market potential is a key factor in the project selection process. The planning and scheduling of each project are inclusive of the commercialization actions required to develop, and ultimately to implement, the project.

Commercialization of a technology or practice includes potential elements depending on the individual situations. In some cases, it is sufficient to transfer the related information to the specific industries which, upon seeing the economic benefits and proven nature of the concept, will readily implement it at an acceptable rate. Other industries might require more tangible evidence of success and may want to see the demonstration unit in operation. In some instances, incentives might be required to stimulate the industrial acceptance of new or existing concepts, e.g., tax credits for energy conservation equipment in the National Energy Act. The results of the Industrial Energy Conservation Program will be closely monitored to establish a measure of its impact and to identify needed improvements in the commercialization process. The current industrial reporting program—that is, direct reporting and reporting through trade associations—provides a vehicle for assessing overall program impact. The specific market penetrations of individual projects will be tracked to get a more specific indication of the program's effectiveness.

In summary, the strategy of the Industrial Energy Conservation Program is to select the mośt energy-conservative techniques that exist today; develop technologies that industry (for various reasons) will not develop on its own; effectively transfer the technologies to the private sector; and stimulate the rapid penetration by the usual marketing practices, i.e., documentation, seminars, films, television spots, trade shows, and, where effective, federal incentives.

## Program Management Strategy

The Assistant Secretary for Conservation and Solar Energy exercises overall management policy and program responsibility for the Office of Industrial Programs through the office director, as shown in Figure 8-3. The industrial energy conservation program functions are performed by two divisions within the

office; staff functions (program planning, budgets, schedule control systems, etc.) are performed under the direction of the deputy director.

The Division of Conservation Research and Development conducts the generic RD&D programs of waste energy recovery and utilization, industrial cogeneration, and alternative materials utilization as well as the RD&D programs for industry-specific technologies of high-temperature processes, low-temperature processes, manufacturing processes, and agriculture and food processes. In addition, the existing but underutilized industry-specific technologies are to be identified and analyzed by this division.

The Division of Conservation Technology Deployment and Monitoring has a diverse role in transferring the existing and new technology products to the industrial sector and accelerating their implementation as well as monitoring the progress of such implementation. The programs of this division include the implementation of specific legislated programs, such as the EPCA. They also include

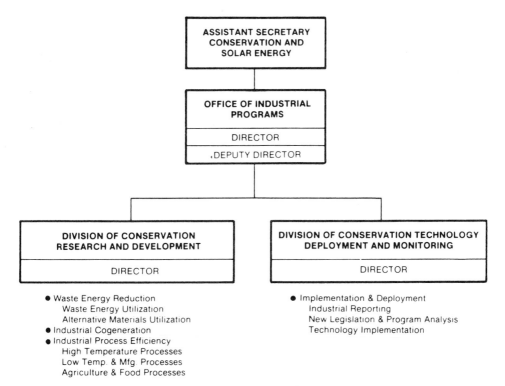

Figure 8-3. Organization chart showing the relationship between the Assistant Secretary for Conservation and Solar Applications and the Office of Industrial Programs.

initiatives encompassing the various techniques of disseminating technology results, exhibits, trade shows, seminars, documentation, films, television spots, and trade association interfacing.

The Office of Industrial Programs will closely coordinate its efforts with other DOE organizations and other federal agencies to ensure maximum integration of related technologies and to avoid redundancy of efforts. Examples of integrated activities with other federal agencies include cofunding projects where common interests exist (such as preheating of glass feed and hyperfiltration of textile dyestuffs—DOE/EPA); support to the Treasury Department on additional tax credits for conservation equipment; and joint efforts of DOE and the Department of Agriculture.

## MAJOR PROGRAM THRUSTS

### Legislative Framework

**Division of Conservation Research and Development.** The Federal Nonnuclear Energy Research and Development Act of 1974 (P.L. 95–577) provides the authority for this division's program functions in the areas of waste energy utilization, alternative materials utilization, industrial cogeneration, high-temperature industrial process efficiency, and low-temperature process efficiency.

**Division of Conservation Technology Deployment and Monitoring.** The National Energy Conservation Policy Act (NECPA), P.L. 95–619, signed in November 1978, amends the Energy Policy and Conservation Act (42 U.S.C. 6341–6346). It requires the expansion of industrial energy-efficiency reporting and initiates reporting on recovered materials for the paper, metals, textile, and rubber industries. It also mandates the establishment of recovered-materials targets for those industries. The NECPA requirements to be implemented by this division include evaluation of the feasibility of performance standards for energy-intensive industrial equipment, specifically, pumps and motors, and the relevance of the second law

of thermodynamics to industrial energy conservation programs.

The Energy Tax Act of 1978 (P.L. 95–618) requires DOE consultation with the Treasury Department in developing tax credit regulations for investments in specially defined energy equipment and recycling equipment.

Conservation technology implementation activities are authorized by: (1) Title III of the Energy Policy and Conservation Act; (2) the DOE Organization Act (P.L. 95–91), which transferred various responsibilities of the FEA and the Department of Commerce to DOE; and (3) the Federal Nonnuclear Energy Research and Development Act of 1974 (P.L. 93–577).

### Specific Program Activities

**Conservation Research, Development, and Demonstration.** The industrial energy conservation RD&D activity focuses on increased end-use efficiency and substitution of abundant fuels for scarce fuels in the processes having wide application across all industry and the processes of the most energy-intensive industries. These RD&D thrusts include waste-energy reduction (the horizontal thrust) and industrial process efficiency (the vertical thrust), which are described in detail in this section.

**Waste-Energy Reduction, Including Cogeneration.** The FY 1981 request for waste-energy reduction, including cogeneration, was $31,800,000. This activity is directed at reducing the energy loss through waste energy of industrial/agricultural processes by improvement in equipment and processes that are common to many industries. Activity areas and some typical projects are illustrated in Table 8-4.

As illustrated in Table 8-4, specific activities include combustion efficiency improvements, industrial cogeneration systems, reduced waste, waste materials as feedstocks and as fuels, waste heat recovery, and effective reuse of recovered waste heat such as in energy-integrated industrial parks. The total potential savings for the specific waste-en-

Table 8-4.  Waste Energy Reduction Activities Areas and Projects.

| ACTIVITY AREA | IMPROVED FUEL CONVERSION | IMPROVED ENERGY USE EFFICIENCY | RECOVERY OF WASTE ENERGY | RE-USE OF RECOVERED WASTE ENERGY |
|---|---|---|---|---|
| Typical Projects | Air-Fuel Ratio Control | Fluidized Bed Waste Heat Boiler | Waste Heat Recuperators | Conversion of Waste Polypropylene to Oil |
| | $O_2$ Enriched Combustion | Reduced Waste in Glass Making | Industrial Heat Pumps | Methane from CO |
| | Industrial Cogeneration Systems | | Waste Tire Reclamation Remanufacturing | Refuse-Derived Fuel in Cement Kilns |

ergy-reduction activities is conservatively estimated as ranging from 6.5 to 9.1 quads annually by the year 1985.

There are numerous current projects in progress in waste-energy reduction forming the nucleus of the projected plans through 1984. These include five high-temperature recuperator demonstrations; three industrial heat pump developments; cogeneration efforts just being initiated by Program Opportunity Notices (PON) and Requests for Proposals (RFP); a parametric study of an energy-integrated industrial park; an air/fuel ratio control being demonstrated; and numerous analytical and laboratory efforts directed toward alternative materials utilization. The projects will be bolstered selectively by additional similar projects focused on other promising industrial applications that are not presently covered and on additional activities based on results of the current efforts. For instance, the air/fuel ratio control system being demonstrated on an industrial boiler can be applicable to multiburner direct-heating operations as well. The principal groups of activities in waste-energy reduction (alternative materials utilization, waste-energy recovery, and utilization and industrial cogeneration) are further described as follows.

The current activities in alternative materials utilization consist of (1) analytical or laboratory-scale experimentation, which will evolve into a significant concentrated effort during the projected 5-year period, and (2) specific development and demonstration projects. Because materials are essential to all industrial operations, the activities in alternative materials utilization are applicable to

many industrial sectors and processes. Materials accumulate energy value as the products progress through the various stages of industrial development.

The alternative materials utilization activity is characterized by the following efforts that have emerged during initial development: substitution and conservation tasks involving shifting feedstock sources from oil and natural gas to coal, wood, or various industrial wastes; size-reduction efficiency improvement, which is basic to most process industries; recycling of industrial wastes and products that have reached the end of their useful lives; concrete systems of blended and low-energy cements; waste material utilization as fuels and as feedstocks; raw material substitutes; and supportive analytical efforts. Specific projects in alternative materials utilization are briefly described below.

The objective of the blended cement program is to facilitate the use of granulated blast furnace slag and coal fly ash as partial replacements for portland cement. The United States production of portland cement presently requires 0.5 quad annually. In other countries, extensive use of slag and fly ash saves significant amounts of energy. This project involves laboratory testing to bolster efforts of the American Society for Testing Materials and other standards and specifying organizations to facilitate greater use of these two materials. Planning has been initiated for a slag granulating and grinding demonstration. Fully implemented (considering economic and institutional factors), the use of slag and fly ash can save 0.3 quad per year by the year 2000.

The Low Energy Cement project is directed

toward the development of a low-energy method for the conversion of limestone to lime used in the manufacture of hydraulic cements. Laboratory process development and feasibility studies are focused on the catalytic decarboxylation of fine-ground limestone to produce a slaked lime slurry. Production of lime in the United States uses the energy equivalent of about 17 million barrels of oil annually. In the conventional process, heat energy is used in a rotary kiln to produce quick lime. The new, nonthermal catalytic decarboxylation process would produce hydrated lime at lower rates of energy use. Approximately 3.5 million barrels of oil equivalent are expected to be saved annually if this process is proven effective and reaches its commercial potential.

The objective of the Refuse-Derived Fuel (RDF) for cement kilns project is to demonstrate the full commercial potential of using RDF for cofiring in energy-intensive cement kilns. Cement clinker (rocklike pieces of unground portland cement) produced in a full-scale plant using RDF has been tested with positive results by a leading cement laboratory. Extension from this short-term, batch operation to a continuous-running plant test demonstration will establish the operating characteristics needed for final commercialization in the industry through information transfer efforts. Test methods will be provided by which any cement company can evaluate for its plants the cofiring of RDF with coal, natural gas, or oil. It is estimated that commercial implementation of this RDF cofiring technology in economically appropriate cement plants would save the energy equivalent of 8 million barrels of oil annually. In addition, this project would provide a needed disposal method for municipal solid waste and industrial packaging material from which RDF is produced. This effort is closely coordinated with the urban waste and municipal systems activities within the Office of Buildings and Community Systems.

The object of the remanufacturing program is to identify durable products whose remanufacture could result in net energy savings and assist, if necessary, in implementing this activity. As an illustration, many automotive parts (e.g., engines, water pumps, distributors, and alternators) are remanufactured on a production-line basis. In most remanufacturing, the greater part of the embodied energy is saved. This saving is significantly greater than if the castings and machined parts were melted for metal recycling. This project will quantify the energy-savings potential of candidate products.

The Waste Lube Oil Recovery project is directed toward the commercial demonstration and implementation of a new process for the recovery of used automotive lubricating oil. The conventional acid/clay re-refining process has become environmentally unacceptable, sharply curtailing the amount of used oil that can be re-refined. Laboratory tests have indicated the technical feasibility of the new two-stage distillation, solvent precipitation process. In this process, all distillate and precipitate fractions of the waste oil are potentially recovered as useful products. Each year, about 650 million gallons of used automotive lube oil are disposed of. If the planned commercial-scale demonstration succeeds in fully implementing the used-oil recovery process, approximately 14 million barrels of oil equivalent per year could be saved.

The goal of the Waste Tire Conversion project is to identify the best contributions to be made by federally supported hardware development in the waste tire area. Analytical studies are being performed on the technical, institutional, economic, and market issues that will be confronted in the eventual demonstration of energy-saving technologies. Concepts involving waste tire reclamation and conversion into energy-intensive materials or fuel, tire reuse, and longer-service-life tires are being evaluated. It is estimated that successful commercialization of technology to convert discarded tires to raw chemical materials would save approximately 26 million barrels of oil equivalent annually.

The COthane Process for Producing Methane from CO project is directed toward the development and demonstration of the COthane process to beneficiate dilute carbon monoxide (CO) waste streams and convert them into high-Btu, pipeline-pressure methane. The successful outcome of technical feasibility studies and laboratory experiments that have been performed would lead to the design and construction of a pilot-scale process development unit. Dilute CO is a substantial volume waste stream now being flared or exhausted from the steel, petroleum cracking, carbon black, and other basic industries. The COthane process would convert dilute CO waste of low calorific value (90–150 Btu/scf) into high-Btu, pipeline-pressure methane in a single, cyclic operation without relying upon hydrogen as an energy source. If the process is successfully commercialized, 0.17 quad of methane per year (the energy equivalent of 29 million barrels of oil) could be produced.

The objective of the Energy Conservation Polymers program is to develop useful energy-conserving polymers and a process for their manufacture for the plastics market. It has previously been shown that CO can be incorporated with ethylene to produce polyketone polymers, and sulfur dioxide ($SO_2$) can be incorporated with ethylene to produce polysulfones. These polymers offer a major potential in conserving valuable ethylene raw material and energy for the polyethylene market. Both CO and $SO_2$ are low cost waste materials.

This project will focus on the following three major tasks: (1) establishment of the process, (2) determination of the processibility and marketability of the substitute copolymers, and (3) design, construction, and operation of a pilot plant for producing the substituted ethylene copolymers. Estimates indicate a conservation of 2 billion pounds of ethylene or 25 percent of the overall United States market by substituting CO and $SO_2$, assuming a 50 percent penetration of the bulk polyethylene market. This savings amounts to more than $280 million per year at the current market price. The savings in energy alone amounts to a substantial 0.1 quad per year equivalent to at least 100 billion cubic feet per year of natural gas.

The objective of the Chemical Production from Waste CO program is to use carbon monoxide, which is presently being wasted and underutilized at many major industrial plants, as a chemical feedstock. The project involves (1) identification of industrial sources

of waste CO, (2) identification of preferred products to be synthesized from this feedstock, (3) economic analysis of the most promising source-product configurations, and (4) demonstration of a commercial-size plant. Of particular interest are those industries using thermo- or electro-reduction processes, such as iron and steel, aluminum, silicon, ferro-alloys, calcium carbide, and elemental phosphorous. In the elemental phosphorous industry, part of the rich CO vent gas is used as fuel for ore drying and calcining; the remainder is unproductively flared. For ore drying, the gas would be replaced with coal. In such applications, the CO can be considered underutilized because its unique chemical properties can provide higher-valued uses. The net energy savings for 34 plants in the year 2000 is estimated to be 0.05 quad.

The Conversion of Plastics to Fuel Oil program involves the development and demonstration of a process to convert waste atactic polypropylene to fuel oil. This process is now operational in a 100-gallon per-hour pilot plant facility. Atactic polypropylene is a waste by-product in the production of polypropylene plastic; United States polypropylene production is growing rapidly and will result in a waste stream of about 200 to 400 million tons per year in 1979. Due to its highly temperature-dependent viscosity and other properties, atactic polypropylene is difficult to handle and is currently land-filled by polypropylene manufacturers. A 10-year search for commercial markets for this waste has been largely unsuccessful. Conversion of this high-energy waste stream to fuel oil could produce a net 0.6 to 1.2 million barrels of fuel oil per year in 1979.

DOE sponsorship was initiated in 1977 for the development, testing, and evaluation of a pilot plant designed to convert atactic polypropylene to fuel oil. The process involves combined thermal and catalytic cracking of atactic polypropylene within a fluidized bed furnace. Successful completion of the pilot plant tests has led to DOE support for an industry commercialization program, including the construction and operation of a commercial-size demonstration plant. Successful conclusion of this project will eliminate a waste-product disposal problem and make a significant contribution to the United States and world oil supply. Exportation of the technology to other countries is likely and will further help to improve the United States balance of payments. United States production of fuel oil from waste polypropylene in the year 2000 is potentially 2 million barrels of oil annually. Currently, a commercial full-scale plant is being designed, and pilot-level research to extend the technology to other high-volume waste plastics has been initiated.

The objective of the Wood Residue Industrial program is to use waste wood from the lumber, timbering, and wood products industries to provide electric power, process steam, and space heat for industry. The program has started with an engineering feasibility study for a forest-residue fueled electric power plant for an industrial park. Planned activities include onsite gasification of lumber wood waste and demonstration of forest residue harvesting/processing economies. The energy potential of annual United States wood wastes has been estimated to be greater than 3 quads.

About 25 percent of the energy that flows into the industrial sector is lost as waste heat. Heat is wasted either when useful thermal energy is allowed to escape with system output streams or when thermal energy is introduced to a system at a higher quality than necessary to perform its intended function. In both cases, the full work potential of the thermal energy is not fully used. This program activity is directed at reducing this waste.

The waste energy recovery and utilization activity is one of the major contributors to industrial energy savings within the time span of this program plan. The effort is directed toward waste heat recovery and utilization and combustion efficiency improvements through the development and use of advanced combustion and heat exchange techniques. The current projects include five high-temperature recuperator development projects, four industrial heat-pump projects, a program of air/fuel ratio control systems, and two oxygen-enriched combustion projects.

These projects are not redundant to efforts that private industry is undertaking on its own. The recuperators under development differ from those currently being manufactured in that they have higher effectiveness, are applicable to thermal ranges and applications not presently covered, and are more cost-effective. The industrial heat pumps operate similarly at higher coefficients of performance and at higher thermal levels than current units. The air/fuel ratio control system features individual burner coordination in conjunction with overall unit control not currently available on the industrial market. The oxygen-enriched combustion is directed to more efficient combustion that is made possible by the use of less costly methods of producing low concentrated oxygen than that which is used by industry today.

As shown later in this plan, the current projects alone will not be adequate to achieve the target energy savings and will be bolstered by second- and third-generation developments that will use first-generation results in their development. Also, the markets for this generic technology are diverse, and proof of success in one application does not necessarily mean that it would be successful in another process. Therefore, during the time frame of this plan, there will be additional activities in the general area of waste energy recovery and utilization that will be initiated to further advance the state-of-the-art and to cover all industrial applications of a given technology. Such areas may include an investigation of corrosion and fouling in waste heat boilers and recuperators as well as advanced heat-transfer concepts that can further improve the economics of heat recovery by reducing the size of heat exchangers, and increase recovery efficiency and operating life. In addition, new areas will be covered, and the currently undiscovered concepts occurring during the period will be investigated.

Specific projects in waste energy recovery and utilization are briefly described below.

Development and demonstrations are underway on two types of *ceramic recuperators*. One is based on built-up, cross-flow, thin-wall unit modules of nominal 850-cfm (cubic feet per minute) capacity. Five early demonstrations of this technique are being completed, having several thousand hours of test time under a variety of severe and corrosive conditions at inlet temperatures of up to 2600°F. Improvements in materials and minimizing leakage have been accomplished. A sixth demonstration, in a steel-mill soaking pit, is being installed. Preheating of inlet combustion air by these units decreases fuel usage by more than 30 percent. These modules are expected to provide "building blocks" for even greater fuel savings when used in very-high-temperature-inlet air burners. Some initial commercial sales of these recuperator modules have been made.

A second ceramic recuperation development is to preheat combustion air in glass furnaces by installing countercurrent special geometry units in the exhaust stack. The exhaust air is especially aggressive in its effect on materials. A major effort was required in selection of the ceramic material. These units should provide greater fuel savings than the presently used checker units due to their higher effective combustion air inlet temperature. Initial technical studies and material component tests are completed, and engineering work and a search for a demonstration site are underway.

Design, construction, and test operation of a full-scale unit was initiated in FY 1980. Using these recuperators, cumulative projected fuel savings through the year 2000 exceeds 15 quads.

Highly efficient, high-temperature metallic recuperators represent a sizable energy conservation potential. DOE is sponsoring the development of a brazed construction, stainless-steel recuperator that makes use of rectangular, offset extended heat-transfer surfaces.

The AiResearch Manufacturing Co., a division of Garrett Corporation, has developed a metallic, counterflow recuperator. It has been demonstrated in commercial service in a gas-fired aluminum remelt furnace. This test unit is installed to recover waste heat from the exhaust stream and to use it to heat inlet combustion air. To date, more than 430 tons of aluminum have been processed in this furnace with a resulting energy savings of more than 32 percent. During this demonstration period, no service or operational difficulties have been encountered. AiResearch is presently pursuing commercialization of this technology with DOE assistance. Demonstration of a heavy-duty unit in a steel mill is being considered.

The projected annual energy savings for this generic conservation technology exceeds 0.3 quad through the year 2000. The technology has good economic potential, showing a projected return on investment in excess of 20 percent.

The performance of existing annular passage, countercurrent flow recuperators (sometimes called *Escher types*) can be improved by providing a *reradiant* insert in the exhaust passage. A demonstration project to evaluate the magnitude of this possible energy savings accurately has progressed to the point that the first two test units were installed on aluminum remelt furnaces in Alabama. The first unit is to evaluate the increased performance obtainable in a conventionally sized unit. The initial tests have revealed the need for more flue gas chemical composition information and improved materials. The second test unit has been fabricated and installed. This second unit is of a smaller size, lower capital cost than a normal-size unit, but is projected to provide equivalent energy savings.

Several studies indicate that many specific industrial processes exist that use 250 psia and above process steam from fossil fuel-fired boilers while at the same time rejecting considerable quantities of low-grade waste heat contained in streams such as nonreturned condensate and process cooling water.

The objective of the *Stirling-cycle heat-pump* project is to evaluate the technical and economic feasibility of using the Stirling cycle operated as a heat pump for these types of industrial heat recovery applications at temperatures generally above 400°F. Other thermodynamic cycles (i.e., Rankine and Brayton) are limited to lower-temperature industrial processes because of the thermal stability of available working fluids or the increasing complexity and decreasing efficiency of the system, or both. Stirling-cycle machines employ gaseous working mediums, such as helium or argon, which are not limited by thermal stability, material compatibility, or toxicity considerations. In addition, Stirling cycles offer the potential for achieving higher system performance. A detailed technical performance and economic assessment study of various Stirling-cycle heat-pump concepts is currently under way. The objective is to define configurations that offer potential for satisfying industrial process needs and for becoming economically feasible commercial products.

The objective of the *Brayton heat pump* program is to determine the technical and economic feasibility of using high-temperature industrial heat pumps to reclaim low-temperature waste heat from industrial processes. Because the Brayton-cycle heat-pump and process operating conditions are noncontaminating, ideal applications are found in the food process industries. Milk drying is a representative process and has been selected for concept demonstration. The high-temperature Brayton-cycle heat pump is ideally suited for energy conservation in many kinds of dryers and ovens. The energy consumed by industrial dryers and baking ovens in the United States exceeds 2 quads per year. Drying conditions are generally similar, and most of the energy supplied (usually as direct-fired natural gas) is exhausted to the atmosphere.

Another objective of this program is to recover the wasted energy from the process—particularly from the exhaust gas leaving the evaporator and dryer—and use it to reduce system energy requirements. For the conditions normally used for milk-and-whey drying, a heating coefficient of performance (COP), a measure of efficiency used for heat pumps, of between 3 and 4 can be obtained with the system. Initial design studies, which are under way, show that the technique can make substantial savings when installed on the evaporator as well as on the dryer and that a variety of prime movers will be economical.

A great deal of energy is unavailable for further use in industry due to its low temperature. The *Rankine cycle–steam recompressor heat-pump* project will take 190°F hot water (or subatmospheric steam) and put it through a heat exchanger to vaporize a fluorinated hydrocarbon working fluid that expands through a special turbine. This turbine will drive a conventional steam compressor to pressurize the subatmospheric vapor to 20 psig (pounds per square inch-gauge) where it can be further used as process steam. The demonstration unit was installed and tested in an acetone-recovery plant in a chemical complex in Virginia. The preliminary design and performance predictions have been completed.

Design approval has been given, and release of long-lead items for procurement has been requested. Installation for the acetone-recovery plant can be made in FY 1980 with a test period to follow.

The commercial equipment developed from this demonstration will have the potential of providing energy savings (an average of more than 1 quad per year between 1981 and the year 2000). This savings is due to the very large amount of process steam in use.

The *electric heat-pump* project was established to improve components and materials to be used in electrically driven, high-COP heat pumps and to conduct a demonstration of individual-scale units in the pulp and paper industry. Development work on compressors, seals, and other components is under way in the laboratory. Especially good results are being achieved in developing and evaluating higher temperature working fluids. The initially chosen demonstration installation in a Kraft paper plant has been deferred, and a number of more appropriate sites have been investigated and their individually desirable operating conditions determined. Following the component development work, a firm site selection will be made, and a pilot-plant demonstration conducted.

A *heat-pump study* has been initiated with a university to study and prepare a report on all the significant parameters that affect industrial heat-pump design and operations. The actual effect of these parameters will be checked by obtaining data from a laboratory model designed, constructed, and operated under the auspices of this project. The results of the study, substantiated by the test operation, will be presented in a report that will be suitable for use as a design manual. This effort was completed in FY 1979.

The use of the *fluidized-bed principle in the design of a waste-heat boiler* is advantageous due to the high heat-transfer rates obtainable and its small physical size. The smaller size plus the increased performance enhances the probability of it being adopted on a retrofit into commercial use.

The current project will first define the potential industrial market for unfired, waste-heat, fluidized-bed boilers, especially as retrofit equipment. Several potential sources of waste-heat recovery and use have been identified and the national energy savings projected. A competitive procurement action, seeking to select at least two contractors to design, build, and test industrial-scale units, was conducted and a contract was negotiated in FY 1980. The project has proceeded to the hardware phase. Payback periods of individual installations based on this technique are projected to be less than 2 years.

The objective of the *air-fuel ratio control* project is to demonstrate the fuel savings attainable by operating multiburner combustors at a low excess air ratio. This requires that the burners operate in parallel and be actively controlled to give identical combustion conditions. When this is achieved, the overall air-fuel ratio can be reduced accurately for all burners, and a significant energy savings can be realized. A prototype unit, operating in the manual mode, has demonstrated fuel savings in excess of 2 percent on an oil-fired boiler. Greater savings are anticipated with automatic control. Following these tests on the oil-fired unit, a similar demonstration is to be made on gas-fired and on coal-fired boilers.

The *enrichment of combustion air by modest increases in the oxygen content* has been demonstrated (on a non-cost-effective basis) to achieve significant savings in fuel requirements by increasing the flame temperature and by not heating additional nitrogen that is subsequently exhausted.

A Program Opportunity Notice competitive procurement action was issued in FY 1978, proposals were received, and selections have been made. The two separate techniques selected for generating and using oxygen-enriched air in multiple types of applications are pressure-swing absorption and spiral-wound membranes. Two contracts resulting from the evaluation were negotiated and design work was initiated in FY 1979 with fabrication, installation, and tests occurring in the subsequent 2 years. Projected payback periods for individual installations using these techniques, based on energy savings alone, are less than 2 years.

Significant activities in waste heat recovery and improved combustion include completeness of demonstration in cross-flow ceramic recuperators, heat pumps, and oxygen enrich-

ment. Figure 8-4 shows major milestones for the Waste Energy Reduction subprogram.

Activities in alternative materials utilization include the start of commercial-scale demonstration of acetylene from coal; commercialization following demonstration of blended cement; completeness of demonstrations of waste plastics conversion to fuel, re-refining of waste oil, paper recycling, waste wood for cogeneration, remanufacturing, reuse of industrial containers and onsite waste-fired process heat; and analyses of evaporative loss controls for chemicals and fuels. (See Figure 8-5.)

Cogeneration of electrical power and process heat in industrial operations is a powerful conservation technique. It has been estimated that the energy saved through additional cogeneration systems could exceed 200,000 BPDE (barrels per day equivalent) by 1985 and 1 million BPDE by the year 2000 with only modest market penetration. The total potential for a fully developed market is perhaps 3 million BPDE or greater by 2000.

At present, industrial cogeneration potential is grossly underutilized. Where cogeneration is applied, it is, in general, poorly applied in terms of achieving large energy savings. Systems must be selected which employ efficient prime movers sized to satisfy plant thermal baseload and wherever possible interconnected with the grid.

The long-range objectives of the industrial cogeneration program are to develop optimized inter-industrial/utility cogeneration activities, to minimize requirement for gas and petroleum fuels, to develop and commercialize improved conversion system technologies, and to generally accelerate the rate of penetration of industrial cogeneration.

To achieve these objectives, the Industrial Cogeneration Program has undertaken a number of activities including:

- Studies to establish cogeneration targets, to define optimum system configurations, and to identify technical and nontechnical obstacles to market acceptance
- Design and development of components required to achieve efficient cogeneration with fuel flexibility

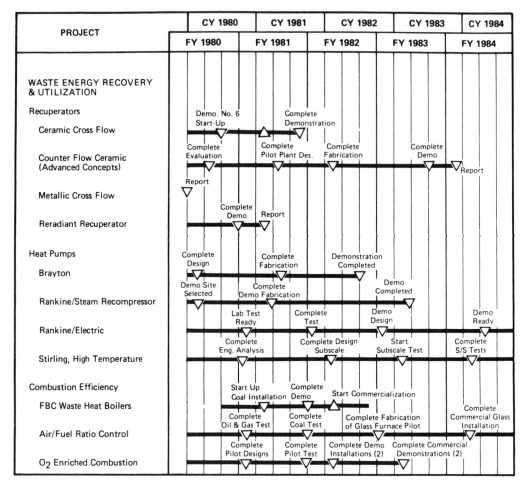

Figure 8-4. Major milestones for the waste energy recovery and utilization activity.

- Demonstrations of efficient cogeneration systems configurations
- Promotion of proven concepts and components to expedite commercialization

Descriptions of specific projects in industrial cogeneration follow.

The objective of the *feasibility study for industrial cogeneration fuel-cell applications* program is to identify and analyze specific industrial sites for the possible application of a 4.8-MW phosphoric acid fuel-cell power plant in a cogenerating mode to generate electric power and hot water in plants where the demand for these energy forms is balanced to the capability of the fuel cell. Four specific industrial sites have been selected. The study is evaluating the potential for conservation and the economic feasibility of the system. A demonstration program plan will be prepared

as one product of this effort. This project is primarily funded by energy technology and supported by IP.

The *cogeneration case-history studies* program is conducted to determine the financial, regulatory, institutional, and environmental factors that influence the original decision to invest and to summarize the industrial experience with the systems to date. Five cases, which represent the following industrial sectors, were selected: food, textiles, pulp and paper, chemical, and petroleum refining. The project was completed in FY 1979.

The exhaust-gas stream of a glass furnace is at a very high temperature and flow rate, so considerable energy is wasted, even with the use of conventional recuperators. A proven technique for increasing the production rate of a glass furnace, with no additional local consumption of fuel, is to electrically "boost" it

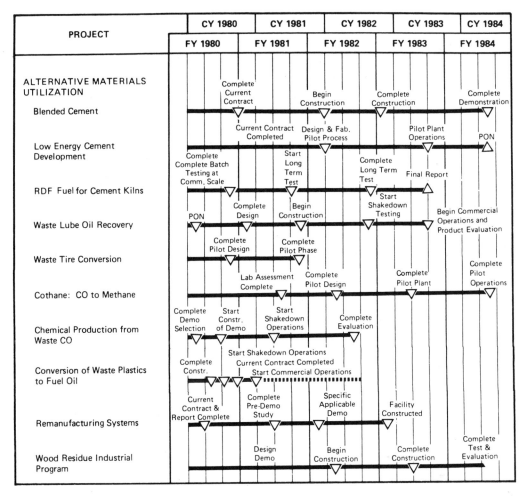

Figure 8-5. Major milestones for the alternative materials utilization activity.

by passing current through electrodes submerged in the molten material tank. The Brayton-cycle glass project is developing a turbomachinery device that is powered by the energy now wasted in the exhaust-gas stream that will generate, on the spot, the electric current for "boosting" the tank.

Engineering and materials studies are under way, component tests have been made, and several small turbomachinery units have been built and tested in a scale-model glass furnace and on a commercial container glass furnace. A major decision is required on the configuration of the full-scale turbomachinery. A full-scale demonstration unit will be prepared and installed in an existing on-line commercial glass furnace and tested to determine performance and economics, and to obtain further design data.

The *industrial cogeneration optimization program* (ICOP) is oriented toward determining optimum conditions for cogeneration systems on site-specific bases. Two contractor teams, headed by TRW and Arthur D. Little, worked on identical efforts that were completed late in FY 1979, after approximately 6 man-years of work. This competitive type of approach resulted in a more comprehensive insight into the optimum conditions for cogeneration systems.

Each contractor collected energy consumption data in five industries—food processing, pulp and paper, textiles, chemicals, and petroleum. They characterized all available and near-term future cogeneration systems considering both topping and bottoming cycles. The use of abundant fuels and the resultant ramifications for these system types and the resulting economics were explored under various energy-pricing scenarios.

The generic information is used together with institutional, regulatory, and environmental constraints to produce preliminary cogeneration system designs on a site-specific basis. The schedule for implementing a system, including design, manufacture, and installation, is to be formulated. The system economics are to be determined and the energy savings to be realized are to be estimated.

The primary objectives of the *cogeneration demonstration* program are to provide near-term, highly visible, and credible hardware demonstrations to increase industry's interest in cogeneration systems and to expedite technology transfer. Emphasis will be placed on those cogeneration concepts that represent a significant advance in applied conversion system technology, process interface, or industry/utility relationship, and can be shown not to be permanently dependent on natural gas or distillate fuel forms. Another objective is to provide industry and DOE with first-hand experience in dealing with institutional cogeneration. These evaluation projects will also contribute to future development of better and more advanced cogeneration systems. The program will consider large and small applications for a wide range of industrial applications. Industries include wood products and pharmaceuticals. Conversion system technologies include externally fired gas turbines, steam turbines, diesels, and fluidized-bed combustion. Coal, wood waste, biomass waste, and residual oil will be used.

For industrial cogeneration during the 5-year period, five demonstrations derived from Program Opportunity Notice I (PON) will be completed for applications of topping cycles in paper, chemicals, and textiles; components will be developed for modular cogeneration systems and two demonstrations will be completed, and PON II will be issued and another five demonstrations will be completed for additionally identified industrial applications.* Major milestones are presented in Figure 8-6.

### Industrial Process Efficiency.

This activity is directed toward increased energy-use efficiency and substitution of

*Roman numerals for the PONs refer to the generation of the technology investigated; e.g., II is second generation.

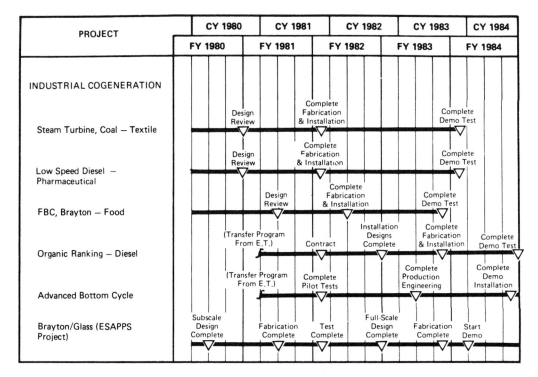

Figure 8-6. Major milestones for the industrial cogeneration program.

Table 8-5.  Industrial Process Efficiency Activities Areas and Typical Projects.

| | HIGH TEMPERATURE PROCESSES | LOW TEMPERATURE AND END PRODUCT PROCESSES | AGRICULTURE & FOOD PROCESSES |
|---|---|---|---|
| ACTIVITY AREA: | Steel, Aluminum, Other Non-Ferrous Metals, Stone, Clay and Glass | Chemicals, Petroleum Refining, General Manufacturing, Pulp and Paper, and Textiles | Agricultural Production and Food Preparation |
| TYPICAL PROJECTS: | • Coal in Aluminum Remelt<br>• Aluminum Cathode Improvement<br>• Direct Reduction Aluminum<br>• Blast Furnace Gasifier<br>• Non-Consumable Anode/ Aluminum<br>• Hot Inspection of Steel Ingots | • Improved Crude Distillation<br>• Refinery Monitor<br>• Critical Fluid Extractions<br>• Post Paint Curing<br>• High Efficiency Arc Weld Power Supply<br>• Cold Corrugation<br>• Hyperfiltration/Dyestuffs<br>• Foam Dyeing/Textiles<br>• Hydropyrolysis<br>• Computer Controls in Paper Processes<br>• High Frequency Hand Tools | • Crop Drying Alternates<br>• Fertilizer Improvements<br>• Improved Irrigation<br>• Food Processes<br>• Sugar Processing<br>• Energy-Integrated Farm Systems |

abundant fuels for scarce fuels in the processes of the most energy-intensive industries, including steel, aluminum, glass, cement, paper, chemicals, petroleum, textiles, agriculture, and food. Activity areas and some typical projects are listed in Table 8-5. The major industries and the estimated current annual energy consumption are listed in Table 8-6.

The targets within each of the major industries are identified by a second law (of thermodynamics) analysis to identify the high-loss points as reflected in Figure 8-7, which uses aluminum as an example. As shown, the large thermodynamic losses occur in the reduction

Table 8-6.  Estimated Current Annual Energy Consumption for Major Industries.

| INDUSTRY | CONSUMPTION (QUADS ANNUALLY) |
|---|---|
| Steel | 3.8 |
| Petroleum Refining | 3.0 |
| Chemicals | 3.0 |
| Paper | 2.5 |
| Agriculture & Food | 2.4 |
| Aluminum | 1.1 |
| Cement | 0.6 |
| Textiles | 0.5 |
| Glass | 0.3 |

of aluminum. Projects are then selected and directed to the high-loss points such as electrolytic smelter modification and direct reduction.

The projects within the industrial process efficiency activity are principally process changes, i.e., total change of the process and replacement of the related equipment. The principal groups of activities within industrial process efficiency (high-temperature processes, low-temperature and end-products processes, and agriculture and food processes) are described in more detail in the following paragraphs.

The high-temperature processes activity is directed toward process efficiency improvements in the most energy-intensive industries using very high-temperature processes. The specific industries addressed are iron and steel, aluminum, other nonferrous metals, and stone, clay, and glass. Each of these industries and related industrial energy conservation projects are discussed below.

Total energy consumption in 1976 by the steel industry was 3.6 quads, including 0.22 quad of oil and 0.60 quad of natural gas. Potentials for energy conservation in the steel industry include: use of furnace off-gas, substitution of abundant fuels for scarce fuels, integrated steelmaking to avoid the inefficien-

## Aluminum

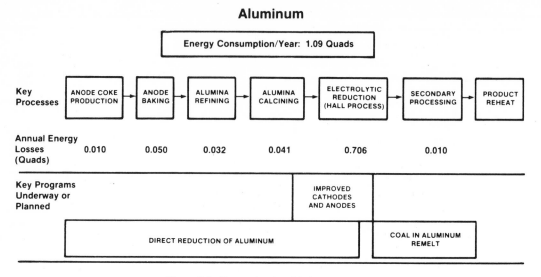

Figure 8-7. Targets for the aluminum industry.

cies of successive heating and cooling operations, improved refractory materials, improved burner designs, use of scrap, and increased efficiency of processes. Current projects include modification of a cupola furnace to eliminate the need for an afterburner, use of obsolete blast furnaces to produce a medium Btu gas, an improved slot forge furnace, and evaluation of electric arc furnaces.

The blast gasifier is a small blast furnace that has coal-oxygen burners attached to the tuyeres (ports through which air is blown into the blast furnace). It produces a medium-Btu gas and some molten iron. A major advantage of this project is that many small, obsolete blast furnaces are standing idle in localities where there are shortages of natural gas. The charge to the furnace includes industry waste products such as coke breeze, mill scale, and BOF slag. The cost of fuel gas produced appears to be competitive with present prices of oil and gas. This technology has the potential for contributing about 0.5 quad per year by the year 2000.

The objective of the *nitrogen-based carburization* project is to demonstrate a new technology that uses a nitrogen-based carburizing atmosphere. The project has two goals: (1) to quantify the energy, cost, and operating benefits of the nitrogen-based atmosphere currently used in ferrous-metal heat-treating, and (2) to develop a commercial facility to demonstrate,

on a full-scale production basis, the advantages of nitrogen-based atmospheres over endothermic atmospheres. Implementation of this technology has the potential of reducing the natural gas resources used for carburizing atmospheres by approximately 75 percent. Potential natural gas savings in heat treating would save approximately 0.01 quad per year.

The largest consumption of petroleum or natural gas in the production of steel occurs in the reheating of slabs to enable hot rolling toward a final flat-rolled product. The present practice within the industry allows hot slabs that have been rolled from an ingot or have been continuously cast to cool to ambient temperature in a slab yard. They are then visually inspected, and obvious surface defects are removed by a manually operated scarfing torch. These slabs are then scheduled, reheated, and further processed. The purpose of the proposed research is to develop a system that inspects the hot slab of steel. This would allow direct rolling (with possible incremental heating) of hot slabs while maintaining a high-quality surface. The energy savings from not having to reheat these slabs is 2 to 3 million Btu per ton, or about 0.25 quad per year.

A large fraction of United States steel production is by electric arc furnace melting. Although modern arc furnaces are rather efficient (75 percent is possible in a well-run operation), the amounts of electricity used are sufficiently large that even a marginal improvement is worth pursuing.

The Massachusetts Institute of Technology, under contract with DOE and in consultation with major electric arc furnace users, has developed analytical models for arc furnace performance that will lead to increased efficiency through improved operating practice. It is anticipated that successful development of the analytical models and their subsequent operation will lead to indicated improvements in electric arc furnace design and operation.

DOE also intends to fund studies on saving energy through improved materials handling practice in electric arc furnace shops, especially with regard to melting direct-reducing iron. Implementation of this technology can save over 0.04 quad per year.

Many metal parts are heat-treated to improve their mechanical or other properties. Heat treatment is generally performed using natural gas or electric heating, and the heating processes are usually very inefficient because of the poor heat-transfer characteristics of the furnaces.

Large amounts of energy could be saved by doing heat-treating in fluidized beds, which offer superior heat-transfer characteristics as compared to conventional furnaces. Several specialized applications have already been developed, but it is desirable to demonstrate the feasibility of retrofitting a major type of conventional heat-treating furnace for fluidized bed-type operation.

Under contract with the Office of Industrial Programs of DOE, Procedyne, Inc., has demonstrated the feasibility of retrofitting a standard integral quench furnace in a commercial heat-treating shop for fluid-bed operation. Implementation of this technology is expected to save 0.03 quad per year.

The aluminum industry consumes 6 percent of the nation's electrical power and is a primary focus of the industrial energy conservation effort in high-temperature processes. There are three principal energy-intensive processes in the production of aluminum: the Bayer process (using natural gas and oil); the Hall-Heroult process (using electricity), and the resmelting process (using natural gas).

Current and future efforts will focus on energy-efficiency improvements in these three areas.

Current projects directed at aluminum include direct reduction, improved anodes, improved cathodes, and the use of coal in aluminum resmelt. The *direct reduction of aluminum* process would eliminate the electricity used in electrolysis and substitute a significantly smaller amount of energy in the form of coke. The basic idea of this project is to reduce a mixture of $Al_2O_3$ and $SiO_2$. (In practice, of course, an aluminum silicate ore containing iron impurities would be blended with bauxite to produce a practical feed with carbon in a shaft furnace.)

The product would be an aluminum-silicon alloy that could be used directly in some applications (e.g., automotive castings), or refined to commercially pure aluminum by a subsequent process. This technology has the potential for saving 0.7 quad annually by the year 2000. The process has as its objective the reduction of energy consumption in aluminum smelting by approximately 50 percent.

All primary aluminum is remelted before being cast in finished shape, and scrap is remelted each time it is recycled. This melting is usually done with natural gas, although distillate oil is sometimes used. Approximately 4000 Btu are required to melt a pound of aluminum by the industry process. Since approximately 2 pounds of metal must be melted for 1 pound of net use, approximately 0.1 quad of fuel is used. This number will increase more rapidly than the increase in primary aluminum usage, due to the emphasis on recycling.

Combustion devices are being designed to use pulverized coal as a fuel, removing the sulfur as part of the combustion process. This will lead to a shift in usage of 0.1 quad per year from natural gas to coal; it will also probably lead to an overall reduction in energy use because of the superior heat-transfer characteristics of coal flames.

Aluminum is produced by electrolysis of a solution of alumina in cryolite at about 1000°C, using a carbon anode made from petroleum coke, which is gradually consumed during the aluminum production process.

If the carbon anode is replaced by a non-consumable, dimensionally stable anode, there are several immediate advantages. First of all, the petroleum now needed to make coke is available for other purposes. Second, the gas evolved is now pure $O_2$, which can be captured and used, rather than a carbon monoxide–carbon dioxide mixture that must be disposed of in an environmentally acceptable manner. Third, the cells can now be sealed, leading to reduced heat losses to the environment.

If the nonconsumable anode is combined with a wettable cathode, very precise control of the cell becomes possible and there are additional benefits from reduction in electricity waste. The potential savings from this innovation, assuming complete adoption, is about 0.1 quad per year.

Aluminum is produced by electrolysis of a solution of alumina in cryolite. This electrolysis is carried out in a Hall cell at temperatures near 1000°C.

The cathodes of Hall cells are made of graphite and are not wetted by the molten aluminum produced. It is desirable to keep the graphite covered with molten aluminum and the layer must be sufficiently thick to ensure complete coverage on a nonwettable surface. This thick layer develops waves under the influence of thermal and magnetic forces, and the electrolyte layer must be kept thick as well to prevent electrical shorts. The large anode-cathode spacing greatly increases electrical losses in the cell.

If a wettable cathode material were used, a very thin metal pad would suffice as a cover, and the anode-cathode distance could be made much smaller. The resulting reduction in cell resistance could result in a savings of 0.2 quad annually by the year 2000.

The approximate savings from complete adoption of such cathodes is about 0.2 quad per year. Since the technology is a low-cost modular retrofit, market penetration can be rapid.

The energy consumed by industries producing other nonferrous metals consumes approximately 0.35 quad per year, almost all in the forms of electricity, natural gas, and petroleum distillates. If new technologies are not introduced, this amount is expected to climb rapidly, because United States high-grade ores are being rapidly exhausted and complying with environmental restrictions is expected to increase the energy cost of these metals.

In several of the remaining metals, most notably copper (which is by a wide margin the most important from the energy conservation standpoint), EPA and OSHA restrictions are forcing major rethinking of traditional technology. DOE's role in ensuring that new technologies are energy-efficient is unclear, and the primary effort up to 1981 has been to analyze the alternatives and determine DOE's role in the development of the new technologies.

Primary production of refined copper in the United States consumes close to 0.25 quad per year. If technology is not radically changed, this amount will steadily increase as environmental requirements become more severe and the grade declines. It is also likely that future environmental requirements cannot be met by traditional plants at any reasonable energy cost.

It appears that the intermediate term will see very extensive changes in copper making technology. A number of new technologies are available in principle, but some have not been tested and others may not be directly applicable to United States ores. At least one new process has been tried in the United States with very disappointing results in respect to both energy and money.

The stone, clay, and glass industries, which consume more than 1 quad per year, have serious and varied problems. For example, the glass industry is heavily dependent upon natural gas and electricity, and in many areas is quite inefficient from the energy standpoint. DOE is focusing attention on how to increase the efficiency of units by developing new heat-recovery technology. The cement industry, on the other hand, can convert almost all its fuel use to coal within existing technology, but there are still opportunities for energy saving through such activities as improved alkali removal and particle size control. DOE is aware of major problems in other segments of these industries (e.g., chemical lime production) but has not yet been asked by the industries in question to take any action.

In support of the International Energy

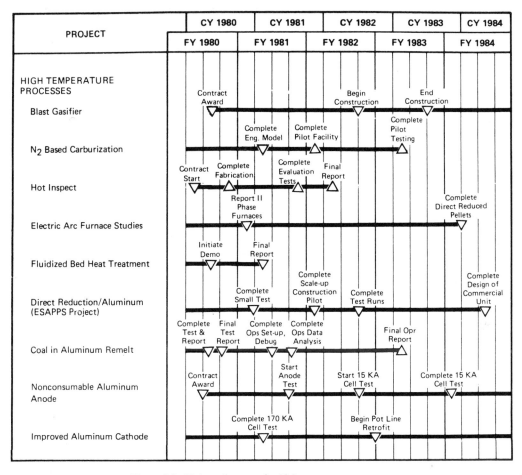

Figure 8-8. Major milestones for high-temperature processes.

Authority, work on the reduction of energy use in cement-making, DOE is funding work on particle size control and improved alkali removal technology. Potential savings are of the order of 0.01 quad per year.

Key projected milestones (Figure 8-8) in high-temperature processes include the completion of demonstrations of blast gasifiers, direct reduction of aluminum, more efficient cupola furnaces, hot inspection and scarfing/steel, and cathode improvement in electrolytic smelting of aluminum.

The low-temperature and end-products processes activity focuses on improved efficiency in the processes and equipment for chemicals and allied products, petroleum refining, textiles, pulp and paper, and general manufacturing. Current energy-use studies indicate that the combined annual consumption of these five categories is over 9 quads or the equivalent of 4.5 million barrels of oil per

day. This amount is exclusive of feedstock fuel values. Major milestones are shown in Figure 8-9.

In the chemical industry, the primary energy-consuming activities are industrial inorganic chemicals (chlor-alkali, industrial gases), plastics and synthetic resins (synthetic rubber and synthetic fibers), drugs, industrial organic chemicals, paints, soaps, and agricultural chemicals. The largest energy-consuming operation is direct heating, which accounts for 0.8 quad annually; next is compression, using 0.75 quad; distillation accounts for 0.6 quad; and evaporation uses 0.4 quad. Feedstocks account for about 1.3 quads annually in the chemical industry.

The major thrusts of the energy conservation program for the chemical industries are the development of new and improved energy-saving processes and improvements and/or innovations in chemical engineering unit

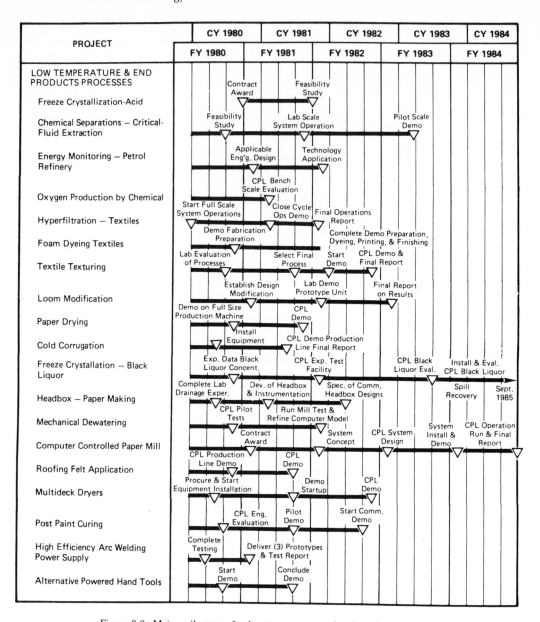

Figure 8-9. Major milestones for low-temperature and end-product processes.

operations such as distillation, evaporation, and heat transfer. Although the chemical industry has historically been aware of energy use in chemical processes, the low cost and high availability of energy, particularly natural gas and petroleum, have placed energy optimization of processes behind product and yield optimization. The federal program in energy conservation will focus on new energy-saving process demonstrations that are beyond the economic reach of many individual companies or those that indicate a high risk compared to return on investment. This federal role ensures the technology transfer of energy-savings processes to small as well as large chemical products companies in the industrial sector.

The objective of the freeze crystallization-acid project is to explore the energy conservation potential of using freeze crystallization, instead of heat evaporation, for concentrating acetic acid. This process, if efficient and economical, could save over 600,000 barrels of oil equivalent per year.

Preliminary evaluation has shown that oxygen at a 90 to 95 percent purity level can be produced by chemical separation from air. This project is aimed at the continuous operation of a laboratory-size apparatus to determine the effects of variables on oxygen production, economics of a full-size operating system, energy conservation potential, and severity of corrosion problems. Potential energy savings are estimated at 0.03 quad in the year 2000.

The chemical separations using critical-fluid extraction project will explore a new process involving liquid extraction using carbon dioxide under pressure near its critical point. The carbon dioxide can be separated from the extracted organic product for recycle by an economical vapor-compression distillation. It is expected that if successful, 75 percent of the energy used today in certain chemical distillation processes could be saved. In the case of ethanol alone, it would amount to 0.01 quad annually. When extended to other chemical and petroleum products as well, the energy conservation implications are very significant.

A rough estimate of energy consumption in the petroleum-refining industry is that 1 out of every 10 barrels of oil is used as fuel for the refinery. At a conservative estimate of 16 million barrels per day throughput in United States refineries, the energy used amounts to 1.6 million barrels per day or 3.2 quads per year. Most of the energy used in a refinery goes into direct heating for the various processing steps: crude distillation, coking, fluid catalytic cracking, catalytic reforming, distillate desulfurization, alkylation, and petrochemical production. The most energy-intensive of these processes are crude distillation and cracking.

The industry consumes products it produces, such as refinery gas, distillates, residual fuel, and coke, as well as purchasing energy in the form of electricity, natural gas, steam, and coal. A unique factor of United States refining as compared to the rest of the world is the significant consumption of natural gas (36 percent of the total consumed).

The major thrusts of the energy conservation RD&D program for petroleum refining are to develop conservation technologies such as process modification, changes in process

conditions, energy optimization models, waste heat recovery, use of alternate fuels, and modifications to equipment and/or unit operations that result in significant energy savings. Descriptions of two projects specific to the petroleum industry follow.

The objective of the energy monitoring–petroleum program is to develop techniques for efficiently conducting detailed energy audits in a petroleum refinery. It is expected that the techniques developed will also have application to some chemical plants. These monitoring techniques, when applied, will determine the plant's operating energy efficiency and will enable identification of areas amenable to modifications for reduced fuel consumption. It is estimated that petroleum refineries consume 1.6 million barrels of crude oil per day to meet a daily throughput of 16 million barrels of crude per day. Thus, a 5 percent improvement in energy consumption through these auditing techniques would conserve over 29 million barrels of crude per year.

The textile industry is extremely diversified, processing a variety of fabrics and fabric blends into a multitude of end products. Its primary customers are the apparel, automobile, and furniture industries. Knit apparel, yarns, and carpets are also produced as final products.

Textile plants range from highly integrated manufacturing complexes that process fibers (natural and synthetic) into finished products, to small nonintegrated contract plants that process goods owned by other producers. There are over 7000 plants distributed over 47 states; however, the major states for the industry are North Carolina, South Carolina, Georgia, and Alabama. The industry is very competitive, and profits, after taxes, typically represent 3 percent or less of sales. It is a labor-intensive industry. The industry uses about 0.5 quad of energy annually, which is roughly split into 50 to 60 percent for the wet processes and the balance for the dry customer.

Dry processes consist of the yarn manufacturing, sizing of yarns, and the knitting and weaving of fabrics. Wet processing consists of the preparation of fabric for dyeing, dyeing, and the predrying and finishing of the fabric.

A breakout of the energy forms used in the textile industry (based on 1974 data) shows that electricity accounts for 28 percent; natural gas, 32 percent; fuel oil, 20 percent; coal, 7 percent; and other, 14 percent. Most of the energy used in the dry processes is electrical energy for motors for machine drives and for air conditioning to maintain proper environmental conditions for the yarn and fabric manufacture. In the wet process area, most of the energy used is thermal energy in the form of steam or heated air. The steam is used for heating hot water for preparing fabrics for dyeing, for heating dye baths, and for predrying fabrics. The natural gas is mainly used for heating air in tenter frames for drying and heat-setting fabrics.

Since the major portion of the energy use in textiles is spent in heating and evaporating water, process improvements that minimized or eliminated the use of water were considered of prime importance for energy conservation. In this regard we have directed our major projects in these areas. Specific projects in the Industrial Energy Conservation Program in textiles will now be described.

The objective of the hyperfiltration-textiles project is to apply high-temperature membrane technology (hyperfiltration) to textile waste streams to reuse the hot water, chemicals, and dyes in a closed-loop system. The system will be applied to a continuous dye range in a textile plant to demonstrate the technical feasibility, economics, and energy savings associated with the use of the membrane technology of hyperfiltration. If this demonstration is successful, potential energy savings are 45 trillion Btu per year as well as an additional cost savings in dyes, chemicals, and water.

Textile fabrics are generally processed through aqueous solutions for preparation, dyeing, and finishing of the fabric. Frequently, these steps consist of alternate fabric wetting and drying during processing, which require a considerable amount of energy to be consumed in thermal drying ovens. The objective of this project is to develop formulations and techniques for applying cleaning chemicals, dyes, and finishing chemical, using foam as a carrier. This results in less water absorption by the fabric and consequently less energy for evaporation, as well as the elimination of most of the hot process water. The estimated savings from this technology amount to 52 trillion Btu per year.

The objective of the proposed loom modification research is the development of simple and economic retrofitting modifications of picking and lay mechanisms of a loom for reduction of their energy consumption. It is expected that over 300,000 looms in the United States can be retrofitted at an estimated energy saving of over 500 million kWh per year.

Many fabrics are composed of synthetic fibers or a blend of natural and synthetic fibers. To make synthetic fibers suitable for fabric use, the fibers are texturized to provide bulk, softness, drape, etc., which are required properties of the fabric. The objective of the dry textile processing-texturing project is to study and develop new techniques for texturing that consume less energy than the present system of gas-fired radiant heaters or resistance-heating elements. The potential energy savings to the industry are 2.5 million barrels of oil equivalent per year.

The pulp and paper industry, like textiles, is also very diversified, producing over 2000 primary products. The industry encompasses the production and sale of pulp derived from wood and other fibrous materials, pulp and paperboard, and certain by-products such as tall oil used in soapmaking and turpentine.

The paper industry is characterized by extensive vertical integration (pulping through papermaking) and is highly capital-intensive. It ranked ninth in capital intensity among all American industries in 1974. Because of this, and because large sums are required for new facilities, the industry is slow to change its practices and processes. Minor process changes are made if they can be retrofitted into existing facilities without large expenditures of fixed capital. Opportunities for change usually arise only when new facilities are added. Even then, however, the process changes must be well documented and demonstrated so that there is minimum risk to production output and quality.

There are various pulping processes used

by the industry, ranging from mechanical (groundwood), semichemical (sulfite), to chemical (Kraft). A particular process is used because of the specific properties and cost of the product best suited to the needs of the end-use application (dissolving pulp for rayon, writing paper, boxboard, etc.). Kraft pulping, however, accounts for over 50 percent of the output and is the leading pulping process.

This industry is very energy-intensive, having used 2.5 quads of energy in 1972 in producing an output of 67 million tons of products. Of the energy consumed, 40 to 45 percent is self-generated from burning the spent pulping liquors (black liquor) and the burning of bark and hogged fuels. The balance of the energy required is purchased. A breakout of this purchased energy shows petroleum 22 percent; natural gas, 21 percent; coal, 12 percent; and electricity, 5 percent. The energy requirements for the pulping process (pulpmaking, chemical recovery, bleaching, and finish preparation) are approximately equal to those required for pulp-drying, papermaking, and paperboard-making.

The program in the pulp-and-paper industry is equally directed toward the pulping process and the papermaking machine. Specific projects of RD&D in papermaking will now be described.

During the wet pressing of sheet on a paper machine, the felt absorbers pick up water from the sheet and become saturated. Vacuum is generally used to subsequently dry the felt. The objective of the paper-drying project is to apply improved techniques of water removal from the felt to improve its wet pickup. The application of a high-velocity airstream, through a mach nozzle, to drive off the water in the felt offers promising potential for improved water removal. This program is expected to improve water removal 5 to 7 percent, resulting in energy savings of 4 to 6 million barrels of oil equivalent per year. The technology can be retrofitted to existing paper machines.

The objective of the cold corrugation program is to demonstrate the use of a room-temperature curable adhesive in the forming of corrugated board. This new method, which replaces the present hot corrugation process, offers major advantages, including a 95 percent reduction in process heat, a 40 percent reduction in electrical drive energy, a 50 percent reduction in corrugator waste, and a 25 percent reduction in new equipment cost. The energy savings, based on present industry practice, would amount to 6 million barrels of oil equivalent per year. The technology can be retrofitted to existing corrugating lines.

The objective of the headbox-papermaking project is to study energy-saving opportunities in the formation of all grades of paper and paperboard. One of these opportunities—the use of high-consistency forming—will be evaluated in the laboratory to determine if the drainage characteristics of the paper sheet is improved. This should reduce the energy required in evaporating moisture from the sheet at the dryer end of the machine. This technology, if successful, could save 2.5 million barrels of oil equivalent per year.

The objective of the mechanical dewatering program is to model the wet end of a paper machine to determine the effect of the various process variables on dewatering paper sheet. Interaction between the computer model and a pilot-size paper machine will be used to refine the computer model. After the pilot machine data and the computer simulations are consistent, a full-size paper machine will be instrumented and operated to verify the computer results. The computer model can then be used to optimize the dewatering process variables and save energy now used in evaporation of this water from the paper sheet. Expected energy savings are 84 trillion Btu annually or 14 million barrels of oil equivalent per year.

The objective of the computer-controlled papermill project is to conserve energy through the application of energy-management automatic control systems to a totally integrated pulp and papermill. The project will demonstrate unit process and central operation control through a hierarchy of automatic control systems. Estimated energy savings are over 9 million barrels of oil equivalent per year.

The roof-felt application project will explore a new process for applying hot asphalt to roofing felt to substitute for the present two-machine hot-dip process. Roofing felt is made on a Fourdrinier machine, after which the dried roll felt is run through a second machine to saturate it with asphalt. The new process involves the application of hot asphalt by spray nozzles, which will apply the asphalt to one side of the partially dried felt to remove the moisture. Subsequently, the second side of the felt will be sprayed to saturate it completely, and this can be done on a single machine. The anticipated energy savings are over 10 trillion Btu per year.

An evaluation of various types of heat exchangers such as heat pipes, heat wheels, fin and tube, etc., and their ease of cleaning will be conducted for application to low-temperature (600° to 800°F) exhaust stacks. This evaluation will consider economics (first cost and operating cost) and energy savings. If the evaluation is positive, a full-size demonstration will be conducted on a building products plant to verify economics and energy savings. Potential applications to other industries such as textiles, pulp and papermills, and forest product mills indicate that annual energy savings of 0.01 quad or more is possible.

The general manufacturing activity covers a variety of manufactured products and industry sectors that in themselves are not one of the major energy consumers; however, in total they represent a significant portion of the energy consumed. It includes transportation equipment, appliance manufacturing, machinery and electrical equipment, and concrete products. Many of these industry sectors do have similar types of manufacturing operations to make their products, so there may be some commonality of processing.

Because of the various types of products manufactured by this grouping of industries, the size of their operations varies greatly from small shops of less than 1000 square feet and 25 or fewer employees to the giant automotive and appliance plants with hundreds of thousands of square feet and thousands of employees. Thus, energy conservation activities and the technology for implementation, as well as capital funds, are generally confined to the larger high-production facilities. However, once the technology is developed by the larger companies, it can be readily disseminated across industry lines by means of the trade associations. For the small shops, however, highly technical and capital-intensive technologies will probably be assimilated very slowly into their operations.

Limited information exists on energy consumption by manufacturing processes in these end-product industries. However, some of the data of these industry sectors are available through trade associations. Several of these industry sectors have manufacturing processes that are common or similar such as metal-forming, machining, painting, heat treatment, oven-curing, welding, etc. The energy consumption of four industry sectors producing metal products such as metal stampings, transportation equipment, machinery, and electrical equipment is shown in Table 8-7.

Since the industries are so varied and cover hundreds of plant operations, the projects in this program area are directed toward the generic processes that have broad application to the industry, or several industries, and are selected on the basis of the energy consumed and the potential of the technology acceptance. Details of significant projects initiated in the consumer products area follow.

Parts that are fabricated and subsequently painted on product finishing lines annually consume over 81 billion cubic feet of natural gas during curing in thermal ovens. This is an inefficient process. The objective of the post-paint curing project is to explore the develop-

Table 8-7.   Energy Consumption in 1985.

| INDUSTRY | TOTAL QUAD | ELECT. QUAD | OIL QUAD | COAL QUAD | N. GAS QUAD | OTHER QUAD |
|---|---|---|---|---|---|---|
| Metal Products | 0.367 | 0.083 | 0.037 | 0.012 | 0.183 | 0.047 |
| Transportation Equipment | 0.347 | 0.094 | 0.040 | 0.046 | 0.131 | 0.040 |
| Machinery | 0.330 | 0.093 | 0.035 | 0.021 | 0.137 | 0.040 |
| Electrical Equipment | 0.228 | 0.081 | 0.023 | 0.012 | 0.085 | 0.030 |
| TOTAL | 1.272 | 0.351 | 0.135 | 0.091 | 0.536 | 0.157 |

ment of radiation-curable coatings (ultraviolet) and their application to post-painted parts. This new process requires only 25 percent of the present energy consumption since the paint formulation is cured without heating. If successful, 60 billion cubic feet of natural gas could be saved each year by the industry.

The objective of the high-efficiency arc weld power-supply project is to improve the energy efficiency of arc welding power supplies through the use of solid-state controls and new inverter technology. This redesign can improve the efficiency of power supplies from the present 55 percent and a power factor of 60 percent to an efficiency of 85 to 90 percent with a power factor near unity. As a result of these improvements, potential energy savings of 42 trillion Btu per year are possible.

The objective of the alternative power for handtools project is to provide theoretical analyses and operational testing of pneumatic, hydraulic, standard electric (60-cycle), and high-frequency electric (180- and 360-cycle) tools for industrial applications to determine the degree of energy savings that is possible. The most efficient and practical tools will be compared to conventional pneumatic tools in a production demonstration to verify energy savings against the theoretical analyses and testing data formerly obtained. Estimated savings in the use of high-frequency electric over pneumatic handtools are in excess of 76 trillion Btu per year.

A study and demonstration of concrete-block curing was completed, which indicated that the use of appropriate insulation on the kilns reduced energy consumption from 35 to 43 percent without affecting quality. The objective of the concrete-pipe demonstration project is to apply a similar technique to concrete-pipe curing. Product differences between pipe and block are such that demonstrations in that industry sector are appropriate to verify energy savings without loss of product quality or production rate. Potential energy savings for this industry segment, assuming a 40 percent reduction in energy consumption, is 1 million barrels of oil equivalent per year. This effort began in FY 1980.

Milestones relating to low-temperature process improvements of the chemical and petroleum industries include completion of demonstrating production of methane from carbon monoxide, ethylene polymer improvements, chlorine production, use of refinery flaring, and a new crude distillation process.

The activities directed to improved energy efficiency in end-product processes (general manufacturing, paper, and textiles) include completion of commercial-scale demonstrations on foam dyeing, closed-cycle hyperfiltration of textile dyestuffs, computer-controlled papermaking processes, nonnatural gas curing of paint, asphalt applications to roofing felt, high-efficiency arc weld power supplies, cutting-tool coating improvements, and improved drying of paper.

The food system is one of the largest and most diverse of our economy. Historically, energy use in the food system has been growing at a greater rate than the population because the United States food system has substituted fossil energy and electrical power for "human energy." Another factor in this energy growth rate has been that more than 75 percent of the food grown on the farm is processed in some degree before being sold to the customer. Recent energy-use estimates indicate that the energy demand for the entire food sector is approximately 6.0 quads per year, and a large percentage of this demand is in the form of natural gas and electricity, two of our more critical energy forms.

The activity in agriculture and food processing focuses on improvements in overall energy use and on use of alternative and abundant fuel forms. The activity is partitioned into 13 segments as follows: integrated farm systems, irrigation, crop drying, alternative farm power systems, fertilizer production, milk and dairy processing, system modeling for agriculture, food-processing efficiency systems, food sterilization, food packaging, sugar processing, citrus processing, and meat processing. Specific projects will now be described.

Energy use on farms is diversified both as to types of energy sources and location of power requirements. The pattern of use is not uniform, and rather large quantities of fuel are used over short periods of time. This diversity and nonuniform pattern of use have both advantages and disadvantages. The primary disadvantage is logistics in supplying such erratic use patterns. However, from the standpoint of using alternative sources such as wind, solar, and biomass, there is an advantage. The integration of alternative sources along with implementation of conservation practices moves some farming enterprises near energy self-sufficiency.

The objective of the energy-integrated farm systems program is to investigate energy production from farm sources such as biomass to investigate the economic applications of solar and wind energy on the farm and to study efficient use of potential energy and non-energy sources in order to promote on-farm integrated livestock/crop-energy production systems. This program will determine the differences between energy-integrated farms and conventional farms in terms of energy used, net profit, and soil and water effectiveness. A particularly significant application can be found in underdeveloped agricultural economies.

Approximately 16 percent of the total energy use on the farm is applied to pumping and distributing irrigation water. A high-priority agriculture and food process subprogram activity is the encouragement and participation in the development of new or improved technologies that will yield greater efficiencies in irrigation pumping plants and irrigation system management while maintaining high levels of productivity.

The primary thrust is to support research in energy-conserving design, development, and commercial demonstrations of improved irrigation systems.

Ten irrigation projects are currently being supported:

- Groundwater irrigation supply system optimization
- Development and evaluation of an ultralow pressure mobile trickle application system
- Analysis of the effect of irrigation scheduling on energy consumption
- Development and demonstration of a variable-speed pump
- Automation of a gated pipe irrigation system

- Low-energy center-pivot sprinkler irrigation system
- Redesign of turbine pump impeller and diffuser using hydrodynamic design techniques
- Irrigated agricultural instrumentation system
- Computerized life-cycle costing irrigation-system pipeline-network design program
- Effects of irrigation-well efficiency on energy requirements

Liquid propane gas is the primary source of energy for crop drying. Approximately 9 percent of the total energy use in agriculture relates to this activity.

Projects are sponsored under the crop-drying activity to improve overall efficiency (currently less than 50 percent) and to encourage the substitution of abundant fuels for liquid propane gas without increasing operating costs. Crop drying—one of the most energy-intensive operations in agriculture—uses, on an almost exclusive basis, liquid propane gas. Moreover, like most heat-treatment operations, drying is only about 50 percent efficient. Five projects are underway:

- Demonstration of the application of electric heat-pump technology to grain drying. (If implemented, this system could save as much as 300,000 barrels of oil equivalent per year.)
- Evaluation of the reduction in energy requirements and improvement in grain quality resulting from a combination of high-temperature and low-temperature drying.
- Development of a new corn-drying system that uses cornstalk residue as a fuel source to dry the grain. (Adoption of this technique could conserve more than 3 million barrels of oil equivalent per year.)
- Performance of a feasibility study to assess the suitability of using a liquid desiccant system for commercial (corn) drying. (This process, if implemented, could conserve the equivalent of 35,000 barrels of oil per year.)

- Development of the use of microwave vacuum-drying technology for moisture removal in grain drying. (At present, the project is concentrating on seed drying, which is extremely critical and delicate, in order to ensure product quality. Present systems rely entirely on LPG, and this alternative would alleviate industry problems.)

Field operations on the farm comprise activities such as seedbed preparation, planting, tillage, and harvesting. Gasoline, diesel fuel, and liquid propane gas (LPG) are all used to power field machines and miscellaneous farm equipment, which are used for field operations. The sum of the energy used for these various field operations is a significant proportion of the total energy required for crop production. The energy used for field operations, though, varies significantly with the crop, soil, and cultural practices. Consumption ranges between 3 and 20 gallons of the energy equivalent of gasoline per year.

If at some juncture the supply of petroleum-based fuels is inadequate to meet the needs of production agriculture, the implications could be disastrous unless alternate equipment systems are available for the farmer. To meet this potential problem, an activity is planned within the alternative farm power systems subprogram to assist in the RD&D of alternate farm equipment systems. The objectives of this activity are as follows:

- Examine more energy-efficient equipment systems technologies for farming
- Assist in the development of fuel alternatives for farm equipment and field machines

Research may include, but is not limited to, unconventional cycles, electric-powered systems, hybrid equipment (employing a combination of two or more energy-conversion/ storage devices), and alternative fuels.

Natural gas is the principal raw-material feedstock for ammonia manufacturing plants, both as a source of hydrogen in the chemical reaction and for process heat in the production cycle. The research planned for the fertilizer production activity includes the development and demonstration of alternative fuels or techniques for heating the tubes in the typical Haber-Bosch process, demonstration of advanced catalyst systems for less energy-intensive reaction of hydrogen and nitrogen to form ammonia, use of alternative methods of producing hydrogen, and development of unique nitrogen-fixation processes to consume less energy.

One project is underway with the Tennessee Valley Authority National Fertilizer Development Center (TVA/NFDC), which is investigating processes for conserving energy in the production of fertilizers in ammoniation-granulation plants under an Interagency Agreement initiated in FY 1976. This 4-year project developed new formulations that reduce moisture added in the granulation process; the formulations use chemical heat of reaction (generated by combination of ammonia and phosphoric acid) as a substitute for natural gas in drying granular fertilizer.

It is estimated that fuel consumption in these plants can be reduced by as much as 65 percent by using the new formulation procedures, resulting in a cumulative energy savings of 5.8 million barrels equivalent per year by 1985, with the first savings occurring in 1979. TVA/NFDC is currently refining a pipe-cross redactor and other melt-type processes, and collecting data from ammoniation-granulation plants on energy requirements for producing fertilizers by conventional processes.

On many modern dairy farms, energy required for cooling milk and heating water for sanitation accounts for up to 75 percent of the electrical energy used. Methods are sought to reduce the energy required for cooling milk and heating water. Investigations are currently under way with DOE support to evaluate the energy savings and market acceptability of sterilization of milk with aseptic packaging. The objective of the milk-and-dairy-processing program is to test and evaluate a "steam infusion, falling-film sterilization" process for fluid milk sterilization. This technique obviates the need for milk to pass over hot metal in the sterilization process, and thereby eliminates the "cooked" tastes obtained from other such sterilization processes. Acceptable sterilized fluid milk potentially offers the significant energy conservation advantages of extended shelf-life, unrefrigerated distribution and storage, reduced travel and marketing procedures, consumer purchasing patterns, and reduced returns of outdated milk. Estimates place the potential energy savings of implementation at 10 million barrels of oil equivalent per year.

The energy use in agricultural production and food processing is being comprehensively modeled by DOE to analyze the options available for use of different technologies. The program, called AGRIMOD, can:

- Identify the impact of possible energy and natural resource constraints on food supply and prices
- Evaluate the effects of policies relating to the imports and exports of food
- Assess the long-term impacts of policies on diets and nutrition
- Analyze the impacts of specific energy conservation policies in planning future programs and assess energy conservation opportunities in the food system

The food-processing industry is diverse; there are a number of processes common to many segments of the industry. Drying is a highly energy-intensive process, which is common but not limited to the prepared animal feeds, wet corn milling, fluid milk, and beet sugar processing industries. In all four of these industries, large amounts of thermal energy are required to evaporate water. Sanitization is another highly energy-intensive process that is common to all the food processing industries. Steam from the boiler is commonly used to maintain sanitation in the food processing plant. Cooling is a third process that is common to many food-processing industries. The meat packing, fluid milk, and frozen fruit and vegetable industries all require refrigeration and freezing to minimize microbial contamination of the food products they process. Specific projects under way include:

- Low-energy concentration of fluid foods and food-processing effluents
- Direct enzymatic extraction of starch

from corn as an energy-saving alternative to production of high-fructose syrup
- Energy use in seafood production and preparation
- Energy conservation in the canning industry
- Advanced dehydration in alfalfa processing to produce a leaf protein concentrate (LPC) for human consumption and improved animal feeds called Pro-Xan that reduces energy use in the conventional alfalfa process by 25 percent, is being demonstrated by Valley Dehydrating Company. The demonstration is supported by the technical staff of the USDA Western Regional Laboratory in Berkeley, California, which developed the process system.

The preservation of foods by canning, freezing, and freeze-drying is very energy-intensive. It was reported in the 1974 Annual Survey of Manufacturers that energy required for the preservation of fruits and vegetables (SIC 203) required the equivalent of 37.6 billion kWh of electricity.

Thermal energy requirements for processing canned fruits and vegetables are relatively large. Canned foods also require a large amount of energy for transport because of the large bulk weight due to the liquid content in the container. Energy use in marketing and end-user storage, though, is minimal. Frozen foods require a relatively small amount of energy in processing, but the requirement of refrigerated energy throughout storage, distribution, marketing, and end-use offsets the energy advantage in processing.

The food sterilization activity is oriented to the acceleration of bench-scale experimentation into new concepts of food sterilization. Bench-scale experimentation will meet either one or both of the following objectives:

- Development of technology that will reduce the energy required for thermal treatments of foods for sterilization
- Development of technology that will assist, along with appropriate packaging, in the reduction of refrigeration

energy required to maintain an acceptable product in distribution and marketing

A project is under way to investigate the preservation of fresh solid foods by gas exchange (GASPAK).

Packaging is a significant component in food-processing operations and in the ultimate food product that we purchase. Packaging serves three basic purposes:

- Containment of the food product
- Maintenance of product integrity
- Assistance in gaining product acceptance by the consumer

A strong relationship exists between the formulation of a food product and the packaging required for the same product. For example, packaging requirements are usually more rigid for liquid foods than for dry solid foods. Based on this relationship, packaging has a significant impact on energy conservation in the formation of food products. The initial thrust of this activity will be to develop a systematic approach to alternate packaging concepts for sterile aseptic packages for fresh foods.

The sugar-processing industry consists of two sectors, the cane sugar industry and the beet-sugar industry. It is estimated that the two industry sectors together consume approximately 15 percent of the energy required for the food-processing industry. This activity is designed to enhance the commercialization of energy-saving technologies (e.g., vacuum pan crystallization and improved concentrating processes) in an extremely low-profit margin industry that is highly conservative and resistant to implementation of untried/unproven technologies. Three projects are underway in beet-sugar processing:

- Demonstration of evaporation by mechanical vapor recompression and falling-film evaporation in the beet-sugar industry
- Feasibility of membrane technology as a means to reduce energy requirements in the beet-sugar industry

- Prelimiting of sugar-beet cassettes to reduce energy in sugar-beet processing

In 1977, it was estimated that 230 million boxes of over 20 billion pounds of fruit were processed by the Florida citrus industry alone. As with many food industries, the processing of this fruit requires high-energy usage to prepare the raw product, remove water, and preserve the final product. Over 50 major processing plants comprise the citrus industry in Florida. These plants have used natural gas for 60 percent of their total energy requirements. The total annual energy consumed has been estimated at 15 trillion Btu, or the equivalent of 2.4 million barrels of fuel oil.

The activity in this area consists of a systems analysis approach to develop improved systems for a growth industry that has suffered from a lack of technological sophistication. To date, two projects have been initiated:

- Energy conservation for citrus-fruit processing
- Evaluation and demonstration of energy-conserving technologies for the citrus-processing industry

In 1975, the meatpacking industry consumed over 100 trillion Btu of energy, or the equivalent of 17 million barrels of oil, more than any other single segment of the food-processing industry. The meatpacking industry is very diverse in sizes of plants, packing processes, and types of products. Despite this diversity, refrigeration, by-product processing, conveyance, and sanitation are energy-intensive operations common to all meatpacking plants.

This subactivity is expected to develop energy consumption data to enhance management prospects of improved energy savings, through adoption and implementation of energy-saving technologies such as heat exchangers, improved recovery systems, modernized steam-boiler controls, etc. The industry has been constrained. The following projects are currently under way:

- Energy conservation in the poultry-processing industry

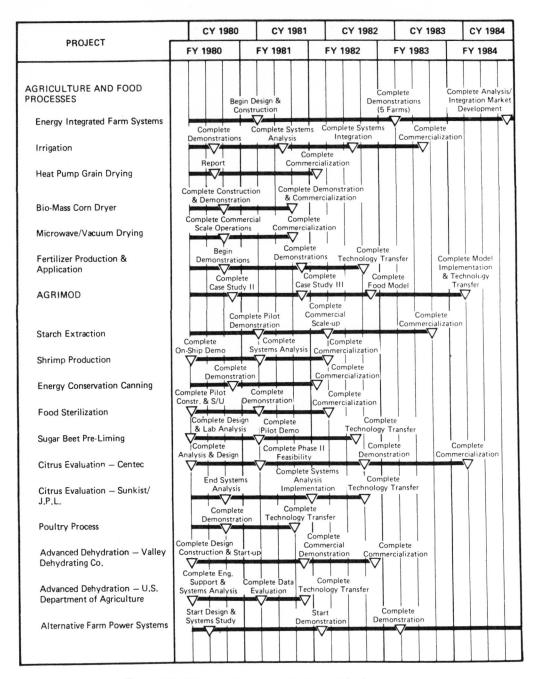

Figure 8-10. Major milestones for agriculture and food processes.

- Food-processing plant systems analysis
- Muscle boning of the unchilled bovine carcass

A wide range of activities will be conducted in improved energy efficiency of agricultural and food processes. Key milestones, shown in Figure 8-10, include completion of five demonstrations of improvements in fertilizer production; demonstrations of milk processing and handling; demonstrations of three different crop-drying systems; demonstrations of alternative power systems for farms and integrated dairy beef, swine, poultry, and grain systems; and two demonstrations of food-packaging systems.

## Conservation Technology Deployment and Monitoring

The new legislation and program analysis activity is directed toward the analysis of proposed new legislation and other programs, and the implementation of new legislative requirements such as those relating to industrial equipment efficiency, industrial reporting, and recovered materials targets. The responsibilities include developing rules, regulations, and analytical studies. Descriptions of specific projects related to this activity follow.

Section 683 of NECPA required DOE to study the relevance of the second law of thermodynamics to the definition and implementation of industrial energy conservation programs. Consultation with the National Bureau of Standards and other appropriate agencies was required. Conservation programs include those authorized by the Energy Policy and Conservation Act (EPCA), the Energy Conservation and Production Act (ECPA), and NECPA, as well as any federal RD&D programs. A report on the results of the study is to be submitted to the Congress.

During FY 1981, it is anticipated that substantial work will be necessary to follow through on recommendations developed as a result of the initial study. The concept of "energy availability" has received substantial attention outside the federal government, and application of the knowledge gained would seem warranted.

Section 374A of EPCA, as amended by NECPA, required that DOE establish targets for use of recovered materials by the metals, paper, textiles, and rubber industries. The materials included are aluminum, copper, lead, zinc, iron, steel, paper and allied products, and textiles and rubber recovered from solid waste. The targets represent the maximum feasible increase in use of energy-saving recovered materials that can be achieved by 1987. In establishing recovered-materials targets, DOE analyzed the technological and economic ability of each industry to meet its target and considered all actions that could be taken to increase the use of recovered materials. Consultation with the Environmental Protection Agency and the affected industries is required. Proposed targets were published in the *Federal Register*, after coordination within and outside DOE.

Removal of constraints to greater- use of recovered materials will be accomplished during FY 1980 and subsequent years. Analysis of the impact of recovered materials use on energy is also to be performed.

Section 342(a) of EPCA, as amended by NECPA, required an evaluation by DOE of industrial pumps and motors in order to (1) determine appropriate standard classifications, and (2) determine the practicability and effects of requiring all or part of the selected classes to meet performance standards establishing minimum levels of energy efficiency. In conducting the evaluation, DOE, among other requirements, must (1) identify significant factors that determine energy efficiency, (2) estimate the potential for improvements in energy efficiency that are technologically feasible and economically justified, and (3) estimate energy savings resulting from labeling rules and efficiency standards. A report on the results of the evaluation will be submitted to the Congress together with recommendations for appropriate legislation.

The National Energy Act also provides DOE with discretionary authority to perform similar analyses with respect to energy-intensive industrial equipment other than pumps and motors, including compressors, fans, blowers, refrigeration equipment, air-conditioning equipment, electric lights, electrolytic equipment, electric arc equipment, steam boilers, ovens, furnaces, kilns, evaporators, and dryers.

After preliminary analyses to determine which equipment to concentrate on, the New Legislation and Program Analysis Branch will, during the 1980s, analyze the costs and benefits of alternative federal programs, both regulatory and nonregulatory, which would be intended to increase equipment efficiency. Congress will be seeking information with respect to the equipment tested as specified in the National Energy Act.

Section 301 of the Energy Tax Act of 1978 provides an additional 10 percent tax credit for investments in certain energy property. The categories of eligible items include specially defined energy property, which consists primarily of waste heat recovery and automatic energy control systems, and recycling equipment. Provision is made for energy-conserving equipment to be added to the list. DOE must provide technical support to the Department of the Treasury in the development of regulations to implement those tax credit provisions. Of particular importance (and specifically addressed in the legislation) is the requirement to provide support in describing performance and quality standards that apply to the determination of eligible equipment.

Each fiscal year, the Office of Industrial Programs has a portfolio review performed to indicate which RD&D programs have environmental considerations associated with their prosecution. Programs involving technology demonstrations that could possibly adversely affect the environment are identified, and a milestone schedule for addressing these constraints is produced. The purpose of this endeavor is to ensure that industry-sponsored RD&D programs can proceed expeditiously and do not incur schedule slippages due to unanticipated environmental problems.

In addition, the Legislation and Program Analysis Branch will represent the office of the DOE Task Force on Industrial New Source Performance Standards and other similar groups.

To make better use of the benefits of several methods for screening proposals, evaluating market potential of competing technologies, and determining the marginal benefits of sponsoring competitive programs, an integrated system is being developed. This incorporates improved data and refinements to the Industrial Sector Technology Use Model (ISTUM) as appropriate to providing a comprehensive planning and budget allocation system.

The Office of Industrial Programs presently uses ISTUM as its primary technology evaluation tool. ISTUM is a sectorial model that compares technologies on an economic basis in the energy marketplace. ISTUM was completed in 1981, and it is proving to be a highly beneficial program-prioritization tool. However, several weaknesses have become apparent. The energy consumption data base built into ISTUM is critical to its operation. The data currently used date to 1974 and must be upgraded. It has also become apparent that ISTUM does not treat retrofit technologies in a completely acceptable manner, and it incorporates no regionalization. Although off-line analysis can be used to overcome these difficulties, it is obvious that the model should be modified and updated.

Development and continuing management of the analytical mechanisms for project selection and comparison are the responsibility of the New Legislation and Program Analysis Branch. This branch will also have the primary responsibility for continuing analysis of current and ongoing programs.

The technologies resulting from the RD&D activities as well as the selected existing but underutilized technologies and practices all must be "commercialized," i.e., widely deployed and implemented by the private industrial end-users. The sequence of activities pertaining to implementation and deployment, spanning from the proving of technological feasibility to the wide-scale implementation by the private sector, is complex for the industrial sector.

The "market" is extremely broad, and, due to the highly fragmented market for the diverse industrial energy conservation technologies, the implementation efforts must be specifically focused. Each market sector must be analyzed with respect to capital availability, asset turnover rates, innovativeness, propensity to accept risks, payback requirements, etc. Each technology product—new or existing but underutilized—must be analyzed to determine its characteristics relative to market acceptance timing, effectiveness of various federal options, rate of return, energy savings, etc., and integrated with the market data to determine match-ups.

Implementation and deployment planning is integral with development and demonstration program planning, and program milestones include those specific events of commercializa-

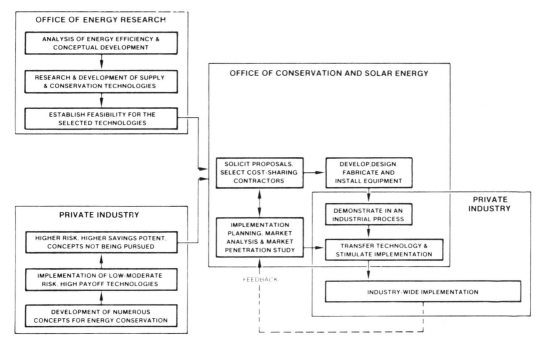

Figure 8-11. Project sequence of industrial energy conservation.

tion as well as hardware demonstrations and testing. Figure 8-11 depicts the sequences related to implementation and deployment of industrial energy technologies. As shown, new and existing industrial energy conservation technologies have three basic origins: private industry (and academia), the Office of Energy Research, and the Office of Conservation and Solar Energy (CS).

The R&D expenditures of the private sector produce many new concepts applicable to energy conservation as well as new product configurations and capacity improvements. The private sector selects appropriate options and generally implements the technologies that have low to moderate risk, high return on investment, and relatively low cost. Many of the remaining conservation concepts are submitted to the government as candidates for cost-shared development and demonstration. Many submittals to DOE relate to existing but underutilized processes that require federal action to implement.

The Office of Energy Research and Fossil Energy conducts R&D on selected supply and conservation technologies and advances them to the point of proving feasibility, usually by a laboratory test. At this stage, the project is

transferred to CS for "commercialization," which in most instances requires demonstrations in industrial environments. The Office of Industrial Programs, within CS, then solicits proposals and selects cost-sharing contractors for demonstrations and, in parallel, starts the commercialization planning sequence.

Next, the designs for the specific application are developed and the necessary equipment is fabricated and installed in a selected industrial end-user facility. A demonstration is then conducted, the technical information is disseminated, and the implementation is stimulated by the usual "marketing" techniques. Feedback on the results of the commercialization efforts is developed to enhance the market planning and analysis for subsequent technologies.

The most effective plan for accelerated industrial implementation of a conservation technology, once proven successful, is to organize a cost-shared contract with a representative end-user company as the demonstrator, the equipment developer (and supplier), and the associated trade association.

A successful demonstration by a reputable company in the industry will have high credibility; the development by a notable equipment supplier will add substantial credence; endorsement and advertising by the related trade association further lend confidence; and cost-sharing by all parties provides a vested interest in making the project successful.

An excellent example of such an arrangement based on a technology developed by private industry is the project for using paint fumes as fuel for curing coil-coated metals. *Coil coating* refers to the prefinishing of metal "coils" (rolls of metal) prior to subsequent formation into products. Typical products of coil-coated materials are aluminum siding, building panels, refrigerator liners, and venetian blinds. The industry doubles in size about every 10 years, and the coil-coating process is, in itself, much more energy-conservative than "post-painting," the spray painting of the finished product after manufacture.

The subject product, however, related to the burning of the paint vapors as a substitute for natural gas used in curing the paint for 95 percent of the industry, which, in 1976 consumed 196 trillion Btu of energy.

The conventional practice is to fire the very large curing ovens with natural gas to drive off the vapors (cure the paint) from the metal surface. EPA regulations require incineration of these fumes before exhausting to the atmosphere, and this requires additional fuel.

Paint curing is one of the most energy-intensive processes in the general manufacturing sector, a contract was awarded to the National Coil Coaters Association (NCCA). As a result of this contract, the NCCA aggressively undertook the challenge to get one of their member companies to propose the application of a new technology (ducting the paint fumes to zone incinerators within the oven so that burning of the fumes serves to cure the paint). This eliminates the need for up to 80 percent of the natural gas to fire the curing ovens as well as the fuel to fire the external incinerator. Thus, a cost-shared contract award was made to the Roll Coater Company, which has since installed and demonstrated the new incineration system. Cost-sharing of over 50 percent was provided by the equipment developer (B&K LTD), the NCCA, and the end-user/demonstrator (Roll Coater Company).

Based on preliminary results, the Roll Coat-

er Company has indicated natural gas savings of 80 percent, a sufficient amount of energy from this one demonstration unit alone to annually heat 8000 average-size homes. The NCCA will take data on this modified oven system to verify the savings and will disseminate the information to their members. In prior preliminary reports, much interest was expressed in obtaining the final results. Since this technology involves retrofit of existing equipment, it is much more attractive to the industry than the complete replacement of their capital equipment.

Fourteen additional systems are already in operation (collectively saving about 3.5 trillion Btu per year), and interest has been expressed by several other companies. Thus, it appears that the technology transfer was taking place before the demonstration was complete.

As previously mentioned, the specific implementation and deployment activities for different technologies differ substantially. Each has to be specifically tailored according to its needs and treated with the optimum combinations of federal options including tax credits, financial incentives, seminars at the demonstration sites, exhibits, documentation, films, plant visits, and focused presentations.

The effectiveness of the implementation and deployment will be measured by tracking the penetration of the specific technology products and feeding the results back into subsequent plans of commercialization for other technology products.

Complete commercialization plans must be developed and implementation programs initiated to transfer existing but underutilized technology and new or emerging technology developed by DOE and others to the private sector. This will require assessments and analyses of such factors as the conservation potential of each technology; the degree of risk and the consequences of failure in both a micro and a macro sense; proprietary aspects and the potential for intra- and inter-industry cooperative projects; the need for and nature of potential regulatory features; development of techniques to ensure that technical information is received by fragmented industry sectors that have little or no research funds; the need for and degree of financial incentives for appropriate technologies; capital requirements and availability; asset turnover rates; and degree of innovativeness by both specific technologies and applicable industry segments with respect to the timing, training, cost, and availability of technical, capital, and human resources.

Analyses of the characteristics and capability of existing industrial institutions and the capital and resource infrastructure will be conducted for each appropriate industrial segment. These will include identification of existing gaps and inadequacies, legal and/or regulatory barriers, and development of planned actions to overcome or modify them.

Marketing plans have been initiated based on completed marketing strategies, including identification and characterization of appropriate industrial segments and target audiences, and development of delivery mechanisms specific to each relevant industrial segment and target audience. Market plan development within the Office of Industrial Programs is closely integrated with the appropriate resource manager within the Office of Commercialization. This activity will include market plans based on the degree of information required, the degree to which the feasibility and benefits of the technology must be proven, and the need for demonstrations and incentives according to the individual situations and characteristics of the target industry segments and audiences.

The ultimate objective of the industrial conservation effort is to effect the transfer of emerging and existing but underutilized energy-efficient technology to the private sector. This element of the office will focus on implementing the actual transfer of technology through such forums as exhibitions, demonstrations, and a wide variety of publications and films.

Within the broad spectrum of implementation and deployment, the technology implementation element is directed toward the following:

- Improving awareness of the nature, extent, and configuration of the energy problems facing the industrial sector and extending knowledge of ways and means to address these problems

- Upgrading existing analytical capabilities, management practices, technological base, and operating procedures relative to energy use in the industrial sector
- Fostering the introduction of emerging and existing but underutilized energy-efficient technologies and practices throughout the industrial community

Implementation and deployment strategy development will encompass a number of projects addressing the energy-intensive industries. Implementation plans for both industry-specific and generic technologies will include pulp-and-paper, iron-and-steel, textiles, aluminum, glass, cement, chemical, petroleum, food, and agricultural industries. This activity is directed toward widespread application of developed technologies on an industrywide basis.

The energy analysis and diagnostic center provides sponsorship to three centers—University of Tennessee, University of Pittsburgh, and Georgia Institute of Technology—to perform energy audits for small- to medium-size industries. The centers conduct detailed energy-use analyses, identify conservation opportunities, perform technical and financial feasibility analyses, and conduct follow-ups on achievements. Since inception of this program, more than 70 audits have been completed, and first-year energy savings of more than 20 times the audit cost have been realized. It is anticipated that the program will continue through FY 1980.

The Office of Industrial Programs periodically sponsors workshops, trade shows, and seminars in an attempt to deploy developed technologies and to further industrial energy conservation practices.

Publications are developed for selected audiences concerning industrial processes. Nine publications are currently in print, and two films have been produced. The publication and audiovisual program is designed to promote industrial energy conservation at all levels.

The purpose of the Energy Impact Scoreboard System (EISS) is to provide the Office of Industrial Programs (OIP) with an accurate and timely measure of regional and national energy savings attributable to the commercial implementation of technologies developed and demonstrated through OIP sponsorship. Once OIP-sponsored technologies have been demonstrated to be technically and economically feasible, those technologies are considered to be commercially available. At that point, through the EISS, each commercial installation of a technology will be noted by location, cost, date of installation, energy consumption/savings, etc., thereby developing a commercial penetration profile for that technology. Analysis of these profile data will enable OIP to evaluate and document its degree of success in developing, demonstrating, and commercializing energy-efficient industrial technologies. The FY 1979 funds will support a pilot data collection and display effort involving two selected technologies. Future efforts will be supported by the Energy Information Administration.

The purpose of the information dissemination project is the preparation of energy conservation technical information products for dissemination to the industrial user community. OIP receives, as a result of RD&D projects, many interim and final reports containing data of use to the industrial community. Much of this information, however, is presented with the government project officer identified as the interested reader. Rarely will the information be presented in a manner that allows the end user (e.g., plant engineers or managers) to evaluate the significance of the results for his own application. Through this project, information is assessed and developed so that the end products will be presented in the most efficient and effective form that will enable the industrial users to assess the appropriateness of a technology to his own need or application. This work includes review of documents; evaluation of technical outputs; preparation of technical and executive briefing reports, technology application reports, technical application manuals, and workshop and seminar reports; and the presentation of workshops, seminars, and symposia.

The Energy Policy and Conservation Act of 1975, as amended by NECPA, requires all manufacturing companies that consume at least 1 trillion Btu to report energy-efficiency progress to DOE. Approximately 1000 companies will report this information, as well as fuel use, either directly or through 50 industry reporting programs. The association reports also include similar information on 2000 voluntary reporting companies that are not mandated by law to report. In addition, identified manufacturing companies within Standard Industry Codes (SICs) 22, 26, 30, and 33 have reported progress beginning in 1979 on the use of recovered materials against the 1987 target of recovered material usage.

As part of the overall reporting program, the Industrial Reporting Program must annually identify those manufacturing companies consuming at least 1 trillion Btu. It must also certify sponsoring organizations' (industry association or third party) reporting systems annually as being in compliance with the act. Upon receipt, reports from all sources are examined for accuracy and completeness. Targets and reporting forms are developed, reviewed, and updated when necessary. Supporting efforts include contributing to rulemaking activities, workshop development, tax-incentive analysis and initiatives, assistance in technology transfer, and technology implementation in the industries with which the programs work.

The program maintains close contact with the industry associations in planning, developing, and conducting workshops, seminars, and other meetings with major industrial corporations to increase acceptance of intensified energy conservation and voluntary participation of companies in the reporting of energy-efficiency achievements.

The following SICs have been identified as the 10 most energy-intensive sectors:

| ENERGY-INTENSIVE RANKING | SIC | INDUSTRY |
|---|---|---|
| 1 | 28 | Chemical and allied products |
| 2 | 33 | Primary metals industries |
| 3 | 29 | Petroleum and coal products |
| 4 | 32 | Stone, clay, and glass products |
| 5 | 26 | Paper and allied products |
| 6 | 20 | Food and kindred products |
| 7 | 34 | Fabricated metal products |
| 8 | 37 | Transportation equipment |
| 9 | 35 | Machinery except electrical |
| 10 | 22 | Textile mill products |

In accordance with the EPCA, overall 1980 conservation targets (1972 base year) for each of the above industries were established. The program assesses the progress of each industry against the target each year in an annual program report to the president and Congress.

Section 375(e) of EPCA, as amended by NECPA, requires DOE to prepare and publish an annual report on the industrial energy-efficiency program established by EPCA. This report includes, as a minimum, a summary of progress made by industry toward meeting industrial energy-efficiency improvement targets, as well as recommendations on how additional improvements might be achieved. In order to be sufficiently comprehensive and to provide appropriate perspective, the annual report also summarizes the total industrial conservation program, and indicates the status of program implementation in response to legislation. The annual report includes a report on the energy-saving recovered materials targets and reporting program established by NECPA.

EPCA requires that approximately 500 manufacturing companies report on energy-efficiency improvement to DOE. With the signing of NECPA in November 1978, the number required to report has about doubled. Each corporation may report either directly to DOE or indirectly through a trade association or other third party, but all must report in conformance with certain requirements established by DOE in response to the legislation.

There has been significant concern on the part of users of the industry reports regarding accuracy and consistency of information. The data validation effort is intended to assess statistical validity and accuracy, and to gauge the degree of conformance with DOE regulations on the part of both company and third-party sponsors.

Major milestones for the Conservation Technology Deployment and Monitoring Subprogram in Figure 8-12.

### Financial History

The Industrial Energy Conservation Program budget history for FY 1979 and FY 1980 is shown in Table 8-8. The FY 1981 budget (operating and capital) is shown in Table 8-9.

Table 8-8.  Budget History of Industrial Energy Conservation.

| SUBPROGRAM ELEMENT | BUDGET APPROPRIATION (DOLLARS IN THOUSANDS) | |
| --- | --- | --- |
| | FY 1979 | FY 1980 |
| Conservation Research and Development | 34,640 | 48,375 |
| Conservation Technology Deployment and Monitoring | 3,160 | 9,800 |
| Personnel Resources | 2,193 | 2,067 |
| TOTAL | 39,993 | 60,242 |

Table 8-9.  FY 1981 Budget Request for the Office of Industrial Programs.

| PROGRAM ELEMENT | FY 1981 REQUEST (DOLLARS IN THOUSANDS) |
| --- | --- |
| Conservation Research and Technology | 51,800 |
| Conservation Technology Deployment and Monitoring | 4,500 |
| Personnel Resources | 2,600 |
| TOTAL | 58,900 |

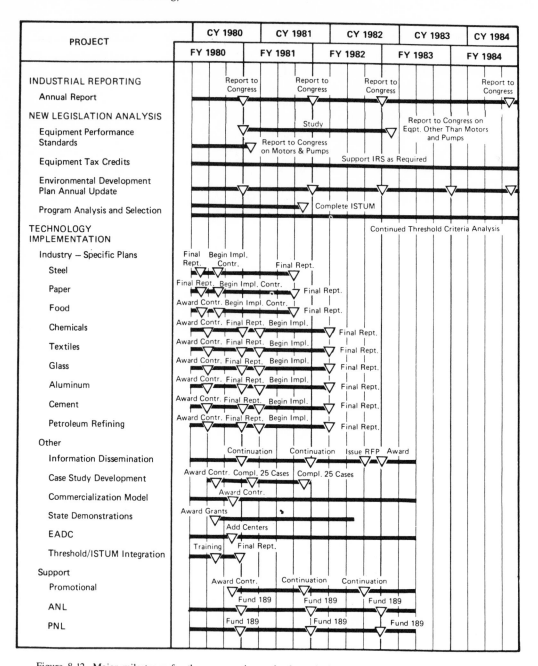

Figure 8-12. Major milestones for the conservation technology deployment and monitoring subprograms.

## PROGRAM IMPACT

### Measures of Effectiveness to Be Monitored

**Program Validation.** The Industrial Energy Conservation Program is a necessary component of the DOE conservation effort. The pro-

gram is directed at the largest energy-consuming sector, which uses 37 percent of the total United States energy. It fills a critical need not adequately covered by regulations and incentives. Federal involvement in industrial energy conservation relates to a variety of factors that change from industry to industry. Energy represents a small fraction of product costs in even the most energy-intensive industries. This fact, coupled with the technical and economic risk in major production facilities, provides little incentive for private-sector pursuit of high-payoff, high-risk energy conservation technology development, especially in the presence of higher priority demands for the limited capital available. Without federal involvement in cost-shared RD&D in industrial energy conservation, a very large, fertile opportunity will be untended.

**Program Options.** The major options that have been examined in formulating the industrial energy conservation program include:

- Regulatory policies
  1. Minimum performance standards for selected equipment
  2. Mandatory energy productivity targets
  3. Energy taxes
  4. Mandatory fuel switching for selected equipment
- Incentive policies
  1. Cost-shared RD&D for new conservation technology
  2. Financial incentives for installing specific technologies
  3. Relaxation of existing policies inhibiting conservation
- Complementary policies
  1. Voluntary energy productivity targets and energy reporting
  2. Technology transfer, technical assistance, and promotional efforts

These options for federal involvement have been compared relative to the following criteria:

- Energy savings impact
- Fuel switching impact
- Technological implementation risk
- Compatibility with other federal mandates (high employment, low inflation, environmental impact, the competitive effect on industry)
- Federal cost of implementation and administration

- Avoidance of displacement of industry initiatives
- Expedience of implementation

The following policy options are found to be consistent with one another and superior to the other options considered:

- *Federal cost-sharing in RD&D.* This program is extremely useful as an incentive for the development and commercial introduction of new, innovative technologies. By making this technology available to all of industry, the program can effect a large energy savings profitable to industry and cost-effective to the government and the nation.
- *Energy taxes.* Taxes levied on industrial use of petroleum and natural gas are a direct means of implementing existing conservation technology. Return of the revenue produced by such taxes to assist industry in accelerating their conservation efforts is basic to a strong federal role in solving the common energy problems of all industry.
- *Mandatory fuel-switching for selected equipment.* If the price of coal falls further below that of imported oil in the future, industry would switch many fuel-burning installations to coal for economic reasons. In the immediate and near term, however, mandatory fuel-switching, now being implemented by the FERC, is the only feasible way to accelerate this outcome.
- *Financial incentives.* Tax credits, loans, and loan guarantees should be modified and administered to provide a strong impetus to investments in existing technology without creating windfall profits.
- *Relaxation of existing policy inhibiting conservation.* Several appropriate federal actions exist that can effectively take the goal of energy conservation into account in policies originally designed for other objectives. These steps as a group appear cost-effective and

necessary for a balanced federal strategy.

- *Complementary programs.* Voluntary energy-efficiency targets, the industrial energy-data-reporting system, technology information dissemination, and the business assistance program are all positive, low-cost contributions to the federal program portfolio. They would appear necessary to enhance the impact of whatever punitive or incentive programs are emphasized in the future.

The activities for the Industrial Energy Conservation Program within the Office of Conservation and Solar Applications consist of cost-shared RD&D and complementary programs with participation with the Internal Revenue Service relative to tax incentives.

Program Effectiveness. The program effectiveness measures to be taken during the next 5 years will include updating and refining the program selection mechanism. The Threshold Analysis Model is being modified to yield a better tool for project selection from a cost-benefit standpoint. A 2-year effort to upgrade the Industrial Sector Technology Use Model (ISTUM) data base is being performed as well. This effort will provide for updating the ISTUM data base, regionalization of the fuel-pricing scenarios, and retrofit technologies. These efforts should provide an opportunity for increased scrutiny in program selection.

Major milestones and out-year funding charts have been developed for each project and reviewed on a regular basis by the Office of Industrial Programs Project Review Board. This will ensure that schedules and financial criteria developed during project selection are met and will provide an opportunity for critical review of any effort not achieving its preestablished goals.

Workshops and seminars with industrial participation will continue to be sponsored to ensure the effectiveness and validity of the overall Industrial Energy Conservation Program. Continuing liaison with the professional societies and trade associations will be maintained through the Industrial Reporting Program.

Activity Measures. Two major areas of activity measure are planned for the next 5 years. The first is development of a conservation technology implementation and development program on an industry-specific basis. Three industry-specific plans were developed in FY 1979 for the food-processing industry, the steel industry, and the pulp-and-paper industry. The second activity area is the development and setting up of an Energy Impact Scoreboard. The first endeavor will provide a measure of the rapidity of adoption of DOE-developed conservation technologies by industry. This will provide a measure of program effectiveness and will identify the technologies and the industries most amenable to energy conservation measures. The Energy Impact Scoreboard will monitor the energy savings being achieved by Office of Industrial Programs technologies. These results will serve as a benchmark for measuring program effectiveness.

## Projected Energy Savings by Specific Subprogram and Fuel Type

The estimated savings for the RD&D activities alone have been developed based on current and projected program efforts. These estimated savings (1.5 quads annually by the year 1985 and 5.5 quads by the year 2000) include additional projects in subsequent years supportive of the targets and building upon the results of the initial work. The estimates are based on the National Energy Act, rising fuel prices, increased industrial consumption, increased market competition for available energy supply, and the effect of uncertainty of supply as stimuli to industrial implementation of increased energy-efficiency technologies.

Table 8-10 shows the estimated savings by subprogram and by fuel type. Table 8-11 lists the total potential savings and estimated actual savings for each activity for the year 1985.

As reflected in Table 8-11, the total estimated savings of the identified technologies within each program area are a fraction of the total potential for the reference year and represent annual savings of an order-of-magnitude greater than program cost. Since each project has been selected partially because it

Table 8-10.   Estimated Savings and Fuel Type.

| SUBPROGRAM | ANNUAL ESTIMATED SAVINGS QUADS | | | | | | | |
| | 1985 | | | | 2000 | | | |
| | GAS | OIL | COAL | OTHER | GAS | OIL | COAL | OTHER |
|---|---|---|---|---|---|---|---|---|
| Waste Energy Reduction | 0.190 | 0.432 | 0.128 | — | 0.738 | 1.378 | 0.165 | — |
| Industrial Cogeneration | 0.008 | 0.035 | 0.107 | 0.080 | 0.033 | 0.151 | 0.464 | 0.352 |
| Industrial Process Efficiency | 0.231 | 0.161 | 0.073 | 0.055 | 0.851 | 0.927 | 0.246 | 0.196 |
| Technology Deployment & Monitoring | N/A | N/A | N/A | N/A | N/A | N/A | N/A | N/A |
| TOTAL | 0.429 | 0.628 | 0.308 | 0.135 | 1.622 | 2.456 | 0.875 | 0.548 |

Table 8-11.   Potential and Estimated Savings.

| ACTIVITY | 1985 QUADS/YEAR | | VALUE OF ESTIMATED ANNUAL SAVINGS @ $3.00/10⁶ BTU |
| | TOTAL POTENTIAL | ESTIMATED SAVINGS | |
|---|---|---|---|
| Waste Heat Recovery & Utilization | 5-7 | 0.63 | $2.25 Billion |
| Alternative Fuels & Feedstocks | 1.5-2.1 | 0.12 | 0.36 Billion |
| Industrial Cogeneration | 2-3 | 0.23 | 0.69 Billion |
| High Temperature Processes | 2-3 | 0.12 | 0.27 Billion |
| Low Temperature Processes | 1.5-2 | 0.18 | 0.36 Billion |
| End Product Processes | 0.2-0.3 | 0.13 | 0.39 Billion |
| Agriculture and Food | 0.4-0.5 | 0.09 | 0.18 Billion |
| TOTAL | 12.6-17.9 | 1.5 | $4.5  Billion |

has high technical risk but also high return on investment (ROI), it is logical to assume that a number of projects could fail. The energy-savings goals of the program have been discounted by about 40 percent to allow for the potential for failure. The key to the success of the overall program is accurate determination of failure and project abandonment when such determination is made, coupled with re-programming to fund new potential opportunities (many such new opportunities have already been identified). The high ROI of a successful project supports rapid market penetration. Considering this and the impacts that could be expected from technologies that are not currently identified but which could be expected to emerge during subsequent time frames, the estimated impacts are considered conservative.

Energy-savings estimates for federal actions on promising but underutilized technologies and the energy impacts from legislated programs have not been determined.

## Accomplishments

The key accomplishments of the Industrial Energy Conservation Program through FY 1980 are:

- Several hundred million yards of textiles have been processed to date using the new energy-conserving foam-dyeing techniques by nine different concerns. The savings from this project today total 2 billion Btu.
- Successful pilot-plant demonstrations have been completed for converting waste polypropylene to fuel oil. Several petrochemical concerns have expressed interest in construction of full-size facilities using this process, and market

penetration appears encouraging. Waste polypropylene now constitutes about 300 million tons annually, which is currently used as landfill.

- A cost-shared development of foam dyeing (foam replacing water in textile finishing operations) has resulted to date in the production of over 300 million yards in nine different mills using the new process. This project, to date, has saved over 300 billion Btu, which at $2.50 per million Btu represents over $150,000 saved for a project with a total federal cost of $662,000.
- A successful demonstration has been completed for fumes-as-fuel in paint curing–coil coatings. The demonstration unit is saving up to 86 percent of the natural gas previously used, and 14 additional systems are in operation.
- A successful demonstration was completed (operating in the manual mode) in FY 1978 for a boiler air/fuel ratio control system, which, for the demo alone, is saving 75 billion Btu annually.
- Eight contracts are in demonstration hardware phases for waste heat recovery. Four are directed at high-temperature recuperators, which are more efficient and cost-effective than conventional units. Three contracts are for industrial heat pumps, which operate at higher thermal levels than existing equipment.
- Three activities directed at advanced aluminum processes are in hardware phases: pilot-scale tests are being conducted for substitution of coal for natural gas in aluminum remelt operations; an electrolytic aluminum smelter has been tested with an inert cathode; and a small-diameter blast furnace was designed, fabricated, and is being tested for direct reduction of aluminum.
- A high-efficiency slot forge furnace has been developed and successfully demonstrated. It features ceramic recuperation, excess air control, slot closing doors, and improved insulation. The furnace achieves savings of nearly 70

percent over conventional units, and expected energy savings of this project are 0.07 quad per year.

- A variety of activities are in progress for increased efficiency in textile processes (dyestuff filtration and waste heat utilization) and in papermaking (high-density headbox and improved black-liquor processes).
- An industrial cogeneration program was initiated in FY 1979, and four contractual efforts have been completed pursuant to demonstration of topping cycles in industrial environments. Designs are complete, and fabrication has been initiated for application of a bottoming cycle system in glass processes.
- Successful pilot-plant demonstrations have been completed for converting waste polypropylene to fuel oil. Several petrochemical companies have expressed interest in proceeding with full-size facilities using this process, which has a potential savings of 2 million barrels of oil annually.
- Design of a vertical-shaft heat exchanger (for preheating glass pellets prior to glass melting) has been completed.
- A parametric study was completed for energy-integrated industrial parks.
- The ammoniation granulation fertilizer "cross pipe" reactor project resulted in commercial implementation. With the support of TVA, operation of 10 plants using the new technology has begun, 8 additional plants are scheduled to implement it, and another 8 plants are considering implementation.

## Project Selection-Analysis of Benefits and Costs

Figure 8-13 shows the sequence of project selection for the Office of Industrial Programs. Incoming proposals are first screened for appropriateness and soundness. Then they are subjected to comparisons to three criteria: return on investment, once the technology is proven successful, must be at least 15 percent; the annual value of energy saved must be at

Proposals

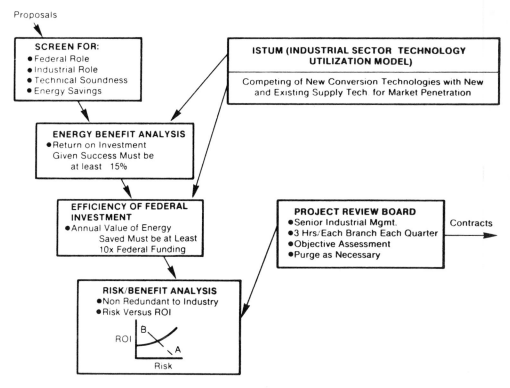

Figure 8-13. Project selection.

least 10 times the total federal cost; and the risk/return on investment at the conclusion of the federal involvement must be within the envelope of acceptable risk/rate of return so that, given success, industry will adopt the technology. The illustration in the lower left center of the figure shows the change in risk and return on investment (from A to B) that federal action should cause. Point A is outside the range that industry would undertake on its own. Point B is within the range of industry acceptability.

Therefore, appropriate candidates for federal involvement would be in the low-ROI, high-risk area (Point A), and the projected result of the federal action would "move" the technology within the bank acceptable to industry on its own. ISTUM inputs on market penetration are used in the analysis as shown. Figure 8-14 shows five examples of actual changes in risk/ROI relationships from projects of the industrial and one projected case (direct reduction of aluminum).

Upon completion of the analyses and comparison with threshold criteria, each candidate project is reviewed by the Project Review Board to assess the relationship to the criteria, cost in relation to budget limits, degree of cost-sharing, proprietary requirements, phase of development, environmental issues, etc., prior to contract development. Subsequent analyses and review are also conducted at the end of each major contract phase prior to proceeding.

Industry, on a nationwide basis, is the beneficiary of the energy savings achieved as a result of the Industrial Energy Conservation Program. The general public is also an indirect beneficiary, as reduced industrial energy consumption will result in increased fuel availability at lower cost than would otherwise be the case.

## ENVIRONMENTAL IMPACT

### Issues

The Office of Industrial Programs sponsors cost-shared RD&D energy conservation pro-

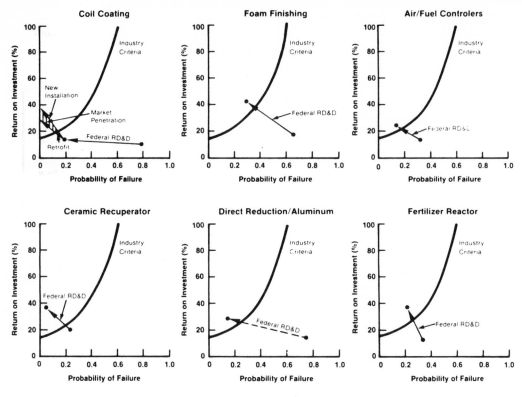

Figure 8-14. Examples of federal actions on risk and return on investment.

grams throughout the energy-intensive indus-trial sector. The programs have both horizon-tal and vertical thrusts because they encom-pass generic as well as industry-specific proc-esses. Because of the disaggregate nature of the industrial energy conservation program portfolio, it is highly probable that many of the sponsored RD&D efforts could have ad-verse environmental effects if not properly conducted. For instance, a major goal of the program is the switching of coal for scarce fossil fuels such as oil and natural gas. Any program of this nature could have an adverse impact on local air quality if proper steps are not incorporated into the RD&D effort. Fur-thermore, it is in the national interest to get these types of technologies developed and im-plemented as rapidly as possible.

## Activities to Resolve Environmental Issues

As indicated, many of the more significant programs, particularly those involving fuel-switching, could have adverse environmental impacts. An example is the Direct Reduction of Aluminum Program, which involves devel-opment of a technology to produce aluminum in a blast furnace as a substitute for the elec-trolytic process. This will involve the sub-stitution of coal for electricity. Although most electricity is produced from coal, the new technology will still be restrained by point and new source emission and all other appli-cable EPA standards. Superior practice dic-tates that all environmental standards be addressed during the technology development rather than on a retrofitted basis. This type of operation minimizes the economic and sched-ule impact of the environmental considera-tions.

The Environmental Development Plan (EDP) system is designed to provide a com-mon basis for planning, managing, and re-viewing the environmental aspects of the various energy conservation technology pro-grams. More specifically, the EDP governs the nature and content of the environmental issues. Development of this plan by carefully reviewing each technology program for possible environmental impacts provides a basis for ensuring that each RD&D effort will have a coordinated timing schedule for required environmental R&D. Careful adherence to the plan ensures that DOE-developed technologies will not have adverse environmental impacts or schedule delays due to unanticipated environmental concerns. The EDP is produced on an annual basis by the Office of Industrial Programs in coordination with the Office of Assistant Secretary for Environment.

## MAJOR PROGRAM ISSUES

The principal program issue is the development of a well-defined federal role for industrial RD&D.

The industrial sector accounts for more than 35 percent of the national energy consumption and is therefore a prime candidate for conservation. Although substantial energy conservation and fuel conversion initiatives are generally being undertaken by industry, significant constraints to achieving levels of energy efficiency exist that are consistent with national energy objectives and minimum national energy costs. These constraints can be generally characterized as follows:

- Economic and financial
- Inertia, i.e., reluctance of industry to move away from that with which they are familiar
- Lack of information, e.g., with respect to emerging developments and existing energy-efficient practices

The actual constraints in the case of any particular industry or corporation are numerous. In addition, they vary substantially from industry to industry. The federal program therefore, must (1) be multifaceted in order to address each of the general constraints; (2) contain elements for which there is a high expectation of payoff in terms of incremental cost-effective energy conservation, and (3) avoid elements directed at factors that are not significant constraints to conservation. Finally, all of this must be accomplished by a program that is relevant for decisionmakers

functioning in response to highly variable conditions and circumstances.

Factors that prevent achievement of more optimal energy use by industry include:

- Energy often accounts for a small portion of the total cost of goods produced, even in most energy-intensive industries.
- R&D is often undertaken by only one or two companies in most energy-intensive industries.
- Corporate investment decisionmaking is not based solely on minimum life-cycle cost considerations, but also on other factors such as maintenance of market share, meeting pollution control regulations, use of processes that are familiar, and maintenance of acceptable debt/equity ratios. There is often little capital for discretionary energy conservation investments, resulting in individual ROI requirements that have no relationship to the depth and economic value of energy supplies.
- The cost of energy actually paid by industry as well as that generally projected in individual corporate ROI calculations, is less than the cost, for example, of imported oil.
- Numerous conservation technologies have perceived high risk associated with their implementation, and industrial decisionmakers tend to be conservative.
- Often, those who specify the equipment and processes for an industrial plant are not those who pay the operating expenses, nor are they concerned with minimizing operating cost.
- Even in industry, there is some reluctance to believe that energy realities have changed.

A federal role seems obvious—even critical—in overcoming such obstacles to improved energy efficiency. Development and demonstration assist in removing risks, or perception of risk, associated with new practices. Also, it is not realistic to assume that industrial energy users have perfect knowledge on energy availability, energy price, and/or what can be done to improve energy efficiency. A well-focused DOE program of commercialization would assist in overcoming industry inertia with respect to energy conservation.

Finally, and most important, programs must be devised that cause industry's investment decisions to be based on a realistic value of the energy resource. In developing and reviewing such programs, care must be taken to avoid getting caught in circular arguments. For example, there are reviewers, including some in industry, who believe that there will be perfect response to changing economics, which suggests that little or no federal presence is necessary. Many of those same reviewers, however, then argue against necessary changes in economics, blocking the condition that could in fact justify less federal involvement. Other reviewers, noting that industry is not responding as would be supposed even on the basis of present energy costs, would force conservation investments without modifying the economics or other basic conditions that have led to the lack of response. Given the complexities surrounding energy use, the reasonable approach seems to be (1) a continuing program to determine the real constraints to conservation and (2) a broad program containing a number of elements geared at removing the constraints.

For these reasons, there is a federal role in sponsoring cost-shared RD&D on high-risk yet high-payoff conservation technologies. The emphasis is on identifying existing but underutilized technologies, developing new energy-saving technologies that are not redundant to the efforts of the industry alone, and using every available means to stimulate the early implementation of these results.

# 9. Transportation Programs

## PROGRAM OBJECTIVES AND STRATEGY

The transportation sector of the American economy is almost totally dependent upon petroleum. In 1978, transportation accounted for more than one-half of the total United States petroleum consumption, which amounts to more than one-quarter of all energy consumed in the United States (see Figure 9-1). Conservation of transportation fuels, therefore, is an important objective of DOE programs. As shown in Figure 9-2, the highway system consumes 78.6 percent of the transportation energy, and the automobile is the major petroleum user in the highway sector.

Until 1950, domestic sources provided virtually all of the petroleum required by this country. Since then, however, demand has exceeded the domestic supply, resulting in a need to import an increasing fraction of the required petroleum; over 40 percent of the petroleum consumed in the United States in 1978 was imported.

The transportation sector's almost total dependence on petroleum and the resulting need for imported oil have a major impact on the nation's balance of payments and result in an already demonstrated vulnerability to petroleum supply disruptions.

### Objectives

Recognizing the problems of increasing oil consumption, depleted domestic supplies, resultant increases in required oil importation, the associated adverse balance-of-payments impact on the economy, and the resultant national defense implications, the Congress and the administration strive to reduce petroleum demand and consumption in this country. This policy is reflected in the overall goals of the Office of Transportation Programs (OTP), which are:

- To reduce the energy consumed by the transportation sector
- To reduce, to the maximum extent possible, transportation's nearly complete dependence on petroleum

Successful development and commercialization of alternative transportation and fuel tech-

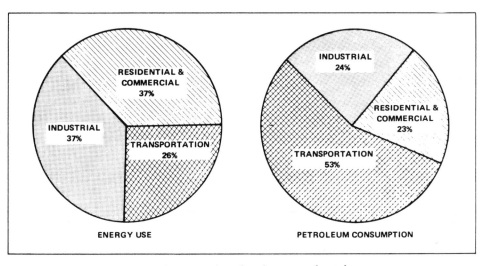

Figure 9-1. Domestic consumption of total energy and petroleum resource.

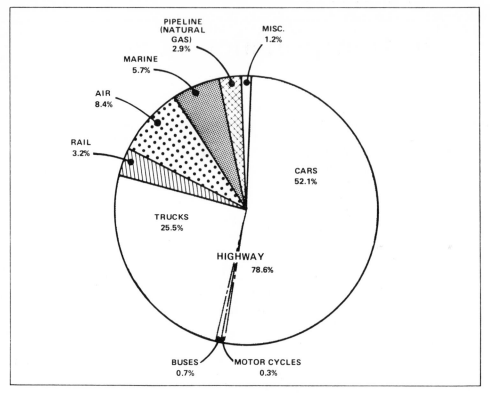

Figure 9-2. Energy use by the transportation sector, 1976.

nologies, together with effective promulgation of educational and informational approaches to encourage greater efficiency in vehicle and systems end-use, are the major paths toward this goal.

As the largest petroleum consumer, the transportation sector can make substantial contributions to solving the petroleum problem.* Savings can be achieved in the near term by an immediate and significant reduction in petroleum consumption, i.e., through improvements in current vehicle technology and in vehicle end-use patterns. Conservation resulting from improved end-use efficiency, although most pronounced in the near term, will continue as longer-term technological developments are commercialized.

The long-term solution to the petroleum problem lies in the widespread availability of vehicles that do not depend on petroleum-based fuels. Independence from petroleum could be achieved through the following options:

- Use of alternative fuels (other than petroleum-derived gasoline or diesel) in near-term and advanced heat engines
- Vehicles powered by electricity

Under the first option, near-term and advanced heat engines will be designed to use alcohols and gasoline-like synthetic fuels, derived from coal or oil shale, which do not require significant processing. The resulting lower grade fuels will provide higher yield per barrel of crude than current gasolines from petroleum. The second option—electric and hybrid vehicles energized exclusively or primarily by electricity generated from coal, nuclear, or hydroelectric power plants—could also provide mobility without the use of petroleum. Since both options show promise of being technically feasible and able to satisfy marketplace requirements, the federal program includes RD&D on both approaches.

OTP has established the following measurable objectives to ensure goal achievement:

- Improve the efficiency of energy use in transportation to achieve a 10 percent

---

*Although this plan is specific to the transportation sector, other DOE program elements are directed at energy conservation and petroleum savings in the industrial, commercial, and residential sectors.

reduction in gasoline use from the levels currently projected for the near term

- Achieve a 25 percent reduction in all forms of transportation energy use from the consumption levels currently projected for the midterm

Each program also has specific objectives that will contribute to attainment of the overall goals. The objectives, controlled by DOE and measurable in some manner, are based upon legislative mandate, industry estimates, technology expectations, market assessments, and budgetary constraints.

## Automotive Technology RD&D

*Advanced Heat Engine Development.* The initial objective of the heat engine program, as conducted by the Environmental Protection Agency in 1971, was to develop heat engines with significantly reduced exhaust emissions. A second objective of high fuel economy was added in 1972. The Automotive Propulsion Research and Development Act of 1978 (Title III, P.L. 95–238) underscored the national importance and priority of these objectives and established a timetable for their achievement by DOE.

The Advanced Heat Engine Development (Ref. 1) portion of Automotive Technology RD&D now places major emphasis on the gas turbine and Stirling engine propulsion systems. Its overall objectives are to achieve the following:

- At least a 30 percent increase in fuel economy over the best 1984 internal combustion engine in a vehicle of equal performance based on EPA test procedures
- Exhaust-gas emissions within the 1983 federal standards
- Engines capable of burning lower grade, non-petroleum-based fuels (higher yield per barrel of crude) without major modifications to the propulsion system
- Suitability for mass production and commercialization in the 1990s

Successful completion of the following objectives specific to the gas turbine project will

contribute to achievement of the overall Advanced Heat Engine Program objectives:

- Provide substantial improvements in fuel economy and manufacturing cost by advancing the technology of low-cost, high-temperature materials, and in the performance of small turbomachinery
- Assist the auto industry in the commercialization of heavy-duty automotive gas turbines
- Acquire in-vehicle/in-revenue service data on advanced gas turbine propulsion systems

Achievement of the following objectives specific to the Stirling engine project will likewise contribute to attainment of the overall Advanced Heat Engine program objectives:

- Technology transfer from Europe to the United States
- Development of computer-design models equivalent or superior to those proprietary models currently used by United Stirling (Sweden)/Philips (Netherlands)
- Development of Stirling engine hardware more conducive to automotive requirements
- Development of technology that will permit solutions to current Stirling engine problems, specifically in the area of low-cost materials and working gas containment

**Vehicle Systems.** The Vehicle Systems Program provides the linkage between advanced heat engine technology development and technology use, leading to accelerated commercialization. Program objectives are:

- Assess and develop vehicle propulsion systems (excluding advanced engines) that contribute to improved vehicle efficiency
- Conduct studies, assessments, service testing, and demonstrations of developed vehicle propulsion systems in actual commercial operations to show technical/economic feasibility and public acceptance and to identify major

socioeconomic problems associated with their introduction

- Provide appropriate data and documentation to facilitate commercialization by private industry and to enhance competition

**Alternative Fuels Utilization.** The Alternative Fuels Utilization Program (AFUP) is directed toward (1) lowering risks associated with development of alternative fuel technologies in the transportation sector and (2) encouraging the use of such fuels. The AFUP has established the following near-term objectives:

*Alcohols*

- To test and evaluate alcohol-gasoline blends, such as 10 percent ethanol–90 percent gasoline, in commercial or government highway-vehicle fleets and ultimately to prove the feasibility and reliability of blends for use in the United States.

*New hydrocarbons*

- To evaluate the various fuels in advanced heat engines. Determination of an optimum fuel is scheduled for mid-FY 1982.
- To achieve and/or evaluate the use of various fuels in present and developmental (for example, stratified charge) internal combustion engines, in order to determine the suitability of these engines for use with new fuels. Evaluation is scheduled for completion during FY 1982.

*Synthetic gasoline and diesel fuels*

- To test and evaluate synthetic gasoline and diesel fuels in production and improved versions of existing engines. Shale-derived diesel studies are scheduled for completion by the third quarter of 1980.

*Advanced fuels*

- Assess advanced and new fuels suggested for possible use in highway vehicles; keep abreast of new developments, inventions, and research programs.
- Select an optimum hybrid fuel by the end of FY 1981; a hybrid fuel demonstration is scheduled for mid-FY 1983.
- Decide upon the most feasible method of hydrogen storage by mid-FY 1983.

*Emergency fuels*

- Determine nonstandard fuels that might be used in existing engines for maintaining the reliability of essential vehicle services in an emergency. A handbook summarizing a study of fuel "extenders" will be published by the end of FY 1980.

**Electric and Hybrid Vehicle RD&D.** An electric vehicle depends solely upon wall-plug electricity. The energy is stored aboard the vehicle in energy storage devices, such as batteries. A hybrid vehicle is fueled from more than one external source of energy, one source being wall-plug electricity (e.g., a vehicle that has both an on-board heat engine and batteries recharged from wall-plug electricity). Over an extended period of time, such as a year, the majority of propulsion energy must be supplied by wall-plug electricity. The Electric and Hybrid Vehicle (EHV) RD&D Program has been established to bring about the commercialization of electric and hybrid vehicles. The passage of the Electric and Hybrid Vehicle RD&D Act of 1976 (P.L. 94–113) established requirements for market demonstrations of these vehicles, financial incentives to industry, and a broadened scope of research and development. The act was amended in 1978 by Title VI of P.L. 95–238, which emphasizes the commercialization process through which the EHV industry, currently in its infancy, can quickly become self-sustaining.

The goal of the EHV RD&D Program is to assure the availability and broad market acceptance of vehicles that depend primarily on externally generated electricity for propulsion energy in order to minimize dependence on imported oil and to ensure continued flexibility in the transportation sector. Specific program objectives are:

- Identify, test, and prove market sections for EHVs
- Develop the necessary EHV support infrastructure
- Provide cash flow to EHV manufacturers
- Develop upgraded vehicles through optimization of off-the-shelf technology
- Assess state-of-the-art in vehicles and technology; establish minimum performance standards
- Facilitate commercialization by 1986 of a cost-competitive 100-mile range urban/suburban electric vehicle
- Facilitate commercialization by 1988 of a cost and performance competitive hybrid vehicle
- Develop a general-purpose hybrid or electric vehicle system

Transportation Systems Utilization. The activities under the Transportation Systems Utilization Program are aimed at accomplishing conservation through (1) the encouragement of energy-conserving equipment acquisition, operation, and maintenance practice; (2) better use of current and future transportation and equipment through such means as increasing load factors, shifting demand to the more efficient modes, and reducing travel and freight transport demands; and (3) development and commercialization of technology for improving the efficiency of the marine, air, railroad, and pipeline modes. The program uses a number of approaches, including education, training, and information transfer; technical assistance; barrier assessment and modification; concept and technology evaluation; and RD&D. The purpose of each of these activities is to motivate the wide range of individuals and organizations, whose decisions affect transportation energy demand, to adopt practices that improve energy efficiency. These groups include consumers;

equipment manufacturers; businesses using transportation equipment and commercial transportation services; public officials who regulate, construct, and operate various elements of the transportation system; private operators of transportation services; and major employers, because of their ability to influence the commuting practices of their employers.

Program activity is pursued with the following objectives:

- *Vehicle performance.* Improved vehicle performance is sought through driver education programs, new-car fuel-economy information dissemination (i.e., *Gas Mileage Guide*), voluntary truck-and-bus fuel-economy programs, and the evaluation and support of new concepts to improve vehicle efficiency.
- *Systems efficiency.* More efficient use of transportation systems is supported by fostering the development and commercialization of improved technology and operating procedures for rail, marine, air, and pipeline systems; intervention in regulatory procedures; technical assistance for state and local agencies; and outreach for private and federal energy conservation programs.
- *Analysis and assessment.* Support for transportation energy conservation program managers is provided through the development and publication of data on the characteristics of the transportation sector and its energy-use patterns, assessment of environmental and sociological impacts of developing transportation conservation technologies, commercialization planning, and program support and analysis.

## Federal Role

In view of the large balance-of-payments impact and the potentially serious consequences of disruptions in our imported petroleum supply, federal involvement in solving the petroleum problem is appropriate in several areas. The transportation industry's investment in existing plants, equipment, production tooling, and methodology is enormous. In order to assist industry in implementing conservation through the introduction and commercialization of new transportation technologies, the federal government can play an active role by:

- Maximizing the potential for early commercialization by direct federal funding and cost-sharing in the R&D of high-risk, high-payoff technologies that might not be otherwise attractive to industry
- Federally funding demonstrations that are designed to favorably introduce industry and the consumer to the newly developed transportation technologies or to innovative means for using existing but underutilized technologies
- Establishing incentives to remove barriers to commercialization, as well as federally supported financial incentives such as planning grants, loan guarantees, and cost-sharing
- Providing education and technical assistance to improve energy efficiency in the use of transportation systems

Through various means of generating awareness and interest on the part of consumers and businesses, it provides them with the information and encouragement necessary to implement end-use energy conservation practices effectively.

In addition to the preceding DOE activities to foster energy conservation, the federal government directs additional programs relating to transportation safety and environmental protection. Although primarily regulatory in nature, many of these efforts by the Department of Transportation and the EPA complement the role of DOE in effecting energy conservation. For example, the net effects of a uniform 55-mph speed limit are a reduction in both automobile-related fatalities and energy use; also, the achievement of the Corporate Average Fuel Economy (CAFE) standards will play an important role in reducing petroleum consumption in this country.

## Federal Strategy

The overall OTP goal is to reduce the energy consumed by the transportation sector and to reduce the transportation sector's nearly complete dependence upon petroleum. Successful development and commercialization of alternative transportation and fuel technologies, together with effective promulgation of educational and informational approaches to encourage greater efficiency in vehicle and systems end-use, are the major paths toward this goal.

The strategy for achieving the Office of Transportation Programs (OTP) goal follows:

- Provide stimuli for the development of new transportation technologies for near-term and midterm R&D. Such stimuli are provided through direct federal government involvement, including cost-shared R&D; this support will reduce the financial exposure to industry associated with the development of new, high-risk, high-payoff technologies.
- Provide encouragement to industry for commercialization of the newly developed technologies through engineering demonstrations and financial incentives.
- Stimulate the demand for newly developed and underutilized technologies through commercial or market demonstrations in normal operational environments.
- Provide informational, educational, and other outreach programs designed to foster public and business awareness of conservation opportunities and encourage adoption of specific methods, techniques, and technology for increasing transportation end-use efficiencies.

## Program Management Strategy

The management policy of OTP is to decentralize the implementation of the various programs as much as possible by delegating authority for conducting specifically assigned tasks to project offices. The major technical management decentralization effort has been to NASA laboratories, e.g., heat engines, electric and hybrid vehicle propulsion systems at the NASA Lewis Research Center; and electric and hybrid vehicle systems at the NASA Jet Propulsion Laboratory. The policy and programmatic decisions are retained in the DOE headquarters program offices. Project management authority is delegated to NASA laboratories, DOE field operations offices, and a few national laboratories. In addition, project tasks are assigned to private-sector contractors and other appropriate government agencies with the necessary facility and personnel capabilities. In general, OTP will use, to the maximum extent possible, the services of nongovernment contractors, industry, and the universities.

## MAJOR PROGRAM THRUSTS

### Legislative Framework

The Energy Reorganization Act of 1974 (P.L. 93–438) established the Energy Research and Development Administration (ERDA). Under that act and the Federal Nonnuclear Energy Research and Development Act of 1974 (P.L. 93–577), ERDA was given responsibility for establishing and executing an energy RD&D program including efforts to provide new and improved energy conservation technologies. The Division of Transportation Energy Conservation (TEC) within ERDA was given responsibility for transportation-related RD&D.

The Federal Energy Administration Act of 1974 (P.L. 93–275) specified that the Federal Energy Administration (FEA) develop and oversee the implementation of energy conservation programs and promote efficiencies in the use of energy resources. The OTP was given responsibility for transportation-related activities within FEA.

On October 1, 1977, when these ERDA and FEA functions were transferred to the DOE under the Department of Energy Organization Act (P.L. 95–91), they were assigned to the Office of Conservation and Solar Applications. As a result of recent reorganization, they are combined in the OTP.

Other legislation pertinent to the conservation of transportation energy is discussed herein.

Automotive Technology RD&D. This program is responsible for conducting automotive RD&D designed to enable early commercialization of new engines, transmissions, and components. The program originated in the EPA and was transferred to ERDA by Section 104(g) of the Energy Reorganization Act of 1974 (P.L. 93–438). In 1977, the program was again transferred when ERDA became part of DOE by virtue of the Department of Energy Organization Act (P.L. 95–91).

The Automotive Propulsion Research and Development Act of 1978 (Title III, P.L. 95–238) continued the authorization for this program. As set forth in Section 302(b), the purposes of this act were:

(1)(A) . . . direct the Department of Energy to make contracts and grants for research and development leading to the development of advanced automobile propulsion systems within 5 years of the date of enactment of this Act, or within the shortest practicable time consistent with appropriate research and development techniques and (B) evaluate and disseminate information with respect to advanced automobile propulsion system technology;

(2) preserve, enhance, and facilitate competition in research, development, and production with respect to existing and alternative automobile propulsion systems; and

(3) supplement, but neither supplant nor duplicate, the automotive propulsion system research and development efforts of private industry.

Further specific requirements of Title III include:

- Establishing new projects and accelerating existing ones
- Intensifying research in key basic areas
- Directing priority attention to fuel-flexible systems

Section 311(a) of the act also amends the National Aeronautics and Space Administration (NASA) Act of 1958 (P.L. 85–568) to facilitate NASA's "unique competence in scientific and engineering systems" toward development of advanced automobile propulsion systems.

Electric and Hybrid Vehicles. The DOE Electric and Hybrid Vehicle (EHV) Program has responsibility for accelerating the commercialization of electric and hybrid vehicles. Responsibility for the EHV Program was given to the Transportation Energy Center when ERDA was established by the Energy Reorganization Act of 1974 (P.L. 93–438). The program received increased emphasis and was modified to reflect requirements mandated by passage of the Electric and Hybrid Vehicle Research, Development, and Demonstration Act of 1976 (P.L. 94–413), as amended in 1978 by Title IV of P.L. 95–238.

Transportation Systems Utilization. Transportation Systems Utilization (TSU) Program is directed toward improving energy efficiency in the use of transportation systems through educational, technical assistance, and demonstration programs designed to encourage improved operational and management practices; through increased load factors and vehicle occupancies, and a shift of travel and freight to more efficient modes; through modification of conservation-inhibiting regulatory practices; and through accelerated use of energy-efficient equipment.

The Energy Policy and Conservation Act (P.L. 94–163), as amended by Section 432 of the Energy Conservation and Production Act (P.L. 94–385), provided for financial assistance to states to carry out state-developed plans for achieving and promoting energy conservation and to encourage the use of renewable resources. It also authorized the FEA to carry out education and technical assistance programs aimed at conserving energy in all economic sectors.

Section 506(b)(1) of the Motor Vehicle Information and Cost Savings Act (P.L. 92–213), as amended by Section 301 of EPCA, directed the FEA to publish data and distribute a booklet (*Gas Mileage Guide*) that provides the EPA fuel-economy rating and related information for new-model-year cars available for sale in the United States. Section 506(a) of P.L. 92–312 further specifies that fuel-economy information labels be affixed to newly manufactured automobiles pursuant to EPA rules.

The TSU Program also participates with other government agencies in the development and support of a comprehensive national energy conservation effort for marine, rail, pipeline, and air transport; and supports energy-related programs affecting more than one mode (for example, intermodal freight operations and alternative fuels).

The TSU Program has been designed to recognize the activities of other government agencies and the private sector. Cooperative projects with the Federal Railroad Administration, U.S. Coast Guard, and the Maritime Administration have been undertaken. The private sector, as represented by organizations such as the Association of American Railroads, the American Gas Association, and the American Waterway operators have also cooperated with TSU initiatives.

## Programs and Milestones

In the sections that follow, the major programs and milestones of each line item activity are described.

Automotive Technology RD&D Program. R&D efforts are underway to promote the development of energy-efficient propulsion systems for automobiles, trucks, and buses. The passenger car, however, is the largest single petroleum user, consuming more than half of the petroleum used by the entire transportation sector. The automobile is therefore the primary target of the automotive technology R&D effort.

*Advanced heat engine development.* Since the inception of the heat engine program, all feasible types of heat engines were studied and evaluated with regard to their potential for low-cost production, versatility in engine size, technical barriers to development, and industry interest in the candidate engine, as well as fuel economy and emissions. Candidate heat engines selected for development were Rankine cycle (steam) systems, gas turbines, diesel engines, and the Stirling cycle system. The Rankine-cycle-system development was terminated in 1974. It was shown that the Rankine-cycle system developed had exceedingly clean exhaust but that efficiency improvements needed for significantly decreased petroleum consumption were not likely, even with continued development. The diesel engine development was terminated in 1976 because at that time industry had initiated its own vigorous proprietary development efforts for light-duty applications of it. Significant emissions, health effects, and combustion technology efforts, however, are being continued by DOE in support of industry's diesel engine development.

Emphasis is now on the development of gas turbine and Stirling-engine propulsion systems because they offer the potential of significantly improved fuel economy, exceedingly low exhaust emissions, and the capability of using virtually any combustible liquid or gaseous fuel. These propulsion systems are also potentially producible at low cost, and are capable of good performance and high reliability.

Industry teams are participating under federal cost-shared contract to work on both gas turbine and Stirling-engine development.

Dynamometer testing of the first experimental engines will begin in late FY 1982. These tests will generate sufficient test data upon which a decision to select a single gas turbine engine contractor will be based. DOE will continue competitive advanced heat engine development with the single Stirling contractor. Convergence to one turbine engine concept may take place by the end of the first quarter of FY 1983 based on the progress toward project goals, engine test data, component rig testing, estimates of engine cost, and program risks. During the evaluation and selection period, both turbine contractors will continue with engine testing and component development. If a single systems contractor is selected, work will continue with engine upgrading and in-vehicle testing until the end of FY 1984, when the technology could be transferred to industry and decisions on mass production and commercialization could be made.

Since the key obstacles to the commercialization of gas-turbine automobiles are competitive fuel economy and cost (as compared to the conventional internal combustion engine), a major emphasis will be placed on the development of ceramic materials and on the improved aerodynamic performance of small turbomachinery. A major barrier to achieving high fuel economy has been the gas turbine's poor part-load efficiency. Low cost and improved efficiency appear to be simultaneously achievable by using ceramic materials in the hog sections of the gas turbine, since ceramics allow the turbine to operate at higher gas temperatures. With low-cost ceramics there would be no need for expensive, high-alloyed, and strategic metals to sustain the higher temperatures.

The following areas are being carefully examined:

- Materials, primarily high-temperature ceramics
- Turbomachinery aerodynamic design techniques
- Heat exchangers, with emphasis on 100,000-mile equivalent life
- Combustion, with emphasis on lowering emissions and fuel flexibility
- Control systems to improve drivability
- High-speed, high-temperature, long-life bearings and seals

Gas turbine vehicles previously developed under DOE contract will continue in use as test beds to demonstrate improvements in gas turbine technology.

Major milestones are shown in Figure 9-3.

*Vehicle systems.* The Vehicle Systems Project deals primarily with development and demonstration of vehicle propulsion systems and subsystems designed to improve vehicle efficiency. Specific project activities include:

- In-service demonstration and evaluation of gas turbine buses for both intercity and urban operations
- Fleet service testing of turbocompound diesels in long-haul trucks
- Development and fleet testing of an organic bottoming cycle using waste heat for long-haul diesel trucks
- Fleet testing of vehicles equipped with controlled-speed accessory drives
- An assessment of light-duty diesel emission control technology

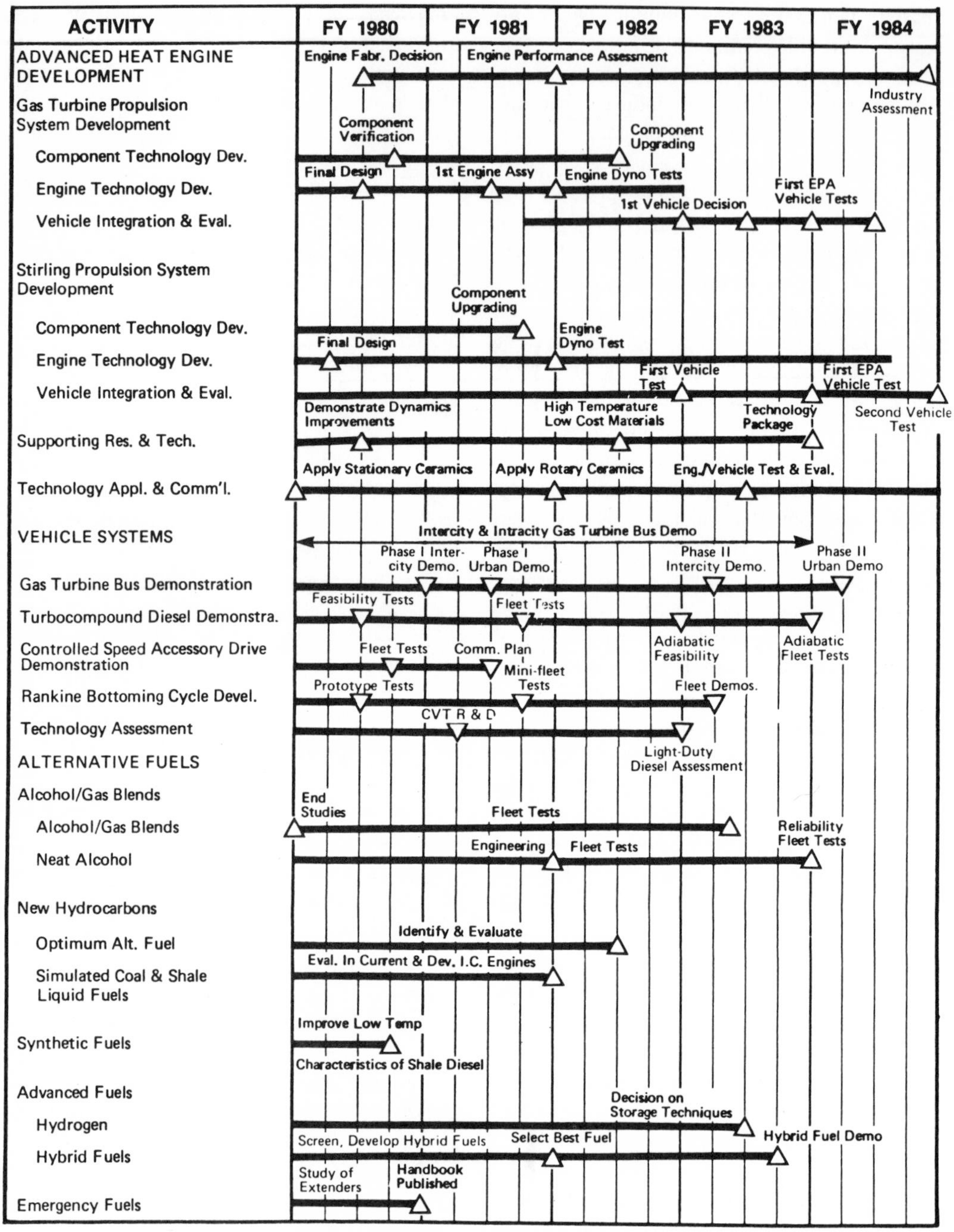

Figure 9-3. Major milestones for the automotive technology RD&D program.

- Advanced transmissions assessments and studies

Major milestones are shown in Figure 9-3.

*Alternative fuel utilization.* The Alternative Fuels Utilization Program (AFUP) provides government participation in R&D designed to find the best means of using alternative fuels in highway vehicles from the standpoint of resource/fuel/engine system optimization. The program will establish links involving direct funding between government, automotive, and fuel industries, through which government resources can be used to accelerate fuels-use R&D.

The systems development portion of the Stirling-engine project will be carried out under contract with the automotive industry, their suppliers, and other high-required technology companies. It will use existing and near-term technology to achieve the project objectives. Fuel-economy improvements will come primarily from the development of a high-required temperature engine, the design and development of an experimental engine optimized for the automobile application, and parallel component development.

Basic supporting research will be conducted on Stirling engines, particularly in critical areas such as seals and materials, and will provide alternative approaches in these areas. A lower-level effort will also be carried out in heat exchangers, controls, and combustors. This work will be conducted by government laboratories, universities, and private companies. Specific research projects include:

- Piston rod seals to retain the high-pressure hydrogen working gas and to prevent oil contamination
- Emphasis on hydrogen containment and low-cost alloys
- Advanced heater heads, coolers, preheaters, and regenerators for higher temperatures, better heat transfer, and greater overall engine efficiency
- Achievement of more efficient and reliable engine power controls
- Component development and test data to validate computer code simulation models
- Development of low-emissions combustion technology through more uniform temperature and fuel distribution
- Initiation of a gas turbine and/or a Stirling commercialization program with the United States auto industry leading toward full line production (500,000 units annually) in the 1990s

AFUP includes the following major activities:

- Identification and evaluation of new hy-

drocarbon fuels based on an understanding of the fuels' production and the design of engines optimized for these fuels. These new fuels include broad-cut and variable composition fuels derived initially from petroleum and eventually from coal, shale, biomass, and other sources.
- Identification and evaluation of alcohol fuels based on an understanding of fuels production and engine design optimized for these fuels. These fuels include methanol, ethanol, higher alcohols, and alcohol blends with hydrocarbon fuels.
- Study of the consequences of using synfuels in standard or improved (for example, stratified charge) versions of existing engine types.
- Assessment of advanced fuels including hydrogen and carbonaceous fuels.
- An assessment of nonstandard fuel use in emergencies.

Major milestones are presented in Figure 9-3.

**Electric and Hybrid Vehicle RD&D.** The strategy selected to meet the EHV Program objectives provides for government assistance to the manufacturing and support industry, stimulation of demand for EHVs in the marketplace, and improvements in vehicle performance, reliability, cost, and acceptability.

Four program elements (i.e., R&D, product engineering, demonstration, and incentives) are coordinated to promote the successful commercialization of electric and hybrid vehicles. In addition, an EHV resource manager has been appointed within the Office of Commercialization. The resource manager will work closely with the EHV Program Office and will be the focal point for promoting the accelerated commercialization of this technology.

Recent actions have been taken to focus the EHV activities toward five major projects: demonstrations, vehicle evaluation and improvements, electric vehicle commercialization, hybrid vehicle commercialization, and advanced vehicle development. Each of these

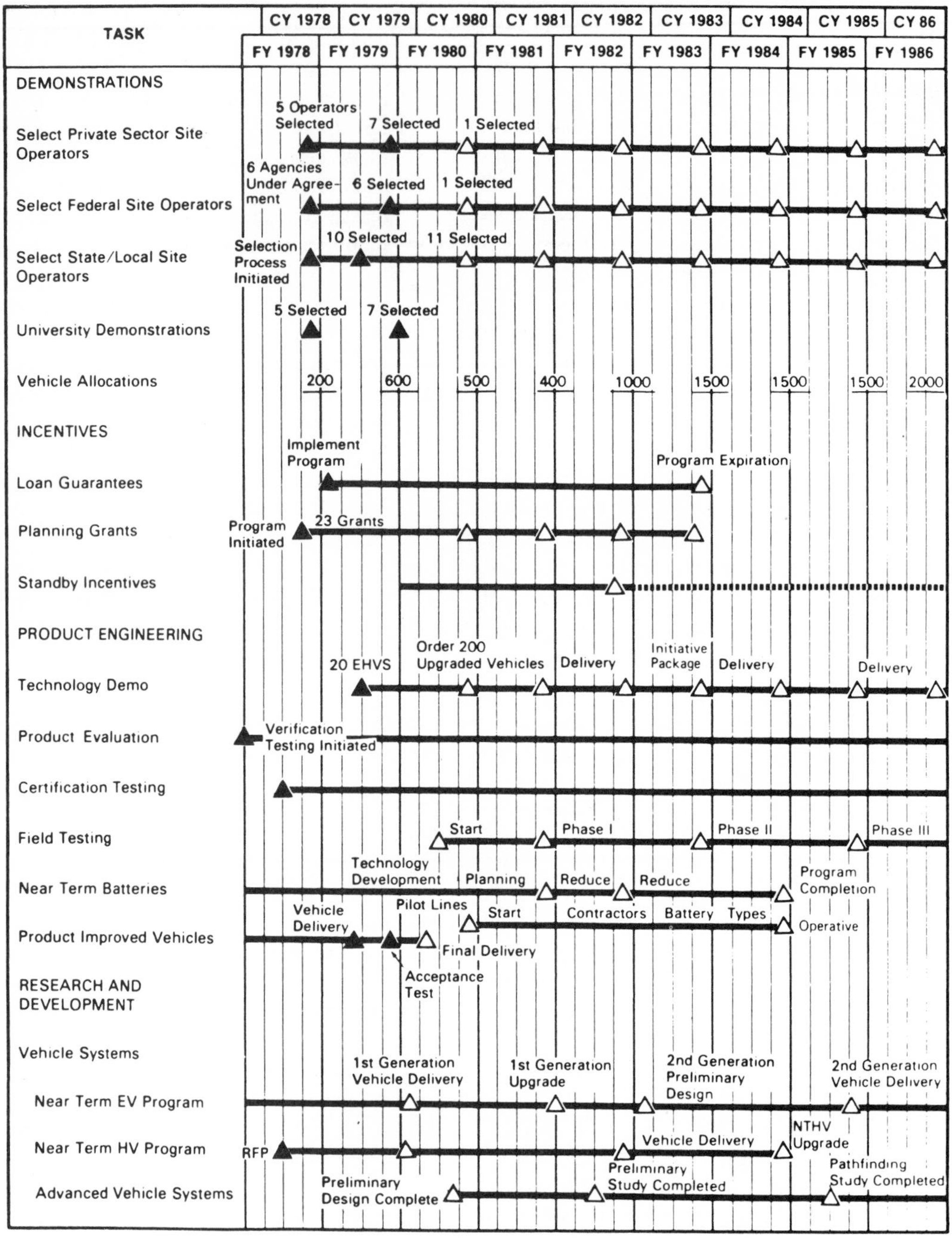

Figure 9-4. Major milestones for the electric and hybrid vehicle program.

projects is targeted toward a set of objectives, which, when collectively met, will result in achievement of the EHV program goal.

The EHV Program milestones are shown in Figure 9-4.

*Research and development.* The EHV R&D plan is to develop, fabricate, demonstrate, and evaluate EHV systems, components, and subsystems in order to confirm that vehicles using these technologies can economically meet commercial and private vehicle requirements. R&D is being conducted in the following three program areas:

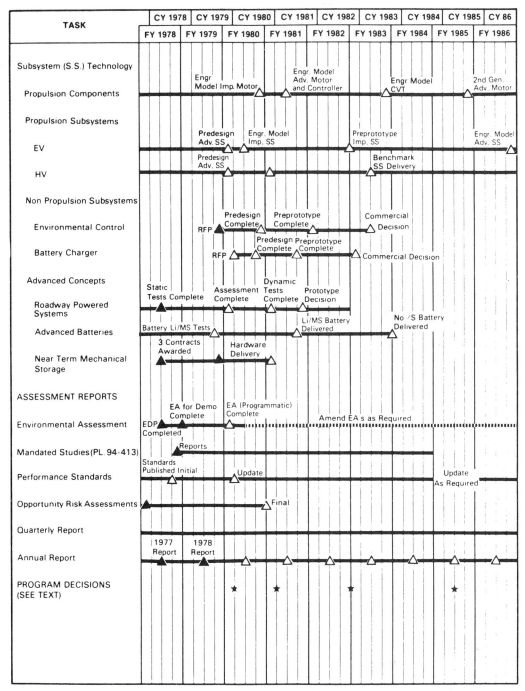

Figure 9-4. Continued.

*Demonstration.* The demonstration project has been specifically prescribed by P.L. 94–413 (as amended by Title VI of P.L. 95–238). It will determine the economic and technological practicability of electric and hybrid vehicles for personal, commercial, and government use in urban and agricultural areas.

The demonstration plan focuses on the market-type demonstrations mandated by the act. The demonstration includes private-sector fleets and dealers, state and local governments, and federal agency fleets. The costs of the demonstration projects are shared by these organizations. The EHVs are production vehicles purchased by these organizations to replace internal-combustion-engine vehicles. The market demonstration vehicles produce market-oriented information on product, performance, maintainability, and general consumer acceptability. It provides a cash flow for the manufacturers to enable them to take advantage of improved technologies resulting from the R&D and product engineering. It also provides the initial elements of a vitally needed support infrastructure. The scheduled placement of cars in the demonstration program is presented in Table 9-1.

*Product engineering.* The product engineering element of the EHV Program provides a bridge between the R&D and demonstration elements and facilitates rapid incorporation of new technology into marketable vehicles. This is done in part by providing engineering and operational data to support modified performance standards reflecting current needs of the marketplace. The product engineering activity will also help the flow of new technology into the marketplace by assuming for the manufacturers some of the risks and costs of engineering testing. Support will be provided to manufacturers to help them improve the reliability, safety, cost, and performance of vehicles offered to demonstration site operations.

Specific activities include:

- developing vehicles that exhibit improved characteristics that could potentially increase the EHV market penetration to levels that will result in significant petroleum savings
- demonstrating these vehicles in engineering field tests and controlled demonstration environments, and iden-

tifying problem areas, strengths, and weaknesses

- providing data to manufacturers on inservice performance, problem areas, reliability, and potential market opportunities
- developing a wide range of near-term batteries suitable for powering intermediate-range vehicles
- providing information to help establish a basis for upgrading the market demonstration performance standards
- working with industry to determine effective ways to incorporate improved technology into marketable vehicles
- facilitating the rapid commercialization of improved electric vehicles in the mid-1980s

*Incentives.* A program of federal loan guarantees, authorized by P.L. 94–413 (as amended), provides for guarantees. Each applicant may receive up to $3 million per project ($6 million per company) with a term not to exceed 15 years to qualified borrowers who lack adequate capital to participate in the production of EHVs. Since the guarantee program was authorized to encourage the commercial production of EHVs, priority is given to those applications that directly relate to projects that will result in early commercial production of EHVs or components. Second-priority consideration is given to projects relating to the development of prototypes, and third-priority consideration is given to R&D projects. Loan-guaranty applications are evaluated from both technical and financial viewpoints; guaranty agreements are negotiated through DOE headquarters.

P.L. 94–413 also provides for planning grants for individuals and small businesses to cover the expense of preparing contract proposals to DOE. Those eligible for planning grants include individuals and small busi-

Table 9-1. Number of Cars Scheduled for Demolition.

| FISCAL YEAR | P.L. 94–413 MANDATED | AGENCY RECOMMENDATION | FUNDS APPROPRIATED |
|---|---|---|---|
| 1978 | 200–400 | 200 | 200 |
| 1979 | 600 | 600 | 600 |
| 1980 | 1,700 | 500 | 500 |
| 1981–84 | 7,500 | 400 (1981) | To be Determined |

nesses with ideas for EHV RD&D and who have the capability to perform in this field, but lack the funds necessary to prepare and submit contract proposals.

P.L. 94–413 further directs that studies be performed to identify barriers impeding the use of EHVs and the application of corrective incentives.

Transportation Systems Utilization. The Transportation Systems Utilization (TSU) Program objectives are accomplished by the following:

- Development, publication, and dissemination of conservation materials
- Public education concerning the need for and methods of conserving transportation energy
- Evaluation and demonstration of energy-conserving transportation components and systems
- Intervention in energy-related procedures of federal and state regulatory agencies
- Development and commercialization of new energy-conserving technologies for rail, marine, air, and pipeline transportation systems
- Promotion of energy-saving revisions and clarifications of regulations and laws
- Promotion of improved equipment acquisition management, operational, and maintenance practices
- Promotion of intermodal shifts of passengers and freight to energy-efficient forms of transportation
- Provision of technical assistance to public agencies and other groups interested in practicing or promoting energy conservation
- Analysis of the transfer of new technology into the entire transportation sector

In addition to providing petroleum savings in a relatively short time frame, it is expected that most of the energy practices brought about by the program will be continuous. Thus, this program complements the longer range technology being developed in the remainder of the OTP effort.

The TSU Program also includes the areas of transportation technology assessment and the evaluation of new concepts in the transportation sector. Major milestones of the program are given in Figure 9-5.

*Vehicle performance*. The New Car Fuel Economy Program, a cooperative effort among DOE, the EPA, and Department of Transportation, increases public awareness of comparative automobile fuel-economy information and thereby encourages the purchase of more fuel-efficient vehicles. The focus of this program is the annual publication and distribution of approximately 15 million copies of the *Gas Mileage Guide* to new car dealers and consumers. The guide gives fuel-economy ratings and interior volumes for new cars and light trucks available in the marketplace.

The following activities are planned:

- Collection and analysis of in-use fuel-economy data
- Characterization of disparity of EPA fuel-economy ratings and actual in-use data
- Fleet characterization (fleet size, type of driving, average fuel economy)
- Identification of those states that have adopted measures for encouraging the purchase of fuel-efficient automobiles

The function of the New Concepts Testing and Evaluation Program is to evaluate, test, and report on new technology that exhibits potential for saving transportation energy and that may also be suitable for further R&D and/or for demonstration. Proposals and concepts may require laboratory and/or field testing or demonstration for evaluation to prove their feasibility for commercialization.

Major aspects of this project include laboratory and limited fleet testing on advanced carburetion systems and other fuel-saving devices, components, or systems culminating in demonstration. Evaluations are culminated by demonstrations on these items that are considered as having potential for saving fuel.

Studies have also been undertaken for the Department of Agriculture to evaluate energy-conserving mobile refrigeration units used in the preservation of perishable agricultural commodities during transit to market. This

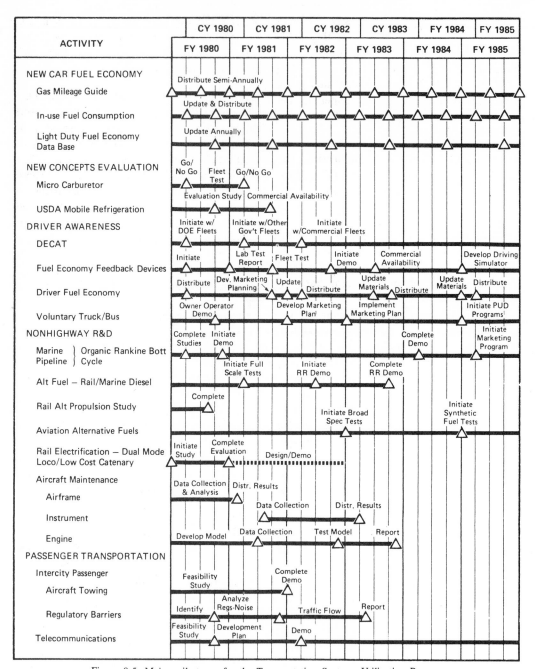

Figure 9-5. Major milestones for the Transportation Systems Utilization Program.

evaluation was completed in 1981. Other studies will be included as appropriate.

The driver awareness effort includes conducting coordinated awareness, motivation, education, and demonstration programs to improve the fuel economy of automobiles in use throughout federal, state, and local governments and the general motoring public. The programs address opportunities to conserve fuel in the motorist's four basic personal economic decision sectors: purchase decisions, driving techniques, vehicle maintenance, and personal transportation planning. In addition, the program includes development and demonstration of a driver feedback device that can assist in improving the fuel economy of the

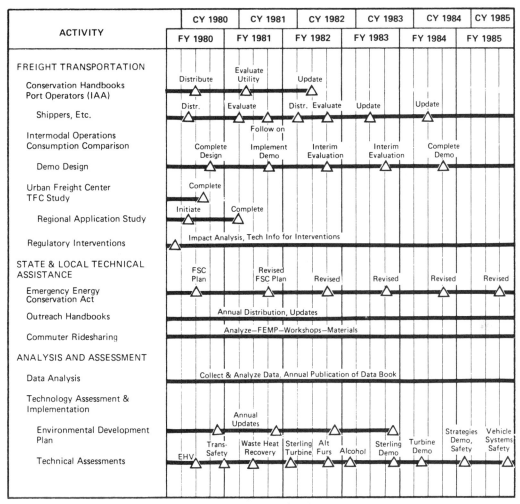

Figure 9-5. Continued.

driver by providing information on cost per trip.

The Driver Energy Conservation Awareness Training Program (DECAT) is an example of the activity in this area. The program has been initiated in DOE fleets and will be initiated in several other government fleets (for example, the U.S. Postal Service).

The Voluntary Truck and Bus Fuel Economy area includes programs to aid the commercial vehicle industry by encouraging and assisting organizations in:

- Developing and offering more fuel-efficient products
- Purchasing fuel-efficient new vehicles and add-on or retrofit components
- Conducting and reporting on fuel-economy testing
- Developing, conducting, and/or supporting driver training and motivation programs specifically directed toward commercial vehicle drivers

Systems Efficiency. Nonhighway Research and Development (NHR&D) includes the air, rail, marine, and pipeline modes. These include liquid, gas, and coal-slurry pipelines; ocean, Great Lakes, coastal, offshore, inland waterway, and recreational marine vessels; freight and passenger line haul and urban railways; and commercial and general aviation.

The program strategy provides for federal assistance to industry for R&D to develop transportation-system capabilities leading to operations that are more efficient and less dependent upon the use of petroleum.

The following description of activities is a partial listing of NHR&D projects categorized by technology application (i.e., air, marine, rail, pipeline):

- Bottoming cycle (or waste heat recovery system) studies are under way to determine applicability to marine, rail, diesel power plants, and pipeline gas turbines. Development and demonstration of bottoming cycle technology in a transportation system is also planned.
- A study of the application of alternative fuels to nonhighway modes has been completed. Laboratory tests to identify and characterize contingent or emergency fuels (in the event of sudden petroleum shortages) and alternative fuels for rail and marine systems are under way with plans for future operational demonstrations.
- Alternative propulsion systems (fuel cell, gas turbine, Stirling engine, and diesel improvements) are being studied for rail application.
- Investigations of rail electrification are under way. They include a feasibility study of a dual mode (diesel or electric powered) locomotive and a low-cost catenary system.

Freight transportation identifies ways to reduce energy consumption in the transportation of freight by all modes through:

- Removing regulatory and institutional barriers
- Shifting freight to more energy-efficient modes
- Encouraging the use of energy-efficient management, operating, and maintenance procedures

Passenger transportation addresses the conservation of transportation fuel in intercity passenger transportation. The objectives are as follows:

- Development and implementation of an

airline operating and maintenance procedures marketing program, including the demonstration of airframe, instrument, and engine procedural improvements
- Initiation of computerized airport gateholding and aircraft towing demonstrations
- Provision of technical support for DOE intervention in regulatory proceedings affecting passenger transportation
- Examinations of the use of telecommunications as a means of conserving energy through improvements in transportation system operations

The primary thrust of state and local technical assistance is to assist state and local government agencies in the development of programs to achieve near-term energy savings by providing the following:

- Information about opportunities for state and local programs to achieve energy conservation in the transportation sector
- Handbooks on driver awareness, ride sharing, truck and bus operations, vehicle purchasing, and other subjects where effective state initiatives are integral to successful commercialization activities
- Services of expert consultants
- Assistance in the commercialization of commuter ride sharing and van pooling through improved effectiveness of federal-employee and private-sector programs

*Analysis and assessment.* The specific function of the Data Analysis Program is to collect, evaluate, and analyze past, current, and forecast data from federal and other sources in interrelated areas: transportation, energy, economics, environment, demographics, and government. These data are used to identify energy relationships, to develop a baseline transportation energy forecast, and to develop a scenario-generating capability and impact identification tool. This tool will provide energy, economic, and environmental information associated with alternative transportation energy technologies.

Annual editions of the *Transportation Energy Conservation Data Book* will be published to provide a comprehensive set of data characterizing all important modes in the nation. Forecasting techniques will be used to assess the potential energy conservation and market penetration of technologies under government development.

*Technology assessment and implementation.* The Technology Assessment and Implementation Program provides technology impact assessment, and implementation and commercialization analyses on the potential of transportation energy conservation technologies (e.g., EHV, alternative fuels, advanced heat engines, etc.) and systems. These analyses, when coordinated and integrated with related activities in other DOE and federal programs, assist in the formulation of DOE energy policy.

In order to establish the conditions under which specific advanced technologies may become successfully implemented in the future, the program is composed of three mutually supportive functions. The first function, technology assessment, examines the future impacts of technologies on the environment in which they are to be operated. Socioeconomic factors are also carefully considered. In an integrated approach, the multi-impact dimensions of transportation technologies are analyzed in a common framework allowing for identification and quantification of critical, or constraining, effects arising from a particular technology. Environmental development planning, electric-vehicle impact assessment, and alternative-fuels impact assessment are included in this effort.

The second function, technology implementation, provides formulation of specific RD&D technology goals and includes formulating plans for the transfer of specialized technical information. This function encompasses the study of near-term technology improvements, as well as strategies and policies associated with their implementation (such as consumer representation, commercialization, technology transfer, demonstration planning, and net energy analysis).

Other program activities include annual updates of the Environmental Development Plan and the EHV impact assessments.

## PROGRAM FUNDING

The budget for FY 1979 and projections for the OTP programs through FY 1981 are outlined in Table 9-2. Yearly budget reviews will take into account current progress in various aspects of the program, and funds will be reallocated as necessary.

All yearly budget requests will be subject to Office of Management and Budget and congressional review. Changes in total budget levels will be reflected in respective program-schedule changes or in alteration of program objectives.

## PROGRAM IMPACTS

### Measures of Effectiveness

The following are examples of general measures of OTP program effectiveness:

Table 9-2. Office of Transportation Programs Multiyear
Budget Projections (Thousands of Dollars).

| PROGRAM | BUDGET AUTHORITY (DOLLARS IN THOUSANDS) | | |
| --- | --- | --- | --- |
| | APPROPRIATION FY 1979 | APPROPRIATION FY 1980 | REQUEST FY 1981 |
| Vehicle Propulsion R&D | 47,800 | 60,500 | 55,900 |
| Electric and Hybrid Vehicle R&D | 37,500 | 41,000 | 42,100 |
| Transportation Systems Utilization | 6,100 | 6,700 | 6,700 |
| Alternative Fuels Utilization | 5,800 | 5,300 | 5,300 |
| Program Direction | 1,949 | 2,923 | 3,000 |
| TOTAL | 99,149 | 116,423 | 113,000 |

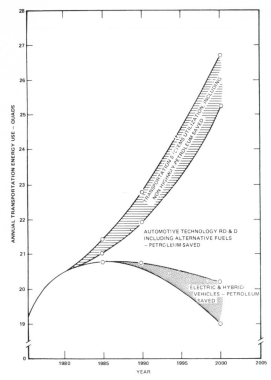

Figure 9-6. Projected annual consumption of transportation energy.

- Industry adoption for production engineering of technologies developed in the automotive technology R&D portion of the OTP program
- EHV demonstrations expanded beyond DOE-supported levels and the market survival of EHV manufacturers.

## Projected Energy Savings

Figure 9-6 outlines the impact of each program in alleviating the nation's energy problem. As shown, in the near term, predominant transportation energy conservation will occur as a result of the TSU Program. From 1985 until the year 2000, the other programs will have significant impact on transportation energy conservation. See Table 9-3.

These savings projections are based on successful implementation of the programs as currently projected, using enhanced funding budgets. Such projections depend upon a consistent set of assumptions of many factors, including market penetration and the national economy. Also, it should be noted that these savings are in addition to those attributable to the achievement of the 27.5 mpg Corporate Average Fuel Economy Standard, which is mandated by law for 1985.

## Analysis of Benefits and Costs

The major transportation programs are expected to save petroleum in the transportation sector as a result of expenditures made by DOE and the private sector (industry and consumers). Two measures of the effectiveness of these expenditures are the cost per million Btu saved to the nation and the benefit-cost ratio to the individual consumer. One very important measure is the shift from resources that exhibit a high depletion rate (quads used per year) to resources with a lower depletion rate.

Table 9-4 shows these values for the three programs. Since the current price of gasoline is about $1.50 per gallon, it is clear that all the OTP aggregate projects can save petroleum at a lower cost than the current selling price of gasoline. Likewise, the benefit-cost

Table 9-3. Projected Annual Transportation Conservation Savings by Each Program Element (Quads).

| PROGRAM ELEMENT | PETROLEUM SAVINGS | | | TOTAL ENERGY SAVINGS | | |
|---|---|---|---|---|---|---|
| | 1985 | 1990 | 2000 | 1985 | 1990 | 2000 |
| Transportation Systems Utilization, including Nonhighway | 0.50 | 0.87 | 1.49 | 0.50 | 0.87 | 1.49 |
| Automotive Technology RD&D, including Alternative Fuels | 0.22 | 1.07 | 4.61 | 0.04 | 0.26 | 2.21 |
| Electric & Hybrid Vehicle RD&D | 1.01 | 0.05 | 1.31 | — | — | — |

Table 9-4. Benefits and Costs of Transportation Programs.

| PROGRAM | $ PER MILLION BTU | BENEFIT/COST TO CONSUMER |
|---|---|---|
| Automotive Technology RD&D | $1.67 | 1.3 |
| Electric & Hybrid Vehicle RD&D | $1.99 | 1.2 |
| Transportation Systems Utilization | $ .58 | 3.6 |

ratios for the OTP projects are positive and will increase in value as the world price of oil increases.

## ENVIRONMENTAL IMPACT

The OTP has developed an Environmental Development Plan (EDP), which identifies the ecosystem, resource, physical environment, health, safety, and socioeconomic issues associated with the DOE RD&D and assessment of transportation technologies. The EDP also addresses several strategy and policy development, and implementation projects. Any federal action that might require an environmental impact statement (EIS) was included. Because technology demonstration and commercialization tend to raise more environmental concerns than other portions of OTP activities, the EDP emphasizes these impacts.

### Issues

Both the Stirling and gas turbine engines being developed under the Automotive Technology RD&D Program will use super alloys (chromium, nickel, cobalt, and tungsten) to withstand higher temperatures required for increased fuel economy. Many of these alloys are generally toxic and will therefore require continued study of any potential detrimental effect on the environment. Other aspects of the heat engine portion of the EDP deal with the hazards of using hydrogen in the Stirling engine and the failure modes of high-speed ceramic components in the gas turbine. Problems with toxicity of the working fluid in the truck bottoming cycle are also being studied.

Considerable effort is under way to determine the effect on the environment of increased or new pollutants associated with the use of alternative fuels such as alcohol, broad-cut petroleum, and synthetic fuels.

The main EDP activity relating to the EHV RD&D Program deals with the propulsion battery systems. In addition to the solid-waste problem of nonrecyclable and nonbiodegradable battery casings, the following considerations are under examination:

- Toxicity of battery component materials
- Shock hazard
- Charging-induced hydrogen generation
- High operational temperature of some battery systems
- Emission of atmospheric pollutants during various manufacturing processes
- Environmental benefits through mobile source reductions

The EDP activity that deals with the Nonhighway Research and Development includes the release of toxic working fluids and the resultant altered pollutant emissions of pipeline bottoming cycle concepts.

Environmental impact activities that are either completed, under way, or planned are listed in Table 9-5.

## MAJOR PROGRAM ISSUES

The impact of President Carter's initiative on basic research in automotive technology is a major issue that remains under study. On May 18, 1979, President Carter met with the heads of the four major automotive manufacturers to discuss a cooperative research program.

To implement the president's initiative, an ad hoc committee on automotive R&D was established, with DOE as a member. The focus of the planning is on basic research and the use of university, industry, and federal laboratories.

## Table 9-5. Status of Environmental Impact Activities.

| OTP PROGRAM | EA COMPLETION DATE | DEIS COMPLETION DATE | EIS COMPLETION DATE | ERD COMPLETION DATE |
|---|---|---|---|---|
| Stirling and Gas Turbine | March 1, 1982 | Jan. 1, 1983 | July 1, 1983 | March 1, 1981 and June 1, 1983 |
| Turbocompound Diesel | Oct. 1, 1979 | July 1, 1980 | Jan. 1, 1981 | Oct. 1, 1980 |
| Accessory Drive | Oct. 1, 1979 | March 1, 1980 | Oct. 1, 1980 | July 1, 1980 |
| Marine Bottoming Cycle | March 1, 1980 | Dec. 1, 1980 | April 1, 1981 | Dec. 1, 1980 |
| Truck Bottoming Cycle | Oct. 1, 1980 | April 1, 1981 | August 1, 1981 | June 1, 1981 |
| Pipeline Bottoming Cycle | Oct. 1, 1981 | June 1, 1982 | Sept. 1, 1982 | June 1, 1982 |
| EHV | April 1, 1979 | May 1, 1980 | Nov. 1, 1980 | Jan. 1, 1980; Jan. 1, 1982; May 1, 1985 |
| Coal-Oil Slurry | June 1, 1980 | April 1, 1981 | Sept. 1, 1981 | March 1, 1981 |
| All Bottoming Cycle Fluids | | | | Feb. 1, 1979 |
| Alcohol Fuels Health and Safety | | | | Oct. 1, 1979 |
| Alcohol Fuels | Oct. 1, 1980 | April 1, 1981 | Oct. 1, 1981 | |
| Alternative Fuels | | | | June 1, 1981 and June 1, 1983 |
| Diesel Alternatives | Jan. 1, 1981 | April 1, 1981 | Jan. 1, 1982 | |
| New Hydrocarbons | Dec. 1, 1982 | June 1, 1983 | Oct. 1, 1983 | |

# 10. State and Local Programs

## PROGRAM OBJECTIVES AND STRATEGY

During President Carter's term in office, Congress acted to reduce United States dependence on increasingly expensive and uncertain supplies of foreign oil by passing four major pieces of energy conservation legislation:

- Energy Policy and Conservation Act (EPCA), P.L. 94–163
- Energy Conservation and Production Act (ECPA), P.L. 94–385
- National Energy Extension Service Act (NEESA), P.L. 95–39
- National Energy Conservation Policy Act (NECPA), P.L. 95–619

Each of these acts recognized the importance of the role of state and local governments in conserving energy. For example, ECPA recognizes that states have a clear understanding of local energy demand requirements and supply options. States can develop and implement laws, policies, and programs tailored to unique economic and geographic conditions, thus minimizing potentially adverse economic and environmental impacts. Furthermore, state and local institutional buildings are themselves prime candidates for energy conservation measures.

Consequently, each of these acts established energy conservation programs that involve the active participation of state and local governments. States are given varying degrees of flexibility in the design and implementation of these conservation programs, since the ability to conserve specific fuels and to develop specific renewable resources varies from one state to another depending on local climate, available energy resources, economic base, population distribution, and institutional structures. Furthermore, the various state legislatures and local governments have the unique authority to pass and implement energy conservation laws such as those that regulate utilities, building practices, and land use.

Not only are states able to develop and implement energy conservation practices that can have an immediate and substantial effect on local energy growth rates, but states also can disseminate energy conservation information internally and serve as collection points for conservation-related information that can benefit other states, other regions, or the country as a whole. While focusing on internal energy problems peculiar to their areas, state and local governments may develop information that can be applied usefully elsewhere.

The CS Office of State and Local (S&L) Programs administers the following programs:

- Schools and Hospitals Grant Program, established by NECPA
- Other Local Government Buildings Grant Program, established by NECPA
- State Energy Conservation Program (SECP), established by EPCA and amended by ECPA and NECPA
- Energy Extension Service (EES), established by NEESA, Title V of the Energy Research and Development Administration Appropriation Authorization Act of 1977
- Emergency Energy Conservation Act (EECA), P.L. 96–102
- Weatherization Assistance Program (WAP), established by ECPA and amended by NECPA

These federal programs are designed to maximize state involvement in the design and implementation of laws, regulations, and programs that either conserve energy directly or that provide information, education, or technical assistance services that encourage increased energy savings.

## Program Goal/Objectives

The overall mission of the Office of S&L Programs is to assist state and local governments in developing and implementing effective programs that increase energy efficiency in state and local energy end-use sectors. To fulfill this mission, S&L has established several goals:

- To support national energy goals through energy conservation programs in which state and local governments play central planning and implementation roles
- To provide states some flexibility to use funds to plan, design, and implement programs to meet their needs
- To provide weatherization assistance to low-income persons and to public or nonprofit schools and hospitals; these groups do not benefit from tax credits and might not make investments in conservation without assistance
- To promote an exchange of ideas, R&D needs, innovations, and accomplishments among local, state, and federal governments
- To disseminate information regarding existing and new technologies developed in DOE sector programs and private industry with significant energy-conserving potential

**Schools and Hospitals Grant Program.** The Schools and Hospitals Grant Program is designed to reduce the growth rate of energy consumption in the nation's schools and hospitals, both public and private nonprofit, through the conduct of Energy Audits (EA) and Technical Assistance (TA) programs, and through the acquisition and installation of Energy Conservation Measures (ECM). EAs assess operational and maintenance changes in buildings to save energy and recommended cost-effective potential retrofit actions. TAs will provide support for detailed engineering analyses of selected buildings to identify energy-saving options. ECPs fund building envelope modifications and the installation of equipment to save energy. Fifty-percent

matching funds are required from the grantees in most phases of the program.

**Other Local Government Buildings Grant Program.** The two-year Other Local Government Buildings Grant Program (FY 1978 and FY 1979), similar in many respects to the Schools and Hospitals Grant Program, is also authorized under NECPA. It provides grants for preliminary energy audits, energy audits, and technical assistance in buildings owned by units of local government and public care institutions. Energy conservation measures (the purchase and installation of energy-saving equipment) are not authorized to be funded out of this program. Eligible buildings include municipally owned or operated facilities (office buildings, libraries, fire stations, etc.) and buildings owned by public care institutions (child care centers, rehabilitation facilities, public health centers). Fifty-percent matching funds are required from the grantees under this program.

Because the S&L programs were established at different times and through different congressional legislation, they contribute to the achievement of overall S&L goals in different ways (e.g., through grants, development of infrastructure, information dissemination, educational services, and technical assistance). Yet together they provide the state and local levels with a range of federal services that either directly or indirectly reduce United States energy consumption.

**State Energy Conservation Program.** The legislation enacting the State Energy Conservation Program (SECP) recognized the responsibility of the federal government to foster and promote comprehensive energy conservation programs and practices by establishing program guidelines, overall coordination, and technical and financial assistance for specific state initiatives in energy conservation. Under the SECP, a state may receive financial and technical assistance to implement either a base plan (EPCA) or supplemental plan (ECPA), or both. To be eligible for financial assistance, the base plan must be designed to achieve scheduled progress toward a specific energy goal for the year

1980 and must include five program measures. These are:

- Mandatory lighting efficiency standards for public buildings
- Programs to promote the availability and use of carpools, vanpools, and public transportation
- Mandatory standards and policies relating to energy efficiency to govern state procurement practices
- Mandatory thermal efficiency standards and insulation requirements for new and renovated buildings
- Right-turn-on-red traffic regulations

To be eligible for financial assistance, each state must include in its supplemental plan three additional required program measures:

- Procedures for carrying out a continuing public education effort to increase significantly public awareness of:
  1. The energy and cost savings that are likely to result from the implementation of energy conservation measures and renewable resource energy measures
  2. Information and other assistance that is or may be available with respect to the planning, financing, installing, or monitoring the effectiveness of measures likely to conserve or improve efficiency in the use of energy, including energy conservation measures and renewable energy measures
- Procedures for ensuring that effective coordination exists among various local, state, and federal energy conservation programs
- Procedures for encouraging and carrying out energy audits with respect to buildings and industrial plants within each state

States also may propose to include in a base plan or supplemental plan any additional approved program measures that they consider appropriate and that contribute to energy conservation.

The SECP requires participating states to develop and implement plans that will reduce projected energy consumption in each state. The overall objective of the program is to achieve a reduction of 5 percent or more in projected 1980 energy consumption. Each state has its own goal which is expressed in terms of a percentage reduction in the amount of energy that would have been consumed in the state in 1980 were it not for the implementation of the state's plan. Based on revised state plans submitted for funding in FY 1978, the states' estimates of energy savings in 1980 total 5.4 quads, which is 6.6 percent of the baseline projection for 1980 energy consumption.

However, since the SECP was implemented later than Congress had planned and since some conservation measures are proving more difficult to implement than first anticipated, the savings expectations of 5.4 quads most likely will not be reached in FY 1980. However, the SECP is still attempting to achieve energy savings amounting to an overall 5 percent reduction in the energy demand projection for 1980. If this objective is reached, energy savings will amount to 4.1 quads in 1980.

Energy Extension Service. The Energy Extension Service (EES) is designed to provide small-scale energy users with access to practical and available energy conservation and renewable resource techniques and opportunities. Thus, it provides a number of outreach services (e.g., energy audits, workshops, hotline services) to residential energy consumers, small businesses, and public institutions and groups that influence energy use (e.g., architects, builders, and bankers).

The states, using existing organizations as delivery mechanisms, provide these services to small-scale energy consumers in order to achieve several objectives:

- Encourage individuals and small establishments to reduce energy consumption and convert to renewable energy sources
- Provide feedback to DOE and other energy policy decisionmakers about barriers to the adoption of energy

conservation measures by small energy users

- Help limit the impact of fuel shortages and price increases on small energy consumers by providing energy-saving information and technical assistance

These EES outreach services are projected to assist small-scale energy consumers to conserve up to 0.04 quad per year after 1980. The EES program will be consolidated with SECP.

**Weatherization Assistance Program.** The overall goal of the Weatherization Assistance Program (WAP) is to weatherize (e.g., insulate, weather-strip, and caulk) the homes of low-income families, particularly those of the elderly and handicapped, in order to reduce energy demand in the residential sector. The program supplies resources to states and certain native American tribal organizations to be used in local programs to improve the thermal efficiency of dwelling units occupied by low-income families. The local programs then purchase and install weatherization materials in eligible dwellings.

The specific objective of WAP is to weatherize 2.7 million homes by 1985. If this objective is reached, energy savings will total 5.4 milion barrels of oil (0.03 quad) per year.

## Federal Role

The role of the federal government in support of state and local energy conservation programs is to provide the federal funding and technical assistance necessary for program implementation. This federal role is carried out in the following specific ways:

- Establishing a national tone or attitude that recognizes the severity of the current energy situation
- Developing state energy management capabilities and a federal/state program delivery mechanism
- Providing a vehicle for disseminating information on new energy conservation technologies
- Supporting research, development, and technology demonstration and transfer

- Providing resources to support program implementation
- Identifying the need for and developing new legislation

## Federal Program Strategy

The key element of the S&L strategy for implementing conservation programs is to provide assistance to appropriate state and local governments as called for by the legislation enacting S&L programs. This assistance is in the form of technical expertise, grants, information exchange, planning and management support, and the development of model legislation.

However, the federal strategy for implementing S&L programs goes beyond providing the assistance explicitly called for in the enabling legislation. There are two further elements of S&L strategy:

- Coordination of S&L activities with other conservation programs and with public-interest groups
- Consolidation of S&L programs into a coherent management structure that is flexible enough to incorporate future program additions or expansion

**Program Coordination.** In order to fulfill its overall mission of assisting state and local governments in developing and implementing effective conservation programs, the Office of S&L Programs coordinates its program activities with a wide range of other conservation programs and also with public-interest groups involved in energy conservation. These coordination activities are the result of two factors: (1) they are mandated by the enabling legislation for S&L programs, or (2) they represent good program management, since they allow S&L to offer the widest possible range of services without duplication.

As an example of legislatively mandated coordination, Section 413 of ECPA requires that regulations promulgated to carry out the WAP must be coordinated with the director of the Community Services Administration (CSA), the secretary of housing and urban development, the secretary of health, education and welfare, the secretary of labor, and the director of the Action Agency. It further requires that regulations setting standards for weatherization materials and energy conservation techniques be coordinated with the secretaries of HUD, HEW, Commerce, and the National Bureau of Standards. This coordination is carried out as an integral part of WAP headquarters management.

The SECP provides several instances of the federal coordination. For example, one of the program measures a state must agree to undertake in order to qualify for financial assistance is to pass a traffic law or regulation that, to the maximum extent practicable consistent with safety, permits a motor vehicle to turn right at a red light after stopping—i.e., a right-turn-on-red law. The Federal Highway Administration (FHWA) is responsible for setting national policy with respect to right-turn-on-red. Therefore, SECP coordinates with FHWA to assure that DOE program criteria and requirements are in concert with those of FHWA even though the authorizing legislation does not specifically mandate such coordination.

Furthermore, two S&L programs, SECP and EES, are designed to serve as mechanisms that coordinate other DOE programs and the programs of other federal agencies with state and local governments. One of the objectives of the SECP is to foster the state-level infrastructure needed to implement current and future energy conservation programs; a major objective of the EES is to foster local-level dissemination of conservation information generated from a wide range of sources including federal R&D programs. Consequently, both these S&L programs depend on other DOE and other federal agency programs for technical support and information, while those programs depend on SECP and/or EES to channel the results of their activities (e.g., a new technology ready for commercialization) to end-users at the local level.

The legislation enacting the EES specifically required that the EES prepare and annually revise a Comprehensive Program and Plan (CPP) for Federal Energy Education, Extension, and Information Activities. The purpose of the CPP is to survey all federal programs that provide energy "outreach" services, to report on considerations important

to their effectiveness, and to provide a plan for their improved coordination. Thus, federal program coordination is a legislatively mandated element of the EES program.

In addition to intergovernmental coordination, the S&L programs are coordinated with state and local public-interest groups, including the National Governors Association, the U.S. Conference of Mayors, the National League of Cities, the National Association of Counties, and the National Conference of State Legislatures, as well as professional associations such as the American Institute of Architects, the National Institute of Building Sciences, and the American Society of Heating, Refrigerating, and Air-Conditioning Engineers. This coordination includes joint conferences and workshops, cooperative efforts to develop technical assistance materials, task forces, and panels to assist and advise the development of program policy regulations and legislative proposals, as well as frequent informal contacts. This type of coordination is intended to involve the largest number of people in a unified approach to a national problem.

Program Consolidation. In recent years the states have begun to administer a growing number of federal energy conservation programs, many of which provide similar services (e.g., technical assistance, information dissemination, and energy audits). As a result, in May 1979, the Carter administration introduced new legislation that would (1) consolidate the functions of SECP, the Supplemental State Energy Conservation Program (SSECP), Energy Extension Service, and the Emergency Energy Conservation Act; (2) provide for a single grant application process; and (3) establish a mechanism for coordinating federal energy conservation efforts at the state and local levels. This proposed legislation, titled the Energy Management Partnership Act (EMPA), currently is before Congress; as soon as it is passed (or some modification passed), regulations will be prepared and implemented.

In general terms, EMPA would provide a "core" program to consolidate the management of federal, state, and local energy conservation activities. Additionally, EMPA would provide for the development and implementation of comprehensive energy management programs at the state level, greater involvement by units of local government, coordination of federal and state activities during periods of energy emergency, and the reduction and simplification of grants management procedures. In addition to providing financial resources to state and local governments to assist them in carrying out existing energy-related responsibilities, EMPA would require additional state activities in areas such as energy consumption forecasts, identification of projected energy supply sources, management plans, facility siting, emergency planning, renewable resources, and assistance to local governments.

If the enabling legislation is passed as currently written, EMPA will operate as a formula grant program.

The S&L program offices have begun to consolidate program management procedures in anticipation of and preparation for EMPA's passage. These management consolidation activities will lead to an S&L program guidance manual for states.

The various S&L program offices have personnel participating in six grant-consolidation task forces, each of which has specific responsibility for coordinating the various aspects of S&L grants management. For instance, the Program Monitoring Task Force is developing a consolidated state reporting monitoring system for all S&L programs. The Information Systems Task Force is developing consolidated information systems needed to meet the requirements of the different program offices. The overall objective of the task force effort is to develop uniform programmatic and administrative policies and procedures in the following areas:

- State grant application process
- Post-award reporting and monitoring
- Evaluation criteria and procedures
- Information systems
- Financial and administrative procedures
- Technical assistance

## Program Management Strategy

As described previously, the role of the federal government in S&L programs is that of facilitator, assisting the states in developing, implementing, and managing their conservation programs through effective DOE administration of technical assistance and financial support. In order to fulfill this role, federal program management responsibilities have been divided between DOE headquarters and the 10 DOE regional offices. Table 10-1 lists the specific responsibilities of each.

## MAJOR PROGRAM THRUSTS

### Legislative Framework

Four pieces of legislation provide the authority for the S&L programs presented in this plan:

- The Energy Policy and Conservation Act of 1975 (EPCA), 42 U.S.C. 6201 et seq., as amended. State Energy Conservation Program (SECP), 42 U.S.C. 6321 et seq. The SECP provides funding and technical assistance to states to develop, modify, and implement state energy conservation plans and supplemental plans. These plans describe the energy conservation activities that states will undertake in order to achieve scheduled progress toward state energy conservation goals.
- The Energy Conservation and Production Act of 1976 (ECPA), 42 U.S.C. 6801 et seq., as amended. Weatherization Assistance Program (WAP), 42 U.S.C. 6891 et seq. The WAP provides grants to states and certain native American tribal organizations for the purchase of materials to be used to weatherize the dwelling units occupied by low-income families, especially the low-income elderly and handicapped.
- The Energy Research and Development Administration Appropriation and Authorization Act of 1977, 42 U.S.C. 7001 et seq. Energy Extension Service (EES). The EES program encourages individuals and small businesses to reduce energy consumption and convert to renewable energy resources. Federal funding is provided to establish state EES programs that offer a variety of

Table 10-1. State and Local Program Management Responsibilities.

| PROGRAM FUNCTION | HEADQUARTERS RESPONSIBILITIES | REGIONAL RESPONSIBILITIES |
|---|---|---|
| POLICY DEVELOPMENT | • Develop, establish, and revise S&L program policies and priorities<br>• Develop and publish program regulations and other guidance materials<br>• Resolve program policy issues | • Assist Headquarters to develop and revise S&L program policies and priorities<br>• Assist Headquarters in development of program regulations and other guidance materials<br>• Identify and raise program policy issues for Headquarters consideration; participate in their resolution |
| PROGRAM COORDINATION | • Coordinate policy, issues, and procedures with other DOE offices and other Federal agencies | • Coordinate policy, issues, and procedures with other DOE offices and Federal agencies at regional level |
| GRANT APPLICATION, PLAN SUBMITTAL AND REVIEW | • Provide guidance to States and regions on preparation, review, and approval of grant applications and plans<br>• Develop the application review process, including standard grant documentation<br>• Oversee application and review process for consistency among regions<br>• Establish administrative review procedures<br>• Prepare State plans for States that do not apply or whose applications are disapproved | • Assist Headquarters to develop applications and plans<br>• Assist States to understand application/plan requirements; review applications and plans; negotiate changes with States; award grants<br>• Assist Headquarters to develop the application review process, including standard grant documentation<br>• Recommend procedural changes and raise major issues for consideration by Headquarters<br>• Conduct administrative review in cases of termination or suspension of State funding and other special circumstances |
| PROGRAM MONITORING | • Provide policy and operational guidance to the regions<br>• Oversee State monitoring for consistency among regions<br>• Make on-site visits as necessary for oversight<br>• Respond to organization, procedural, and substantive problems identified by regional offices through grant monitoring<br>• Review regional reports and discuss issues/problems with regional offices<br>• Assist DOE Inspector General in providing procedures for State audits of their grants | • Maintain official State grant records; provide copies of approved plans and all State reports to Headquarters<br>• Monitor State progress as compared to State plans; resolve program implementation problems with States<br>• Make on-site visits to States to review program progress and problems<br>• Identify issues for consideration by Headquarters<br>• Report to Headquarters on status of State issues/problems<br>• Establish timetables for State audits of their grants; assist States in understanding audit procedures<br>• Arrange audits by Federal personnel as necessary<br>• Provide Headquarters with advance warning of the potential need to initiate suspension or termination procedures |
| PROGRAM EVALUATION | • Conduct sample surveys; analyze data as necessary<br>• Prepare required evaluation reports for the DOE Secretary, the President, and the Congress<br>• Initiate program and procedural changes to improve effectiveness | • Validate State data collection procedures and data collected<br>• Review and comment on reports to the DOE Secretary, the President, and the Congress<br>• Recommend to Headquarters changes which would increase program effectiveness |

*(continued on next page)*

## Table 10-1. (Continued)

| PROGRAM FUNCTION | HEADQUARTERS RESPONSIBILITIES | REGIONAL RESPONSIBILITIES |
|---|---|---|
| TECHNICAL ASSISTANCE | • Identify national technical assistance resources and capabilities<br>• Support regional offices in handling State TA requests<br>• Hold conferences and workshops of national interest | • Identify regional and State technical assistance resources and capabilities<br>• Respond to State TA needs and requests<br>• Hold conferences and workshops of regional interest |

energy conservation services, including energy audits, specialized training, technical information, workshops, and hotline services.

• The National Energy Conservation Policy Act (NECPA) of 1978, 92 Stat. 3206 et seq. Schools and Hospitals Grant Program and Other Local Government Buildings Grant Program, 92 Stat. 3238 et seq. These programs are designed to assist in identifying and supporting adoption of energy conservation measures in schools, hospitals, and local government buildings. During Phase I of the program, grants are provided to states to conduct and/or administer energy audits in eligible buildings. During Phase II, grants are provided to institutions for technical assistance programs and energy conservation measures, as appropriate.

In addition, Title II of the Emergency Energy Conservation Act provides for the development of state plans to meet fuels curtailment targets in the event of an energy emergency.

## Current Programs and Milestones

The following sections describe in detail the current status and future milestones of each S&L program.

Schools and Hospitals Grant Program and Other Local Government Buildings Grant Program. During FY 1979, the Schools and Hospitals Grant Program and Other Local Government Buildings Grant Program published proposed and final rules and regulations related to Phase I (preliminary energy audits and energy audits) and Phase II (technical assistance and energy conservation

measures) program activities. Prior to September 30, 1979, approximately 25 state plans for activities were received at headquarters for review, and 108 grants for preliminary energy audits and energy audits were awarded. By February 1, 1980, DOE had approved 45 state plans.

Energy savings resulting from the two programs will begin to be effected during FY 1980 and beyond. During FY 1980–1984, based on anticipated funding levels for grants, more than 58,000 energy audits will have been funded, more than 45,500 technical assistance grants awarded, and more than 14,700 energy conservation measures in eligible institutional buildings sponsored. Milestones are shown in Figure 10-1.

State Energy Conservation Program (SECP). The SECP has been in operation since 1976, and a total of $171 million in financial assistance has been provided to 55 participating states and territories. However, since 1978 was the first year of actual implementation of state energy conservation plans, many program measures are just beginning to reach the stage where results can be observed.

Due in large measure to the implementation of the SECP, State Energy Offices (SEO) have been established in each of the 50 states and six participating territories (District of Columbia, Puerto Rico, the Virgin Islands, American Samoa, Guam, and the Northern Mariana Islands). These SEOs have developed state energy conservation plans and have begun to serve as the infrastructure needed to implement a wide range of conservation programs.

Each of the state plans includes the eight mandatory conservation measures described in EPCA and ECPA. The status of each of the required program measures was as follows:

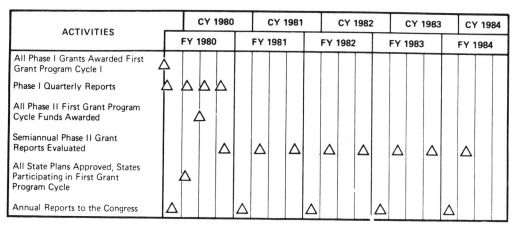

| ACTIVITIES | CY 1980 | | CY 1981 | | CY 1982 | | CY 1983 | | CY 1984 |
| --- | --- | --- | --- | --- | --- | --- | --- | --- | --- |
| | FY 1980 | | FY 1981 | | FY 1982 | | FY 1983 | | FY 1984 |
| All Phase I Grants Awarded First Grant Program Cycle I | △ | | | | | | | | |
| Phase I Quarterly Reports | △ | △ △ △ | | | | | | | |
| All Phase II First Grant Program Cycle Funds Awarded | | △ | | | | | | | |
| Semiannual Phase II Grant Reports Evaluated | | △ | △ | △ | △ | △ | △ | △ | △ |
| All State Plans Approved, States Participating in First Grant Program Cycle | | △ | | | | | | | |
| Annual Reports to the Congress | △ | | △ | | △ | | △ | | △ |

Figure 10-1. Major milestones for the Schools and Hospitals Grant Program and Other Local Government Buildings Grant Program.

- *Thermal efficiency standards.* Twenty-nine states and four territories have either adopted thermal energy efficiency standards for new and renovated buildings or have passed legislation requiring the development of standards consistent with SECP measure requirements. All states not currently in compliance with the minimum requirements have submitted action plans, including milestones, which will result in full compliance in 1980.
- *Lighting efficiency standards.* The status of adoption of the lighting efficiency standards is essentially the same as for the thermal efficiency standards.
- *Government procurement policies.* The government procurement policies program requirement has been met by all states.
- *Carpool/vanpool public transportation.* This program requirement has been met by all states.
- *Right-turn-on-red (RTOR).* All states and territories have a right-turn-on-red law that meets current guidelines.
- *Energy audits.* The energy-audit requirement has been met by all states.
- *Public education.* The public-education requirement has been met by all states.
- *Intergovernmental coordination.* The intergovernmental coordination requirement has been met by all states.

DOE is working closely with those states that

have not met all requirements to establish action plans with milestones to assure full compliance in 1980.

In addition to the eight required program measures, states are implementing state-specific conservation activities in the different end-use sectors: transportation, commercial/industrial buildings and processes, residential buildings, agriculture, and utilities. Many states are issuing statewide executive orders and providing technical assistance to local governments to undertake comprehensive energy management. Several states are proposing tax credits and other incentives to encourage individuals to take energy-conserving actions. Most states are including energy-conserving techniques in their driver education programs and are developing other public-education campaigns related to energy conservation.

Furthermore, many states have found that SECP funds provide the possibility of critical leverage for the state energy office in achieving cooperation from other state offices, such as in purchasing and transportation, and that it has been possible to establish good working relationships to achieve the goals of the program. The SECP also has led to the development of education and information programs designed to enhance citizen comprehension of energy conservation benefits, and the creation of state capability to respond to energy emergency situations.

During FY 1980, the SECP was in a transition phase between the current program and

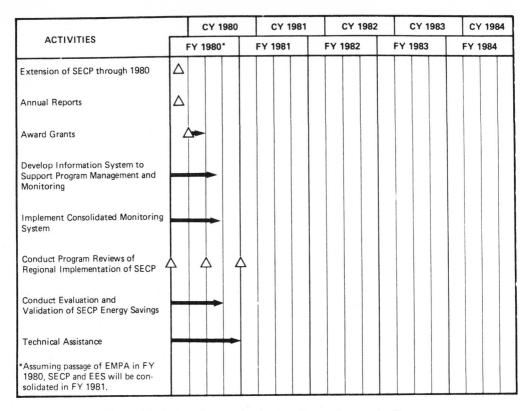

Figure 10-2. Major milestones for the State Energy Conservation Program.

the new consolidated program. Therefore, major program milestones will be related to the extension of the SECP through FY 1980 (the program is only legislated through FY 1979), the development of consolidated management, monitoring, and information systems preparatory to EMPA, and the evaluation and validation of currently reported energy savings. Figure 10-2 presents major SECP milestones through FY 1980.

Energy Extension Service. The Energy Extension Service (EES) was initiated in August 1977 on a 10-state pilot basis to give personalized information and technical assistance on energy conservation to small-scale users of energy. With federal funds, the 10 pilot states designed and implemented practical energy-savings programs such as self-help solar workshops, energy audits for homeowners and small businesses, and energy management services for local governments and hospitals.

The EES pilot program has been the subject of intense scrutiny over the last several years. In March 1979, the DOE submitted to the

Congress a report evaluating the program. The report, based on field observations and surveys of both program clients and comparison groups not receiving services under the program, showed that EES clients were highly satisfied with the services they had received. In addition, clients made and planned more energy conservation improvements and saved more energy than did people who had not received services under the program.

The actions of EES clients have been dollar-efficient as well as energy-efficient. The value of the energy saved as a direct result of EES services produced a 30 percent return on investment for homeowners and over 100 percent for small businesses.

The EES also has had a multiplier effect in that EES clients have precipitated action by other people who were not served directly by the program. For example, the New Mexico EES trained people from other states to build solar greenhouses, who in turn have worked with community residents in their states to construct solar greenhouses.

As a result of the successful 2-year pilot

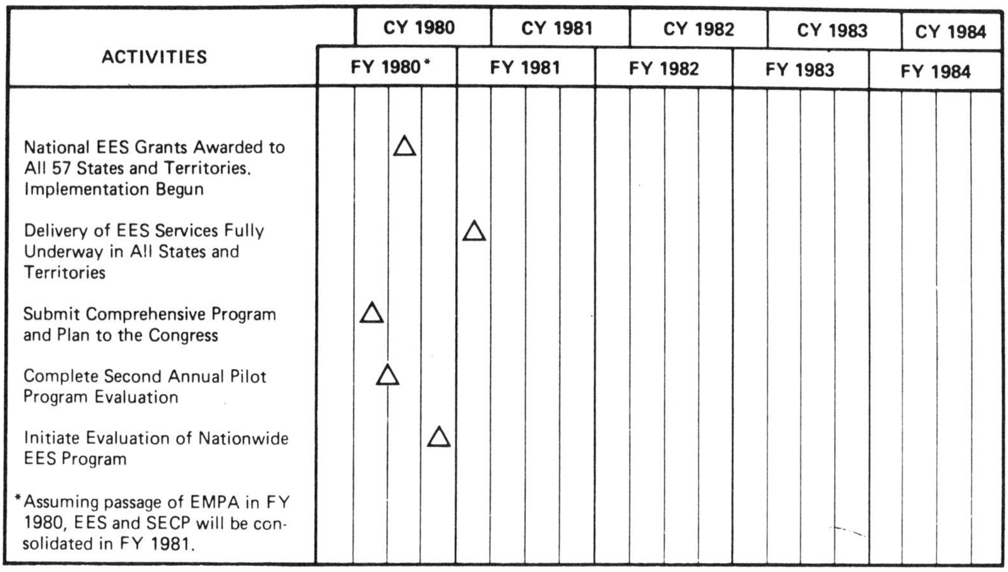

| ACTIVITIES | CY 1980 | | CY 1981 | | CY 1982 | | CY 1983 | | CY 1984 | |
|---|---|---|---|---|---|---|---|---|---|---|
| | FY 1980* | | FY 1981 | | FY 1982 | | FY 1983 | | FY 1984 | |
| National EES Grants Awarded to All 57 States and Territories. Implementation Begun | | △ | | | | | | | | |
| Delivery of EES Services Fully Underway in All States and Territories | | | △ | | | | | | | |
| Submit Comprehensive Program and Plan to the Congress | △ | | | | | | | | | |
| Complete Second Annual Pilot Program Evaluation | △ | | | | | | | | | |
| Initiate Evaluation of Nationwide EES Program | | △ | | | | | | | | |
| *Assuming passage of EMPA in FY 1980, EES and SECP will be consolidated in FY 1981. | | | | | | | | | | |

Figure 10-3. Major milestones for the Energy Extension Service.

program, the EES is expanding in FY 1980 to serve 57 states and territories. During FY 1980, the EES participated in S&L management consolidation activities in anticipation of merging with the SECP. The major EES milestones for FY 1980 are shown in Figure 10-3.

In 1979, DOE issued three sets of amendments to the Weatherization Assistance Program (WAP) regulations, which made the administration of the program at the state and local levels simpler and more flexible. In January, DOE issued amendments that liberalized categories of allowable expenditures and made possible the payment with grant funds of certain previously ineligible costs. In May, DOE issued amendments that increased the maximum expenditure per dwelling unit and expanded the population eligible for assistance under the program by raising the income eligibility level. In August, the "Retro-Tech" rulemaking was issued in order to simplify the energy-audit procedure for home weatherization.

During the same period, WAP awarded nearly $200 million in grants to the eligible states and native American tribal organizations. Of those funds, over $15 million was earmarked for training and technical assistance. Furthermore, as of June 30, 1979, approximately 174,000 low-income homes had been weatherized with WAP funds, resulting

in an annual energy savings of over 0.5 million barrels of oil equivalent.

ECPA authorized WAP through FY 1979, and NECPA extended it through FY 1980; congressional action is expected to further extend the program. The major WAP milestones for FY 1980–1984 are shown in Figure 10-4.

## PROGRAM FUNDING

Each S&L program described herein has been established by recent federal legislation related to energy policy and conservation. The funding level for these programs also has been established by Congress. Table 10-2 summarizes the funding for each S&L program for FY 1979 through FY 1981.

## PROGRAM IMPACTS

### Measures of Effectiveness to Be Monitored

As discussed previously, S&L programs have two basic goals:

- To assist states in conserving energy directly by providing the funding and technical assistance necessary to improve the energy efficiency of homes,

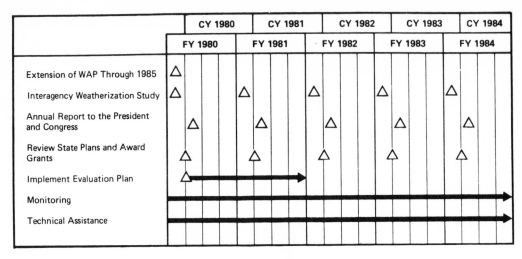

Figure 10-4. Major milestones for the Weatherization Assistance Program.

commercial buildings, schools, hospitals, etc.

- To encourage further energy savings over the long term by helping to build state and local level infrastructure (energy offices, educational programs, information centers) that assist a wide range of energy users.

The effectiveness of programs that directly affect energy use can be measured quantitatively—e.g., the number of homes weatherized via the WAP multiplied by average energy savings per home results in a total amount of energy saved. Total energy savings over the life of the individual homes also can be computed. The same kind of calculations can be made for capital improvements that improve the energy efficiency of schools, hospitals, and public buildings.

However, the energy-savings impacts of programs that serve to develop conservation

## Table 10-2. Funding for the State and Local Programs.

| | BUDGET AUTHORITY (DOLLARS IN THOUSANDS) | | |
|---|---|---|---|
| PROGRAM | APPROPRIATION FY 1979 | APPROPRIATION FY 1980 | REQUEST FY 1981 |
| Schools and Hospitals Grant Program | 100,100 | 143,750 | 202,500 |
| Other Local Government Buildings Grant Program | 7,300 | 17,700 | 0 |
| Energy Management Partnership Act | 0 | 0 | 151,625 |
| Energy Policy and Conservation Grant Program | 47,800 | 37,800 | 0 |
| Energy Conservation and Production Grant Program | 10,000 | 10,000 | 0 |
| Energy Extension Service Program | 15,000 | 25,000 | 0 |
| Emergency Energy Conservation Program | 0 | 0[1] | 4,072[2] |
| Weatherization Assistance Program | 198,950 | 198,950 | 198,950 |
| Program Direction | 2,980 | 7,340 | 11,437 |
| TOTAL | 382,130 | 440,540 | 568,584 |

1. FY 1980 supplemental request is $14,072,000.

2. State implementation grants to be included under EMPA.

infrastructure at the state and local levels often cannot be measured directly or separated completely from the impacts of other federal, state, local, or individual actions.

Consequently, although all S&L programs attempt to measure the effectiveness of their programs, the measures of effectiveness vary from program to program. For instance, the SECP measures the degree to which states achieve state-specific conservation goals; in their savings figures, states often include the savings results from other S&L programs (e.g., WAP). Thus, some of the savings resulting from EES, WAP, the Schools and Hospitals Grant Program, and Other Local Government Buildings Grant Program are included as part of the savings claimed by SECP.

The EES measures the effectiveness of its program through sample surveys of its clientele and comparison groups. The evaluation estimates the incremental amount of energy saved by EES clients' conservation activities and then compares it to the energy saved by the actions of the comparison group. However, this methodology does not include further savings that result from the "ripple" effect of EES programs, since it is difficult to measure the number of people whose activities are influenced by EES clients who have received educational services, information, or technical assistance from the program.

The effectiveness of WAP, the Schools and Hospitals Grant Program, and Other Local Government Buildings Grant Program are easier to quantify accurately because they provide funds to increase the energy efficiency of individual buildings, and energy savings can

be measured directly. However, the last two programs encourage other savings that are not so easily measured, since they pay for energy audits that may lead to conservation activities not necessarily funded.

## Projected Energy Savings

Applying the measures of effectiveness described above, each of the S&L programs has estimated expected energy savings resulting from its program activities (Table 10-3). Figures cannot be added, since the SECP savings include some of those from the other S&L programs. Furthermore, all programs assume that savings effected in one year carry over through the following years (e.g., a house weatherized in FY 1980 will continue to save energy through FY 1984). In this sense, energy savings are cumulative.

## Accomplishments to Date

Since the S&L programs were enacted at different times, some have begun to effect energy savings, while others have not. For instance, SECP was implemented in 1977, and at the end of 1978 the states reported the estimated energy savings resulting from their state energy plans. The EES estimated 1978 energy savings as a part of a program evaluation submitted to the Congress; the WAP estimated that energy savings through 1978 would be a function of the number of homes weatherized. Since the Schools and Hospitals Grant Program and Other Local Government Buildings Grant Program only recently began to disburse grants to the states, energy savings

Table 10-3. Projected State and Local Energy Savings.

| PROGRAM | ANNUAL SAVINGS (QUADS) | |
|---|---|---|
| | FY 1980 | FY 1984 |
| Energy Policy and Conservation Grant Program and Energy Conservation and Production Grant Program | 4.15 | 5.8 |
| Energy Extension Service | 0.04 | 0.29 |
| Weatherization Assistance Program | 0.006 | 0.03 |
| Schools and Hospitals Grant Program and Other Local Government Buildings Grant Program | 0.01 | 0.12 |
| Emergency Energy Conservation Program | 0.0 | 0.0 |

## Table 10-4. State and Local Energy Savings Through 1978.

| PROGRAM | ENERGY SAVINGS THROUGH 1978 (QUADS) |
|---|---|
| Energy Policy and Conservation Grant Program and Energy Conservation and Production Grant Program | 0.07 |
| Energy Extension Service | 0.0024–0.0048 |
| Weatherization Assistance Program | 0.0059 |
| Schools and Hospitals Grant Program* | 0 |
| Other Local Government Buildings Grant Program* | 0 |

*Program authorization not received until November 9, 1978.

due to these programs have not yet begun to accrue. Table 10-4 shows estimated energy savings to date resulting from each S&L program.

## Analysis of Benefits and Costs

Benefit-cost analyses can be conducted at almost any time in a program's life cycle, yet the most reliable analyses are those based on factual data resulting from program experience. However, in the S&L program area, only a limited amount of data has been accumulated. Consequently, it is difficult to differentiate, based on program experience, the possible causes of fuel savings (e.g., differentiating savings resulting from a physical change such as weatherproofing from savings caused by adjustments such as concurrently lowering thermostat settings).

Obviously, the direct benefits from S&L programs are the dollar savings resulting from the purchase of less energy resources than would have been purchased if the conservation effort had not been made. Secondary benefits expected are less easily identified and include such considerations as the effects of spending more federal funds in the state and local areas (e.g., more local jobs, more disposable income, more state income-tax revenue). Secondary benefits will not be discussed further here.

In this benefit-cost analysis, dollar savings are compared to the expected costs of S&L programs, and estimates are expressed in terms of expected payback periods. Costs for WAP are accumulated over 8 to 10 years and compared with the expected savings after 1985. Costs for the Schools and Hospitals Grant Program are totaled for a 4-year period and compared with the expected savings per year after 1981. Costs for the Other Local Government Buildings Grant Program are totaled for a 2-year period.

Because of the differences in time between the expenditure of program funds (costs) and the accrual of benefits, this analysis could have applied "present-value" calculations; however, this has not been done because current data are only approximations. Consequently, it should be noted that benefits and payback period calculations are somewhat overstated, with the greatest effect applicable to WAP. Note also that the benefits and costs of EMPA have not been addressed due to the lack of data.

## Weatherization Assistance Program. A

major goal of this program is to save energy currently wasted in older, less energy-efficient housing by weatherizing 2.7 million homes by 1985, resulting in approximately 0.03 quad of energy savings per year thereafter. If it were assumed that these energy savings could be credited entirely to petroleum products, foreign oil imports (at $20 per barrel) would be reduced by as much as 5.4 million barrels per year, resulting in $108 million savings each year after 1985.

Total WAP funding to accomplish this goal could reach $1.5 billion by 1985, assuming continued funding at almost $200 million each year from FY 1980 through FY 1985. Assuming savings of $108 million per year, the payback period for these costs would be approximately 14 years.

Schools and Hospitals Grant Program. It is estimated that there are more than 200,000 school and hospital buildings throughout the United States totaling about 7.5 billion gross square feet of floor area and using about 2.32 quadrillion Btu annually. Schools and hospitals both are experiencing difficulty coping with increased fuel prices largely because their buildings were constructed when energy was cheap and energy efficiency was not a paramount concern. It is anticipated that savings of conservatively 0.12 quad annually will result at the conclusion of the planned 3-year program. Assuming that this savings could be credited entirely to petroleum products at 180 million barrels of oil equivalent per quad, an estimated 22 million barrels of oil can be saved each year after 1982.

Authorized funding levels for this program total approximately $900 million. Payback of this amount could begin within just 2 years of implementation or, assuming that only 45 percent could be credited to equivalent petroleum savings, the payback period would still be only slightly more than 4 years.

Other Local Government Buildings Grant Program. There is little information available on local government buildings, but it is estimated that local governments may own or operate buildings totaling 1.4 billion gross square feet of floor-space area and use about 0.25 quadrillion Btu annually. This program differs from the Schools and Hospitals Grant Program in that grants can be used for audits and technical assistance, but not for energy conservation measures. The expected energy savings is correspondingly less, perhaps a maximum of 0.025 quad per year. At 180 million barrels of oil equivalent per quad, assuming that the savings can be credited entirely to petroleum products, an estimated 4.5 million barrels of oil can be saved each year.

Authorized funding for this program totals $65 million over a 2-year period. Payback of this amount could occur between 8.6 months and 1.8 years, depending on the assumption of whether all or 45 percent of the savings can be credited to equivalent petroleum products.

## ENVIRONMENTAL IMPACT

Environmental assessments have been completed for each current S&L program and published in accordance with procedures established by the National Environmental Policy Act (NEPA). Program regulations have been reviewed by the Environmental Protection Agency (EPA), and no significant environmental impacts are anticipated. In fact, overall impacts are judged to be beneficial.

Passage of EMPA is expected to necessitate a programmatic environmental impact statement prior to publication of final guidelines as well as individual state environmental assessments prior to funding. In addition, since state energy conservation plans will include energy supply options, site-specific environmental assessments and impact statements may be required. NEPA procedures related to EMPA will be worked out concurrently with development of program regulations.

## MAJOR PROGRAM ISSUES

Over the next five years, the development of the various conservation programs managed by S&L will be affected by the resolution of a number of current issues, which are discussed below.

### Energy Management Partnership Act

The most important issues currently facing S&L are related to the timing and final form of EMPA, since the legislation ultimately enacted by Congress will change the structure of the EPCA Grant Program, the ECPA Grant Program, and EES by combining them under a single management system. Consequently, as EMPA moves through the legislative process during the coming months, the following issues will have to be resolved:

- The EMPA legislation has been submitted to the Congress, but the legislative schedule is still uncertain. The speed of the legislative process will affect the time frame for program implementation as well as SECP and EES budget projections and headquarters and regional staffing actions.

- The funding cycles and grant processes for the two affected S&L programs are not integrated. When EMPA is enacted and the two programs combined, the funding cycles and grant processes will be consolidated. Preparatory work in these areas is under way.

## Regional Offices

In order to properly implement, coordinate, and evaluate the S&L conservation programs, each DOE regional office must fulfill specific state and local responsibilities established by DOE headquarters. These responsibilities include the processing of grants and administrative paperwork for each program and the provision of technical assistance to the states.

Regional offices are understaffed relative to the responsibilities assigned to them. For example, with implementation of the Schools and Hospitals Grant Program and the Other Local Government Buildings Grant Program, regional offices are being given the added responsibility of processing and administering an additional 42,700 grants. The resources provided to them in FY 1980 (approximately 121 positions for all programs under state and local management) are inadequate to manage the grant programs. Consequently, their ability to function effectively will be affected by the number of additional staff positions allotted to them in the future.

# 11. Multisector Programs

## PROGRAM OBJECTIVES AND STRATEGY

The oil embargo of 1973 signaled a fundamental change in the ability of the industrialized nations to chart their own economic destinies and to guarantee the economic security of their citizens. In the United States, the embargo led to nationwide shortages of petroleum, a $60 billion decrease in GNP, more rapid inflation, and large balance-of-payment deficits that continue to plague the economy today. The recent crisis in Iran, with the resultant cutbacks in oil production and the accompanying increases in the prices of imported oil and gas, has aggravated the economic conditions in this country and spotlighted United States vulnerability to energy shortages.

The Energy-Related Inventors Program was instituted in 1974, and in 1977 Congress authorized the establishment of a grants program in small-scale, energy-related technologies within the Office of Conservation of the Energy Research and Development Administration (ERDA). In the National Energy Plan of 1977, President Carter stated, "A new Office of Small Scale Technology is also proposed, in order to tap more fully the potential of individual inventors and small business firms." In his June 20, 1979, address on energy issues, the president advocated drastic measures to reverse dependence on imported energy forms by rapid development and commercialization of technologies using renewable resources and those using other nonrenewable resources (e.g., coal) in abundance. In addition, the president established that a goal of 20 percent (18 quads) of the nation's energy use is to be supplied by renewable resources in the year 2000.

The president's plan indicated that a balanced energy program should promote conservation and the use of nonconventional resources, and seek out the talents of individual inventors and small businesses, as well as large organizations and institutions.

In order to pursue these plans, the DOE established a $1.3 million regional pilot program for a Small Scale Appropriate Energy Technology Grants Program. A national program began and the Office of Inventions and Small Scale Technology (OISST) was created in FY 1979. This office also operates the inventor's program. In addition, multisector programs will include the Energy Conversion Technology Program in FY 1981.

The OISST is the only program office within DOE that has individuals and small businesses as its principal constituents. The office includes programs for all conservation, renewable and nonnuclear energy technologies. Through these programs, OISST can offer encouragement to thousands of individual innovators and small businesses. The office can disseminate information about outstanding projects that could be useful to other program offices in DOE or be widely accepted by the public. This information transfer provides a valuable service that has previously been unavailable.

### Objectives

In order to properly manage the Office of Inventions and Small Scale Technology during 1980 and beyond, a set of objectives has been identified. The objectives have been divided into two sets: those that can be accomplished with the current staffing level and those additional objectives that can only be achieved with an enhanced staff level. With the current level of staffing, the following Appropriate Technology Small Grants program objectives can be achieved:

175

- Write revised regulations.
- Institute operating guidelines for both the headquarters and regional operations.
- Develop applicable material for the required congressional reports.
- Establish and operate the decisionmaking structure to clearly define the headquarters and regional roles.
- Manage the next solicitation for grants.
- Create applicable budget documents and supporting material.
- Construct appropriate operating plans in accordance with the CS Division's guidance.
- Identify the reporting requirements necessary to manage the programs.
- Work closely with state agencies in the operation of the programs.

To achieve important program objectives, the enhanced staff level will be required. The following objectives would then be applicable:

- Produce and use a program evaluation methodology to improve operations.
- Develop a product and service marketing strategy and an implementation methodology, and conduct an operational test of the approach.
- Design and implement an information collection, categorization, and dissemination strategy and the accompanying procedures for use, both domestically and internationally.
- Provide policy guidance.
- Organize and implement a procedure for interfacing OSST with other federal agencies and the private sector.

Long-range objectives are:

- To create a sheltered environment wherein, with federal assistance, independent inventors and small businesses are provided improved opportunities to make significant contributions toward solving our nation's energy problems.
- To provide a system of means and incentives for individuals and small businesses to develop, exercise, and improve their capabilities for invention and ultimate innovation.

Short-range objectives are:

- To continue to improve the efficiency and effectiveness of the evaluation process by developing and maintaining statements on the current state-of-the-art in specific technologies that are prevalent among inventions submitted. Knowledge of the current state-of-the-art in a readily accessible manner will not only reduce the costs of evaluation but will also enhance the quality of submissions by individual innovators.
- To enhance the benefits of the evaluation process by referring to other groups and programs those promising inventions that do not meet the criteria for recommendation to DOE under the Energy-Related Inventions Program. Examples of such programs and groups are Appropriate Technology, Consumer Products and Technology, University Innovation Centers, SBA assistance programs, and state, local government, or regional development programs.
- To improve the timeliness and quality of the one-time support provided to inventors whose inventions have been recommended to the DOE. These improvements are required primarily in recognizing and planning for the business-related aspects of producing and marketing new energy-related inventions.
- To foster and support, where appropriate, selected innovation centers that can provide local assistance to inventors and innovators by direct interaction. These innovation centers not only stimulate the creativity of inventors and provide initial assistance that improves the quality of submissions to the NBS but also provide business advice and assistance, thus directly enhancing accomplishment of the first three objectives.

## Federal Role in Appropriate Technology

The federal government's role is to assist and encourage the use of appropriate technologies by administering the Small Grants Program. It is also necessary for the federal government to assume the decisive role of taking the next "step" in providing impetus to the initiatives. Federal initiatives will be directed toward:

- Providing financial assistance to encourage development of small-scale technology projects that might otherwise be considered too risky by the private sector
- Educating both the public and private sectors on the use and benefits of appropriate small-scale technologies

## Federal Program Strategy for Appropriate Technology

The federal program strategy will be aimed at encouraging the use of small-scale appropriate technologies whenever possible. In order to realize these goals, OSST will continue to update the regulations; coordinate the solicitations; maintain applicable budget documents and operating plan; and specify reporting requirements required for program management under the current level of staffing.

## Program Management Strategy

To achieve the objectives, much work needs to be accomplished. Inherent and important to the overall dynamics of this plan are the roles played by headquarters and the DOE regional offices. The general activities that are to be accomplished under current staffing are described as follows:

- *Operation of the Appropriate Energy Technology Small Grants Program (AETSGP)*. Operate the AETSGP by announcing regional solicitations, performing evaluations of grant applications, awarding grants, evaluating grant results, and developing information and appropriate distribution procedures and marketing strategies.
- *Operation of the Energy-Related Inventions Program*. Operate the program through evaluation of the inventions and support needed to assist inventors to develop their energy technologies.
- *Conservation and solar operating plans*. Develop and maintain required divisional annual and multiple-year operating plans.
- *Program reporting requirements*. Specify administrative requirements necessary for program operation and management. This effort consists of (1) designing an appropriate reporting system to be used by the regional and headquarters staffs and (2) providing the necessary guidelines.
- *Congressional reporting*. Provide appropriate material for the annual congressional reports.
- *Administration*. Design and implement management procedures uniquely necessary and applicable to this program.
- *Staffing requirements*. Develop specific functions for each position and draft position descriptions in support of office staffing.
- *Regulations development*. Prescribe amendments required to the existing program regulations as operational changes are required and suggested.
- *Budget documentation*. Create and maintain budget documents and supporting materials.
- *State participation*. In order to reach out to individuals and small businesses, a federal program needs to work as closely as possible with the "local" level. State governments have and will continue to play an important role in this area. State energy offices and other agencies will work in "partnership" with the regional DOE/OSST in the following areas: program grant solicitation, application technical reviews, peer review of proposals, announcement of grant awards, grant review and administration, commercialization activities,

information development and dissemination, and program procedures and management development.

- *Interaction with the Energy-Related Inventions Program (ERIP).* It is envisioned that the coordination with this program will require a two-way exchange of data. Some of the concept developments that are funded in the program will possibly be candidates for funding as energy-related inventions. These ideas must be made to ERIP, thus providing .original grantees access to this program. For the reverse process to occur, one must be aware of the inventions that might be candidates for demonstration or prototype model development grants under the program. To facilitate this exchange of information, a procedure will be established that will ensure the proper exchange.

In order to achieve other important program objectives, it is necessary to accomplish several activities. First, methodologies must be developed and implemented to evaluate both individual grants and the overall program. A method will be developed for evaluating each individual program activity, program solicitation, technical review, peer review, grant negotiation, information dissemination, commercialization, and grant monitoring.

In developing the methodologies, the following criteria, as a minimum, will be included as the basis for evaluation: energy saved over the next 2 to 5 years; meeting the plan's objectives on schedule; achieving energy-saving results in all levels of income and in all regions of the nation; achieving maximum use of funds; and achieving the maximum coverage for information and dissemination.

Development of the methodologies will incorporate a follow-up mechanism to ensure that all evaluated program activities are changed as deemed necessary. For example, the regulations are updated as better ways of doing business are identified.

Particular attention will be given to the evaluation of the transition-to-renewables process or the commercialization effort. This process is especially critical since it is the process that gets the results of the grants into the marketplace and to the using public.

Marketing strategies must be designed and implemented. The development effort will produce a marketing strategy that will be circulated to regions for review and input. It will serve as the guide for accomplishing the next step after original grants have been completed. This effort will include development of an implementation plan. The plan will contain such major features as procedures for identifying barriers and incentives; business plan outlines; a marketing decisionmaking procedure; and government, and nongovernment sources of funding.

Literature must be developed for all projects, and for the program in general; collection and dissemination procedures will be specified; and information packages for all types of users will be designed, e.g., potential users of the products or services, DOE management, the general public, grantees, state and local governments, Congress, and other federal agencies.

Many activities and projects involve program interests of numerous offices and divisions within the DOE. For example, OISST grants have been awarded to research and/or demonstrate the following technologies: direct solar heat, alcohol fuels, small-scale hydropower, weatherization, building conservation, electric vehicles, heat storage methane utilization, geothermal, and wind energy. The results of these projects will probably be of interest to other offices within the department such as Solar Applications, Geothermal Energy Program, Transportation Programs, Solar Energy Programs, Small-Scale Hydropower Programs, Buildings and Community Systems, Industrial Conservation, Weatherization Assistance Programs, Energy Extension Service, Energy Storage Systems, and Wind Energy Electric Systems.

It is the intent of the OISST to develop efficient working relationships with all DOE offices involving similar and specific program functions. For example, results of OISST project grants in all solar-related technologies will be distributed for review to each "solar" office within DOE according to specific program interest. More importantly, it is highly probable that additional funding or assistance may be warranted. In such a case, the appropriate DOE program office will be contacted by the OISST. This coordination will be effected at both the headquarters and regional office levels.

For reasons similar to those prompting intraagency coordination, the OISST will undertake a coordination effort with a number of federal and private-sector organizations. As the value of application of technologies to the energy crisis becomes more and more apparent, increasing numbers of agencies are showing an interest in developing programs accordingly. Therefore, one objective of OISST must be to assist in program development and to coordinate program management and interagency understanding and effectiveness.

Already initiated but in the development stage are the following relationships:

- The National Park Service has made a commitment to enhance the public's awareness of energy and natural resources through education, exhibits, seminars, conferences, etc., within the existing facilities of the nation's parks system. Pilot planning programs are under way in two national parks. As results are evaluated, the program is expected to be expanded to other parks on the federal, state, and local levels.
- The Action Agency has a well-established network of programs that aid less advantaged sectors of the world's societies.
- The Small Business Administration (SBA) oversees problems that can have a direct impact on the development of technologies. The OSST intends to coordinate its programs and projects with the new SBA energy loan program, and the Senior Core of Retired Executives. Additionally, the SBA is instituting business development centers that can be of assistance in bringing its technologies to the marketplace.

The development and application of tech-

nologies is of international interest. There must be an information exchange between both developed and lesser developed nations. Although some channels already exist, more need to be opened. Organizations currently involved include VITA, Action, Appropriate Technology International, AID, and NATO. The OISST will participate in the future development of this international dialogue.

The native Americans still represent the most socially and economically disadvantaged group of United States society. Through application of small-scale technologies, they can lessen the impact of meeting their energy conservation and production requirements, develop personal and cultural self-reliance, and enhance minority small-business development opportunities. The OISST intends to pursue these objectives by assisting the native American community through coordination efforts involving many federal agencies (HUD, HEW, BIA, DOL, etc.) and numerous nonprofit organizations established to assist native Americans such as United Indian Planners Association and American Indian Development Administration.

Government has limited funding capabilities for individuals and small businesses. The OISST intends to leverage its abilities by coordinating with local and national foundations in the development, acceptance, and use of conservation and renewable technologies.

The National Center for Appropriate Technology has many objectives that correspond to program interests of the OISST. These regional and national interests will be reviewed, and mutually cooperative actions will be planned. Some areas of common interest are regional coordination, low-income groups, project review, and information dissemination.

Many other opportunities exist for effective interagency coordination and will be explored. For example, interests in small-scale appropriate energy technology are developing within the National Science Foundation and the Departments of Agriculture, Commerce, Housing and Urban Development, and Health, Education and Welfare.

Efforts toward such coordination were initiated in FY 1979. Much more will be accomplished as the OISST begins to expand its staff and develop management capabilities. Substantial gains have been achieved in this important field of interagency coordination.

A developing national phenomenon in the area of energy awareness is the proliferation of local energy fairs and exhibitions. As the public demands solutions to the energy dilemma, such functions are proving to be very effective educational events. It is not uncommon for each event to spawn several more in widely diverse locations. People want to know what they can do to help themselves and their communities.

Several events have been supported by the DOE Conservation and Solar Applications Division. Many thousand individuals around the nation learned about those technologies that are appropriate solutions to their local energy needs. Consequently, a growing demand exists for more fairs and special events to demonstrate the application of needed technologies. In FY 1980, the OISST will begin an active participatory role in enhancing public energy awareness by assisting in the sponsorship of such events. The intent will be to provide "seed" money to local groups with which they can approach other agencies and organizations for additional support and technical assistance in producing an energy fair or event. Many types of events will be explored: Appropriate Technology Fairs, State and County Fairs, participation in exhibitions, seminars, conferences, training sessions, etc.

If the OISST is to effectively coordinate the development and application of technologies within DOE and among other agencies and organizations, the Assistant Secretary for Conservation and Solar Applications must assume an active role. This role requires involvement of the assistant secretary in pulling together the resources of the department to encourage other offices and divisions to adopt the philosophy that the United States must find its own solutions to the energy crisis. Government can assist but cannot provide a panacea. The assistant secretary must also support the OISST in its approach to other agencies and organizations by sponsoring high-level planning and organization meetings with federal, state, and private-sector groups.

The 10 DOE regional offices administer and oversee operations of various projects within their respective jurisdictions. The primary responsibilities of each regional office are:

- Management of fiscal outlay
- Management support
- Project management

Close liaison is maintained between the regional offices and headquarters to resolve outstanding issues and contingencies that may arise periodically.

In addition to coordinating the efforts of the regional offices, headquarters identifies projects that require additional technical expertise and attempts to furnish such assistance by drawing on the resources of national laboratories, if so required. Headquarters also encourages the use of simplified procedures in all aspects of the program. These procedures should provide easier access to the individual or small-business person not accustomed to working with the federal government.

The following sequential procedures have been established for operation of the Small Grants Program:

- Announce regional solicitations.
- Perform evaluations of grant applications.
- Award grants.
- Evaluate results of grants (grant review and administration).
- Collect and disseminate information.
- Implement marketing strategies.

Regional Solicitations. Program funds are allocated by region and are weighted according to geographical distribution and population distribution. DOE regional offices are responsible for conducting these solicitations.

OISST tries to notify all interested persons of implementation of the program in their regions. Proposals are solicited through program announcements in the Commerce Business Daily, newspapers, and trade and technical publications. Announcements are also sent to state and local governments and to associations and groups that have expressed interest in the programs to DOE.

Straightforward procedures for grant application have been established to ensure that all applicants receive equal consideration. Applications are evaluated by people familiar with state, local, and regional requirements and resources to ensure that the projects selected for funding are responsive to local needs and concerns.

The eligibility requirements for participation in the program are varied, and grants may be awarded to:

- Individuals.
- Local nonprofit organizations and institutions, including corporations, trusts, foundations, trade associations or institutions entitled to exemptions noted in the Internal Revenue Code, section 501(c)(3), or others not organized for profit and which provide no net earnings to any private shareholder or individual, including educational institutions.
- State or interstate agencies or instrumentalities.
- Local units of government, including county, municipality, city, town, township, local public authority, special district agencies or instrumentalities, and regional or intrastate entities. (Regional or intrastate entities include intrastate districts, councils of government, and sponsor group representative organizations.)
- Indian tribes and nations, regional or village corporations, or other legally established groups of communities of native Americans.
- Small businesses that are organized for profit, independently owned and operated, not dominant in the field of operations in which they are submitting proposals, and have fewer than 100 employees.

Each appropriate technology proposal must fall into one of the following categories:

- Concept development project grants are awarded for the development of an idea, concept, or investigative finding.

Concept development projects could describe new concepts of energy sources or develop and describe new applications or uses of old energy-related procedures or systems. Grants for concept development projects will not exceed $10,000 for any single program participant. Grants in this category are made for the definition of a concept and do not involve fabrication, development, or demonstration activities.

- Development project grants are awarded for studies, investigations, models, hardware development, experimental tests, or operational tests that systematically use or apply investigative findings and/or theories and produce useful products. A proposal in this category should examine a specific problem and develop a solution or test a specific technology system under controlled laboratory or other experimental conditions. Development grants may not support manufacturing or production engineering or full-scale operational testing. Grants for development projects will not exceed $50,000 for any single program participant.

- Demonstration project grants are awarded for the testing of a technology or system under actual operational conditions to show that commercial application of the technology or system is technically, economically, or environmentally feasible. Grants for demonstration projects will not exceed $50,000 for any single program participant.

Any project, regardless of type, must be completed within 24 months from the beginning of funding. Also, the project must be conducted in the region in which the grant is awarded. The total amount of money made available to any participant in the program, including affiliates, consultants, or staff, will not exceed $50,000 over any 2-year period. There are no minimum limits.

Evaluation of Proposals. Evaluation of proposals is based solely on the content of the grant applications. After DOE prescreens the proposals, applicants may not be contacted or interviewed by anyone involved in the proposal evaluation process until completion of the selection procedure. Therefore, the application must reflect the ability of the applicant to satisfy the general requirements of this program and be as complete, detailed, and thorough as desired. The technical, state, and DOE evaluators have a large number of proposals to review in a very short time, so proposals should be concise but provide sufficient information to permit adequate evaluation.

The proposal evaluation process consists of four steps: prescreening, technical and feasibility evaluation, state peer review, and the DOE selection process.

Prescreening involves a review of the applications by the DOE regional office to determine whether they contain sufficient technical, cost, and other information. If proposals do not meet minimum requirements, a letter will be sent to the applicant, indicating the reason(s) why the proposal cannot be considered.

A technical evaluation panel, composed of citizens of a given region and recognized for their expertise in energy-related appropriate technology, performs a technical/feasibility evaluation of the proposals. Each proposal is reviewed to determine the following (listed in their descending order of relative importance):

- If it is technically/procedurally feasible
- If the results of the project can be measured or evaluated
- If potential environmental impacts are addressed
- If the proposal can be carried out with the funds being sought

The technical evaluation includes a review of the qualifications and capabilities of the key individuals involved in the project. In addition, the project's organization and management is reviewed to determine if the project can be completed within a reasonable time schedule, and if the project's goals and requirements are consistent with the proposed budget.

The third step involves a review of the proposal by a panel of state peer reviewers, composed of five or more citizens of the representative state. Each represents a cross section of the state's economic, social, and cultural interests. The peer reviewers use the technical evaluation panel results and perform independent reviews of the proposals. Criteria used by the state peer reviewers include the following items that affect the specific state, community, or both:

- Potential impact on the needs of the community or state, including energy awareness and education
- The energy resource involved and its importance to the state
- The expected energy savings/production and its importance to the state
- Local institutional barriers (e.g., environmental or zoning regulations) that may substantially affect the project and the potential of the project to overcome those barriers
- The likelihood of commercialization, expanded use, market potential, and cost-benefit value
- Potential environmental impacts
- Innovative nature of the project compared to similar work being done in the state
- The extent of work required beyond the funding period to complete the project
- The extent to which local resources, materials, and labor will be used
- Adequacy of the business aspects of the proposal

The state peer reviewers then prepare recommendations regarding proposals within their state. Each state's recommendations are sent to the DOE regional office for the final DOE selection process.

The DOE regional office will select a DOE selection panel of technically knowledgeable DOE personnel. After receipt of the recommendations from the state peer reviewers, the DOE selection panel reviews the recommendations and performs a final evaluation of the proposals, taking into account the following program policy factors to establish a regional mix of projects:

- Regional distribution, including geography, population, and climate
- Project type distribution, including a diversity of methods, approaches, and technologies
- Diversity of participants
- The best overall use of the funds available

The criteria are of equal importance, and the DOE panel places no special emphasis on any individual criterion.

After confirming or modifying each state's rank-ordered list of proposals, the DOE selection panel produces a regional list of proposals, ranked in order of excellence. This list is forwarded to the grant selection official who makes the final decisions on grant award.

On the basis of the DOE selection panel's rank ordering, the selection official selects the proposals to maximize the following:

- The distribution of grants throughout all states in each region
- The distribution of projects in a variety of geographic, climatic, and density situations
- The distribution of projects among a variety of renewable energy resource measures and/or energy conservation measures
- The distribution of grants among the types of eligible program participants

**Award of Grants.** On completion of the DOE selection process, successful applications are further examined to determine the nature and extent to which funding would be necessary to conduct the project. The total amount of financial support made available to any participant in the Small Grants Program may not exceed $50,000 during any 2-year period. There are no minimum limitations.

Cost-sharing or in-kind contributions from sources other than DOE are encouraged to support the projects funded in this program. Cost-sharing considerations, however, are not to be a factor in evaluating applications.

Payments to all grant recipients are made in two or more stages. In practice, an advance payment of 60 percent of the grant award is initially made, and the balance is made subsequently upon request.

Upon selection for funding, the applicant and the DOE enter into a grant agreement. This agreement contains certain required provisions dealing with reports and records, allowable expenses, accounting practices, publication and publicity, copyrights, patents, discrimination, conflict of interest, insurance, safety, changes, resolution of disputes, and other standard government and contract requirements. Details of these conditions are made available to applicants.

**Grant Review and Administration.** Periodic oral and/or written progress reports may be required of individual grantees. In addition, all grantees are required to submit a final written report within a stipulated period following the completion of the grant period.

To ensure proper use of government money, grantees are required to account for expenditures on the projects in a manner acceptable to DOE. All records of funded projects are subject to audit by DOE.

The grantee must allow DOE personnel to visit the project site at reasonable times to obtain progress information. Sharing and distributing information developed in the individual projects is a major goal of the program. Consequently, the results of projects are made public record. During and after the project, grantees may be asked to make the project open to the public during appropriate hours. DOE monitors completed projects on a regular basis for a specified period of time during which the project must be suitably maintained.

## Energy-Related Inventions Programs

The Energy-Related Inventions Program encourages innovation by reducing uncertainties and risks in the development of energy technologies. Small businesses and independent inventors with promising energy-related inventions may obtain federal help through this program in developing and commercializing their inventions.

The National Bureau of Standards (NBS) evaluates the inventions and recommends the most promising to DOE for support. DOE then determines whether the inventions should be supported and how much support they should receive. DOE expects to support almost all of the inventions recommended by NBS.

**Evaluation of Inventions.** A new concept, device, product, material, or industrial process may be submitted for evaluation as an energy-related invention. The invention is first reviewed to determine if it is acceptable for evaluation. An invention may be rejected at this stage, for instance, if it is not energy related; if it deals with the production or use of nuclear energy; if the description and basis for claims are not clear and complete; or if it contains obvious technical flaws.

If the invention passes the initial test, it goes through a first-stage evaluation in which staff evaluators, other government scientists or engineers, or outside consultants and contractors provide technical opinions about the invention. A staff engineer then reviews the opinions, and a decision is made about the invention's potential. If the invention is rated "promising," it moves to the second-stage evaluation.

A more in-depth analysis is conducted in the second-stage evaluation, and a formal report is prepared. If the initial "promising" rating is confirmed, the results are forwarded to the Office of Energy-Related Inventions with a recommendation for government support. No testing is performed in either evaluation.

**Type of Assistance.** Support to each invention is decided on the basis of individual merit and need, but is provided on a one-time-only basis. Usually, support is provided by a grant award, but assistance has included contracts as well as testing of the invention at a DOE facility.

Some inventions are beyond the development stage and require market surveys or the preparation of business plans. Such support may be provided through a grant or through arrangements with nonprofit technology innovation centers.

DOE funds cannot be provided for capital costs, such as production tooling. The Small Business Administration, however, has agreed to give special attention to loan requests from inventors recommended by this program.

## Energy Conversion Technology

**Subprogram Description.** The Energy Conversion Technology (ECT) Subprogram is designed to develop and promote the use of an expanded technology base and advanced concepts in energy conversion and utilization. The purpose of the ECT Subprogram is to ensure a continuing flow of applied scientific advanced concepts to meet the conservation and Solar Energy Program goals of improved energy productivity.

The ECT Subprogram supports work in three functional categories:

- *Applied research.* Achievement of fuller scientific knowledge for specific conservation energy requirements.
- *Exploratory development.* Investigation of innovations in particular energy conservation technologies.
- *Technology development.* Achievement of feasibility and evaluation of potential energy conservation concepts.

ECT Subprogram interacts with other Conservation Subprograms by sponsoring advanced research and technology development in support of those subprograms. The ECT Subprogram also provides interfaces with the Office of Basic Energy Sciences and advanced technology subprograms in the Solar Energy, Fossil Energy, and Nuclear Energy Programs.

The ECT Subprogram is executed according to the following guidelines:

- Joint participation by universities, government laboratories, and nongovernment R&D organization
- Coordination with other R&D programs in government, industry, and foreign countries

- Cooperative activities with potential user industries
- Prompt dissemination and utilization of results

In addition, participation by nongovernment representatives in planning, review, and evaluation is promoted.

ECT Objectives and Goals. The objectives of the ECT Subprogram are:

- To develop a technology base to support future development of energy conversion and utilization systems with major improvements in energy productivity and capability for alternate fuel utilization
- To develop advanced concepts in energy conversion and utilization technologies to achieve feasibility and gauge conservation potential

The term *technology base* refers to a body of knowledge and expertise drawn upon by industry in developing improved products. The base is developed and expanded by dissemination and utilization of the results of basic research, applied research, and exploratory development.

An "advanced concept" is an idea or plan embodying an innovative approach to the design of a component or system. An advanced concept offers potential for significant improvement over current designs.

The ECT Subprogram establishes continuing R&D activities in project areas that encompass the principal technology needs in energy conversion and utilization processes and equipment. ECT program elements and projects have been defined as follows:

The strategy being followed involves activation of new projects as project plans are developed and as management and financial resources become available.

## MAJOR PROGRAM THRUSTS

### Legislative Framework

The Appropriate Energy Technology Small Grants Program was established by Section 112 of P.L. 95-39, June 3, 1977, as part of the Authorization and Appropriations of ERDA. The granting procedures are governed by Section 9 of the Federal Nonnuclear Energy Research and Development Act of 1974 (P.L. 93-577). The criteria for reporting the progress of the program are specified in the Federal Nonnuclear Energy Research and Development Act of 1974, as amended by P.L. 95-238, Title II, Sec. 206(b), Feb. 25, 1978, 92 Stat. 61, which added subsection (a)(4). The program regulations are specified in the *Federal Register*, Vol. 43, No. 153, Tuesday, August 8, 1978.

The Energy-Related Inventions Program was established to implement Section 14 of the Federal Nonnuclear Energy Research and Development Act of 1974 (P.L. 93-577), which directed the National Bureau of Standards (NBS) to evaluate promising energy-related inventions, particularly those submitted by individuals and small companies for the purpose of obtaining direct grants from ERDA. ERDA's responsibilities under this legislation have been assumed by the DOE.

### Programs and Milestones

Appropriate Technology. To provide a basis for the Small Grants Program, DOE

| PROGRAM ELEMENT | PROJECT |
| --- | --- |
| Energy conversion technology | Open-cycle power systems |
| | Closed-cycle power systems |
| | Boilers and furnaces |
| Energy utilization technology | Materials |
| | Physical processes |
| | Chemical processes |

conducted a pilot program in Federal Region IX, which comprises Arizona, California, Hawaii, Nevada, American Samoa, Guam, and the Pacific Trust Territories. The response to this pilot program was overwhelming: more than 1100 proposals for grants were received, and 108 proposals totaling $1.3 million were funded. The pilot program was initiated in September 1977 and was implemented to provide monetary grants for small-scale, energy-related projects for the following:

- Idea development for concepts-demonstrating potential
- Concepts testing for projects that have gone beyond the idea-development phase and are ready for testing
- Demonstration to develop projects that, having been tested, now must be proven through actual use

Prior to award and grants, it was ascertained that the technologies funded would make best use of available renewable energy sources, conserve nonrenewable resources, depend largely on human labor, and emphasize use of local materials and labor skills.

The project should also meet the following criteria:

- Satisfy local needs.
- Increase community energy understanding and self-reliance.
- Be environmentally sound.
- Result in durable recyclable systems and/or products.

A variety of projects covering a wide spectrum of technologies received funding. Projects included solar-powered devices, heat-recovery systems, applications of biomass and wind energy, integrated systems, and an array of miscellaneous devices for energy conservation/conversion. Also funded were a number of projects that included courses and manuals of data or instruction. In addition, projects concentrating on exhibitions, guided self-help projects, and student education received appropriate grants.

Major milestones for this program are as follows:

- Open solicitation of the FY 1981 grants program—2/15/81
- Announce awards for the FY 1981 grants for all regions—8/30/81
- Complete negotiation and granting of monies—9/30/81

**Energy-Related Inventions.** Since April 1975, NBS has received more than 12,600 evaluation requests and has completed action on all but 800, which are in process. Approximately half of all requests received are found acceptable for evaluation. About 1 in 40, or 2.5% of those inventions that are evaluated, are recommended to DOE for support.

The National Bureau of Standards has recommended 121 inventions to DOE. Awards include 55 grants and two contracts averaging almost $70,000 and totaling $3.8 million. In addition, 4 inventors have received assistance other than direct financial support. Support is imminent for 6 more inventors, and negotiations are continuing with another 45. In nine cases, there has been no feasible way for DOE to provide support.

Inventions address almost all nonnuclear energy-related technologies. Of those inventions recommended to DOE, about 30% pertain to industrial processes, 22% to buildings and structures, 14% to solar and other natural sources, 14% to combustion engines and automotive systems, 8% to fossil fuel production, with the remaining 12% in other categories.

About one-fourth of the inventions are near the prototype stage and are ready to enter the market. Many of these exhibit potential for significant energy savings, social benefits, and near-term return on investments.

## PROGRAM FUNDING

Currently, the Multi-Sector Programs are funded at $19.0 million from all sources. The Appropriate Technology Program accounts for $12 million of this total, while the Energy-Related Inventions Program accounts for $6.4 million, including funds from the Departmental Administration Budget. The remaining $635,000 is for the personnel resources in program direction.

Requests for FY 1981 amount to $31.6 million from all sources with the addition of the Energy Conversion Technology program. This program will add $11 million to the budget. The appropriate technology request totals $14.8 million (including program direction) and requests for the Energy-Related Inventions Program amount to $5.8 million (including Departmental Administration funds). Details are shown in Table 11-1.

## PROGRAM IMPACTS

### Measures of Effectiveness to Be Monitored

Since an important task to be accomplished in FY 1980 is the development and application of a program evaluation methodology for all segments of the Appropriate Technology Small Grants Program, the following criteria will be incorporated into the evaluation methodology:

- The bottom line in judging how effective the program is functioning is the payback in energy saved over the next 2 to 5 years.
- The objective of this plan must be met in the required time frame in order to meet the program goals.
- Another measure of program success is whether the program achieves results in all levels of income and in all regions of the nation.
- The program must place money in a balanced manner in all technologies. To achieve the best total results, however,

it may be necessary to concentrate time, money, and effort in specific pilot areas for the marketing and commercialization phase.

However, unless additional staff are acquired, this very important task cannot be accomplished.

An evaluation methodology is also being developed for the Energy-Related Inventions Program during FY 1980. Since this program deals with unique individuals—inventors— particular attention will be given to their special needs in trying to develop innovative ideas and bring them to the marketplace.

### Projected Energy Savings by Specific Programs and Fuel Types

Since the Small Scale Appropriate Energy Technology Grants Program has only recently been announced nationally, the only experience developed to date is from the Region IX pilot program. Most of the data from the pilot program have been developed at the Lawrence Berkeley Laboratory. Of 20 selected small-scale projects, early analysis indicates that, for the dollars granted, the return on investment (ROI) could be as high as 500 percent based on the dollar value of fuels saved (gas and electricity).

In one possible scenario, it is estimated that, if 13 percent of the proposals deemed worthy of grant support are funded and if commercialization efforts are pursued, yields as high as 0.06 quad of savings in United States energy consumption may be evident in

## Table 11-1. Multisector Budget.

| ACTIVITIES | BUDGET AUTHORITY (DOLLARS IN THOUSANDS) | | |
| --- | --- | --- | --- |
| | APPROPRIATION FY 1979 | APPROPRIATION FY 1980 | REQUEST FY 1981 |
| Appropriate Technology | 8,000 | 12,000 | 14,100 |
| Energy-Related Inventions Program* | 2,000 | 4,200 | 3,400 |
| Energy Conversion Technology | 0 | 0 | 11,000 |
| Personnel Resources | 243 | 635 | 700 |
| TOTAL | 10,243 | 16,835 | 29,200 |

*An additional $2.2 million is available for this program in FY 1980 and $2.4 million in FY 1981 under the Departmental Administration budget.

1985, 0.12 quad in 1990, and 0.12 quad in 2000.

To demonstrate the magnitude of these energy savings, the 0.06 quad of energy saved in 1985, if applied to the residential sector, would be sufficient for space heating and cooling of over 467,000 single-family detached homes. This would be equivalent to savings of approximately 29,200 bbls.

## Accomplishments

Major accomplishments have been the implementation of the Region IX pilot program and the implementation of the program in FY 1979. However, these results need to be analyzed before any meaningful conclusions can be drawn.

## Analysis of Benefits and Costs

Preliminary analysis indicates that the benefits will exceed the cost by a ratio of at least 5 to 1 for the Appropriate Technology Small Grants Program. A study is now under way to examine the cost benefits for the Inventors Program.

## ENVIRONMENTAL IMPACTS

One of the primary requirements of the Small Scale Appropriate Energy Technology Grants Program and the Inventor's Program is that the projects be environmentally sound. Therefore, no adverse environmental impacts have resulted from the projects funded to date. The Energy Conversion Technology program has not yet been funded within CS.

## MAJOR PROGRAM ISSUES

The lack of adequately trained and experienced staff is the most inhibiting factor to achieving the objectives of this program. The current staff is overwhelmed with many quick response actions, correspondence, and general administrative details.

# 12. Energy Impact Assistance

The goal of the Energy Impact Assistance Program is to provide technical and financial assistance to states, regions, and local governments to foster the development and implementation of programs to prevent and mitigate the potential adverse economic, social, and environmental impacts resulting from major energy development, and to provide for the financing of public facilities and services.

The strategy for grants achieving this goal is to provide federal assistance to areas impacted by increased employment in the mining, processing, and transportation of coal and uranium. The grants are available for planning and for the acquisition of site development of land to be used for the construction of housing and public facilities.

To be eligible for assistance under this program, an area must:

- Have had an 8 percent increase in coal or uranium-related employment in the most recent year, or

- Have an 8 percent increase in such employment projected for each of the next 3 years, and have an existing or projected shortage of housing or public facilities and services resulting from this employment increase

To attain the specific program goal, it is necessary to determine the appropriate federal role.

Redevelopment of domestic energy resources entails costs at the local level that cannot readily and in a timely fashion be internalized in the cost of the energy developed and marketed in the country to replace imported fuel oil. The federal role is designed to compensate for the inequity of benefits and costs not occurring at the same location.

The specific accomplishment to date is that $70 million has been approved for expenditure in local areas to overcome impacts associated with the rapid buildup of population due to energy projects.

# 13. Energy Information Campaign

The objective of the Energy Information Campaign is to increase DOE effectiveness as a source of reliable consistent information for the nation's consumers on energy conservation. To meet this objective, a major paid advertising campaign is being developed to be initiated in FY 1981. The campaign, which is to be national in scope, will reach 95 percent of the American population an average of at least 100 times during the year. It will utilize print and electronic media including major magazines, newspapers, and network and local television and radio. It is augmented by an aggressive public relations and media event program that multiplies the effectiveness of the campaign by extending its audience and focusing attention on its objectives. It is complementary to the DOE energy extension service and residential conservation service and is supported by DOE publications, exhibits, speakers' bureau, and various other outreach activities. It is also designed to be complementary to, rather than competitive with, the energy conservation messages currently being promoted by private industry. It is intended to encourage private industry to accelerate and extend its energy messages by adding a sense of urgency and credibility to the national energy conservation effort.

This program is similar to those already in place in a number of energy-consuming nations around the world. In total, 11 countries have paid advertising programs designed to encourage energy conservation. Most of these programs have been in place at least 5 years.

In addition to its ongoing public affairs activities and conservation outreach programs, DOE proposes to undertake a highly specialized paid advertising and communications program that will incorporate the optimum advertising techniques required to encourage the American public to understand the national energy situation and to conserve energy in every form. Tie-ins with private-sector advertising and promotion activities are contemplated, similar to those used in DOE's previous paid advertising programs: Energy Cost of Ownership, Low Cost/No Cost, and Fuel Oil Marketing. The FY 1981 budget request for this campaign is $50 million. Funding at this level is needed to bring this message before the American public through national exposure at an effective level.

More specifically, it is projected that the following amounts will be required for the purposes indicated:

| | |
|---|---:|
| Production of ads and commercials for television and print | $ 1,500 |
| Purchase of media (television, radio, print) | 44,000 |
| Supporting program (exhibits, special brochures, media events) | 4,000 |
| Market research | 500 |
| Total | $50,000 |

# PART III
# SOLAR ENERGY

# 14. Introduction

Part III, Solar Energy, presents an overview of the U.S Department of Energy Solar Energy Program in the Office of the Assistant Secretary for Conservation and Solar Energy. Solar energy R&D activities, which were formerly the responsibility of the Assistant Secretary for Energy Technology and the Assistant Secretary for Conservation and Solar Applications, have been merged under Conservation and Renewable Energy. They are now the responsibility of the Deputy Assistant Secretary for Conservation and Renewable Energy.

This chapter reviews the status of these solar programs and presents 1981–1982 activities. The eight solar technologies are:

- Biomass energy systems
- Photovoltaic energy systems
- Wind energy conversion systems
- Solar thermal power
- Ocean systems
- Agricultural and industrial process heat
- Active solar heating and cooling
- Passive and hybrid solar heating and cooling

## BUDGET SUMMARY

The budget request for the Office of Solar Applications was prepared along functional lines (e.g., market analysis, market test, and applications). The budget request of the Office of Solar, Geothermal, Electric, and Storage Systems, hereafter referred to as the Office of Solar Technology, was prepared along technology lines (e.g., photovoltaic systems). In order to be comprehensive and present the solar budget clearly, the entire Solar Program is described herein by both technology and function.

The combined FY 1981 solar energy budget request was $652 million, which is an increase of $56 million from the FY 1980 appropriation. A budget summary for these projects is presented in Table 14-1 for Solar Technology and in Table 14-2 for Solar Applications.

Details of the Solar Program by technology, including budget activities for the Office of Solar Technology, are presented in Chapter 16; a functional crosscut including proposed FY 1981 activities for the Office of Solar Ap-

## Table 14-1. Office of Solar Technology: Summary of FY 1981 Budget.

| PROGRAM ELEMENTS | BUDGET AUTHORITY (DOLLARS IN THOUSANDS*) | | |
| --- | --- | --- | --- |
| | APPROPRIATION FY 1979 | APPROPRIATION FY 1980 | ESTIMATE FY 1981 |
| OFFICE OF SOLAR TECHNOLOGY | | | |
| Biomass Energy Systems | | | |
| Technology Support | 16,406 | 16,000 | 22,000 |
| Production Systems | 5,085 | 5,700 | 11,600 |
| Conversion Technology | 14,695 | 24,500 | 11,500 |
| Research and Exploratory Development | 5,213 | 8,300 | 14,400 |
| Support and Other | 1,001 | 1,500 | 3,500 |
| Total Biomass Energy Systems | 42,400 | 56,000 | 63,000 |

## Table 14-1. (Continued)

| PROGRAM ELEMENTS | BUDGET AUTHORITY (DOLLARS IN THOUSANDS*) | | |
|---|---|---|---|
| | APPROPRIATION FY 1979 | APPROPRIATION FY 1980 | ESTIMATE FY 1981 |
| OFFICE OF SOLAR TECHNOLOGY (Continued) | | | |
| Photovoltaic Energy Systems | | | |
| Advanced Research and Development | 33,087 | 47,000 | 50,000 |
| Technology Development | 40,546 | 61,000 | 60,000 |
| Systems, Engineering, and Standards | 6,492 | 14,500 | 15,000 |
| Tests and Applications | 23,675 | 24,500 | 15,000 |
| Total Photovoltaic Energy Systems | 103,800 | 147,000 | 140,000 |
| Wind Energy Conversion Systems | | | |
| Implementation & Market Development | 4,530 | 2,008 | 0 |
| Research and Analysis | 6,466 | 8,739 | 7,700 |
| Wind Characteristics | 4,447 | 5,401 | 6,200 |
| Technology Development | 9,925 | 13,181 | 17,100 |
| Engineering Development | 34,187 | 34,071 | 49,000 |
| Total Wind Energy Conversion Systems | 59,555 | 63,400 | 80,000 |
| Solar Thermal Power Systems | | | |
| Central Receiver Systems | 53,034 | 65,000 | 37,250 |
| Distributed Receiver Systems | 31,822 | 34,000 | 46,250 |
| Advanced Technology | 13,444 | 22,000 | 34,000 |
| Total Solar Thermal Power Systems | 98,300 | 121,000 | 117,500 |
| Ocean Systems | | | |
| Project Management | 3,728 | 4,655 | 4,520 |
| Definition Planning | 3,000 | 2,953 | 2,900 |
| Technology Development | 11,550 | 9,359 | 8,200 |
| Engineering Test and Evaluation | 19,247 | 19,546 | 20,000 |
| Advanced Research and Development | 3,620 | 3,487 | 3,380 |
| Total Ocean Systems | 41,145 | 40,000** | 39,000 |
| Solar Information Systems | 0 | 0 | 1,400 |
| International Solar Program | 0 | 0 | 11,000 |
| Solar Energy Research Institute Facility | 3,000 | 6,900 | 10,000 |
| Technology Support & Utilization | 6,700 | 3,100† | 0 |
| Program Direction | 3,390 | 3,688 | 4,000 |
| TOTAL SOLAR TECHNOLOGY | 358,290 | 441,088 | 465,900 |

*Includes Operating Expenses, Capital Equipment, and Construction
**Does not include pending supplemental authorization
†Includes purchase of scientific computer at SERI

### Table 14-2. Office of Solar Applications: Summary of FY 1981 Budget.

| PROGRAM ELEMENTS | BUDGET AUTHORITY (DOLLARS IN THOUSANDS*) | | |
|---|---|---|---|
| | APPROPRIATION FY 1979 | APPROPRIATION FY 1980 | ESTIMATE FY 1981 |
| OFFICE OF SOLAR APPLICATIONS | | | |
| Market Analysis | 2,713 | 6,000 | 7,300 |
| Systems Development | 41,000 | 53,000 | 56,000 |
| Market Test and Commercial Applications | | | |
| Buildings | 55,000 | 36,750 | 18,000 |
| Agricultural/Industrial Process Heat | 11,000 | 14,000 | 20,800† |
| Photovoltaics | 15,000 | 10,000 | 35,200 |
| Federal Buildings | 25,668 | 11,750 | 2,000 |
| Subtotal | 106,668 | 72,500 | 76,000 |
| Market Development and Training | 2,800 | 20,500 | 39,500 |
| International Solar Program | — | — | 4,000 |
| Program Direction | 2,751 | 3,077 | 3,300 |
| TOTAL SOLAR APPLICATIONS | 155,932 | 155,077 | 186,100 |

*Includes Operating Expenses, Capital Equipment, and Construction
†Includes wood commercialization

plications is provided in Chapter 17. The left column of the two budget summary tables identifies the appropriate section of the document that describes activities supported by specific budget elements. Other solar activities are supported by the Office of the Assistant Secretary for Environment and the Office of Energy Research, as well as the Divisions of Electric Energy Systems in the Office of the Assistant Secretary for Resource Applications and Energy Storage Systems in Conservation and Renewable Energy.

# 15. Overview of the Solar Program and Strategy

## THE NATIONAL SOLAR GOAL

The renewed interest in solar energy has resulted from recognition of its importance as an energy resource, both now and in the future. Solar energy includes energy from sunlight, wind, biomass, hydroelectricity, and the oceans (i.e., thermal differences, waves, etc.). These resources are either inexhaustible or renewable, and they could satisfy most of the United States energy needs in the long term, i.e., until sometime in the next century. Moreover, the resources are sufficiently diverse geographically that almost all areas of the United States have an abundance of one or more of the solar resources. Chapter 20, Solar Resources, discusses the nature, size, and diversity of the solar resource base.

Perhaps even more important, solar resources represent an energy supply that is secure against oil import disruptions, possible slowdowns in the use of nuclear power, or coal production. Furthermore, the price of solar energy is not subject to foreign manipulation. The solar resources also represent very flexible energy-supply options. Many of the resources can be used both in small-scale decentralized applications and as part of larger centralized systems. Any environmental problems associated with solar energy use are viewed as minor compared to the problems associated with other energy resources. Consequently, there is strong justification for a serious examination of the solar energy options.

For these reasons, President Carter initiated the Domestic Policy Review of Solar Energy on Sunday, May 3, 1978. The key results of this Domestic Policy Review (DPR) were that solar energy has great potential as a United States energy supply, that it has significant advantages over other energy resources (such as environmental quality and oil import replacements), and that significant energy contributions could be made by the year 2000.

Subsequently President Carter established the national goal for solar energy in his National Solar Message of June 20, 1979.

We have a great potential and a great opportunity to expand dramatically the contribution of solar energy between now and the end of this century. I am today establishing for our country an ambitious and very important goal for solar and renewable sources of energy. It is a challenge to our country and to our ingenuity.

We should commit ourselves to a national goal of meeting one-fifth (20 percent) of our energy needs with solar and renewable resources by the end of this century. This goal sets a high standard against which we can collectively measure our progress in reducing our dependence on oil imports and securing our country's energy future. It will require that all of us examine carefully the potential solar and renewable technologies hold for our country and invest in these systems wherever we can.

The goal* is admittedly ambitious because, historically, new energy sources have typically taken 30 to 50 years to make significant energy contributions. Solar energy currently contributes about 5.0 quads of primary energy to the total national consumption of 78.0 quads. This contribution comes primarily from hydropower and biomass use (e.g., wood combustion). Obviously the solar goal

---

*For the purposes of the goal, *solar energy* includes all direct forms (e.g., insolation) and indirect forms (biomass, wind, hydropower, and waves) of solar energy.

## Table 15-1. Estimated Solar Technology Contributions.[1]

| | | 2000 | | |
| | | BASE CASE | MAXIMUM | TECHNICAL |
| SOLAR TECHNOLOGY | 1977 | AT $32/BBL | PRACTICAL | LIMIT |
|---|---|---|---|---|
| Active Heating and Cooling | Small | 1.3 | 2.0 | 3.8 |
| Passive Heating and Cooling | Small | 0.3 | 1.0 | 1.7 |
| Industrial and Agricultural[2] | — | 1.4 | 2.6 | 3.5 |
| Biomass | 1.8 | 4.4 | 5.4 | 7.0 |
| Photovoltaic Systems | — | 0.2 | 1.0 | 2.5 |
| Wind Systems | — | 0.9 | 1.7 | 3.0 |
| Solar Thermal Power | — | 0.2 | 0.4 | 1.5 |
| Ocean Thermal | — | — | 0.1 | 1.0 |
| Hydro | 2.4 | 4.0 | 4.3 | 4.5 |
| (High Head) | (2.4) | (3.5) | (3.5) | (3.5) |
| (Low Head) | (Small) | (0.5) | (0.8) | (1.0) |
| TOTAL QUADS | 4.2 | 12.7 | 18.5 | 28.5 |

1. The estimates in this table represent the amount of conventional energy that can be displaced by solar systems, rather than the amount of energy actually delivered by solar systems.
2. Includes process heat, on site electricity, and heating and hot water.

Source: Solar Energy Domestic Policy Review, Response Memorandum, February 1979.

cannot be achieved without a well-designed program of federal activities, which is carefully coordinated with work being done by states, localities, and individuals.

The DPR used three broad scenarios to examine the potential for solar energy: (1) a base case that represented a continuation of current policies and assumed that oil prices would rise smoothly to the equivalent of $32 per barrel in the year 2000†; (2) a maximum practical case where federal and state policies are used to accelerate solar use; and (3) a technical limits case that represented the maximum quantities of solar that could be installed, given the limits of production and turnover in capital stock. Table 15-1 summarizes the DPR estimates. The maximum practical case corresponds most closely with the 20 percent goal.

The DPR analysis goal assumed that the United States would need 95.0 quads of energy by the year 2000; this suggests that meeting the goal would require the production of 19.0 quads of energy from solar sources, which is an increase in the energy now supplied from solar sources by a factor of almost

4.* While the essential contribution of each solar technology may change as more careful analyses are performed, the contributions shown provide a useful working model for designing the national program. It is assumed that the use of hydroelectric energy supplies, mainly from small-scale hydro,* will grow by about 2 percent per year for the next two decades.† Larger increases in the use of this attractive energy source are not expected because, at the rates shown, the United States would be close to saturating its exploitable hydroelectric resource.

The table indicates that biomass will contribute more to United States energy supplies

†Today's world oil price is already approaching levels assumed by the Domestic Policy Review for the year 2000.

*The DPR recognized that uncertainty exists in making projections about the future and that the rate of solar adoption depends on the costs of solar technologies, comparable costs of conventional fuels, government policies, and personal decisions to be made by millions of consumers.

*Small-scale hydropower commercialization is the responsibility of the Office of the Assistant Secretary for Resource Applications and is not discussed herein.

than any other solar technology. The biomass contribution increases by approximately 5 percent annually to a total of 5.4 quads by the year 2000. More than half of the increased use of solar energy hoped for in the year 2000 will need to come from technologies now making negligible contributions to national energy supplies—some of the technologies are not supplying any commercial energy at the present time. Confidence in meeting the specific goals for these technologies must necessarily be lower.

This is one example of how solar technologies could meet the 20 percent goal for the year 2000. Work is now under way to determine the best mix of conservation and solar technologies to meet the goal. The solar energy strategy described next is designed to accomplish this goal.

## NATIONAL SOLAR STRATEGY

The national solar strategy must take into account the enormous complexities of energy markets, the diversity of solar resources, regional need, and environmental factors and be able to accommodate rapidly changing conventional energy prices.

The range of market needs for energy services is quite broad, including high-temperature process heat, low-temperature space heat, mechanical power, electricity, fuels for transportation, and chemical feedstocks. The magnitude of these market needs varies considerably by region, as do the price and availability of competing fossil fuels. The regions also differ in the extent of their institutional infrastructure development—another determinant of the ease with which solar technologies can penetrate the markets.

Technologies to tap the different solar resources for use by end-use markets are available now or are under development. Other technologies convert either direct or indirect solar energy into energy forms to meet basic consumer needs. The solar resources include direct sunlight (insolation) as well as indirect resources such as biomass, hydropower, wind, and ocean energy. These are illustrated in Figure 15-1, which shows how various solar resources and technologies can be linked

to end-use energy demands. In view of the diversity of potential markets and solar resources, there is clearly a need to support the development and assessment of a broad slate of solar technologies with which they can be matched. The potential for large international markets for such solar technologies heightens the value of a broad-based technology support program.

It must be emphasized that technologies that utilize all of the solar resources have been technically proven. One key difference among the technologies, however, is their closeness to economic feasibility under current market conditions. Some technologies are cost-competitive today in specific markets; for example, wood combustion for heating, passive solar heating of homes, and photovoltaic electricity generation for remote markets. Other technologies are sufficiently close to being competitive in cost that removal of market imperfections and realization of economies of large-scale production will move them into the marketplace.

Private industry and the public have so far been slow to adopt the currently available solar technologies, even though many of the technologies are reliable and cost-effective in certain applications; this demonstrates that certain present market conditions do not favor it. The private sector will move heavily into marketing solar energy systems only when the profit incentive is strong and immediate. A number of economic, social, and market barriers are inhibiting the rate of solar adoption.

Therefore, if solar energy is to attain President Carter's goal for the year 2000, the barriers must be identified and overcome as soon as possible. Furthermore, a balanced solar strategy is required. It must meet the diverse market and regional needs and address the different problems faced by solar technologies that are in different stages of development, commercialization, and market use.

The DPR called attention to five major barriers to solar energy that must be addressed in any solar strategy. These included:

- Financial barriers faced by users and small producers

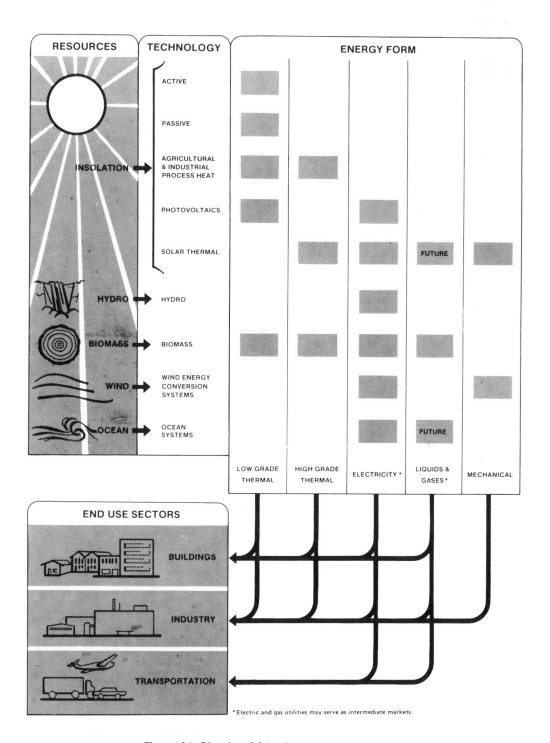

Figure 15-1. Diversity of Solar Resources and Applications.

- Limited public awareness of, and confidence in, solar technologies
- Subsidies to conventional energy sources through federal and state policies and market imperfections
- Inadequate and inappropriate funding of solar technology R&D
- Lack of action by or coordination with state and local governments

These DPR findings helped to formulate the National Energy Policy and the solar strategy. President Carter identified several specific initiatives to speed the use of solar energy, including:

- Incremental cost pricing of natural gas
- Review of electricity pricing policies
- Deregulation of domestic natural gas and oil prices
- Solar Energy Development Bank
- Synthetic Fuel Corporation
- Exemption of gasohol from federal gasoline taxes
- Tax credits for passive buildings, agricultural and industrial process heat systems, and efficient wood stoves
- An expanded set of federally funded solar technology demonstrations and program initiatives, including information and training programs, warranty programs, mortgage and energy-cost analysis programs, solar building codes, a federal gasohol fleet, agricultural projects, and the Federal Buildings Program
- Efforts to expand international solar markets

More information on the Domestic Policy Review impacts is provided in Chapter 6.

President Carter also announced the formation of a standing Subcommittee on Solar Energy under the cabinet-level Energy Coordinating Council. It will be the responsibility of this new committee to ensure that all parts of the administration are working together to accelerate solar energy development. It should be noted that at least 10 federal agencies have significant responsibilities in the federal solar program. This chapter discusses work taking place in departments other the DOE when that work is funded through the two offices submitting this budget request.

## DOE SOLAR PROGRAM

### Role of DOE in the Solar Program

DOE is the lead manager of the national solar program. Its role is threefold: policy analysis and coordination (such as assessment and recommendation of new incentives); regulatory (such as rulemaking on utility rates for solar backup charges); and development, implementation, and administration of federally sponsored R&D and commercialization programs. Thus, the overall DOE solar effort includes work sponsored by the Office of Energy Research, the Assistant Secretaries for Resource Applications, Environment and Policy, and Evaluation, as well as the Economic Regulatory Administration. However, since the conservation and solar energy programs represent more than 95 percent of the DOE-sponsored work in solar energy, and since the purpose of this document is to explain the FY 1981 solar programs within the Office of Conservation and Solar Energy in DOE, the other efforts will not be discussed here.

The major factors influencing the DOE solar strategy to meet President Carter's 20 percent goal in the year 2000 include:

- Cost of meeting this goal
- Expected reduction of oil imports
- Environmental benefits from solar
- Equitable treatment of various regions of the country with differing energy needs and resources
- Diversity of market needs for energy services
- Coordination with other national energy policies

These factors result in a solar energy strategy, which, although complex, is responsive to the complexities and rapid evolution of the United States energy situation.

Briefly, the DOE solar R&D strategy consists of the following five steps:

- Characterizing market needs
- Performing basic, applied, and developmental R&D as necessary to define the costs of solar technologies in early development stages
- Identifying and selecting cost-competitive applications for solar technologies and appropriate federal roles in development
- Performing R&D to reduce costs of selected solar technologies in advanced development stages
- Providing technology transfer support

Characterization of market needs includes activities that consider the effects of conservation on growth and energy demands, identify environmental needs, examine the price sensitivity of various end-use energy services, determine reliability requirements, and so forth.

Cost-definition R&D consists of basic and applied research and studies that are designed to identify and define what the current and potential future costs of new solar technologies might be in their early development stages—for example, advanced biomass techniques, such as algae production and biophotolysis, or ocean-systems research, such as that on salinity gradients and waves. It also includes activities designed to identify the appropriate federal role in further development of specific technologies.

Identification of cost-competitive applications includes all activities that seek to define market sectors and market-service demands for solar technologies, including the identification of international opportunities. These activities will identify the initial and subsequent cost-competitive markets to which specific solar technology development should be directed.

Cost-reduction R&D is performed on selected advanced solar technologies such as wind, photovoltaic, solar thermal, biomass conversion, and active cooling systems. It is designed to develop components and systems that will reduce the costs of the selected solar technologies to a point where they can penetrate initial markets and where economies of mass production will enable larger market penetration to occur.

Technology transfer activities include information, training, test-marketing, and limited demonstration projects designed to move solar technologies that have approached the competitive cost range into the marketplace. Technologies such as active and passive heating of buildings are in this phase of development.

This overall solar strategy is coordinated with national energy policy and is designed to meet the national solar goal with cost-effective and reliable substitutes for conventional fuels, which meet the spectrum of market and individual needs. The specific strategies associated with individual technology programs are described in Chapter 16.

## Organization of the Solar Energy Program

Implementation of the Solar Energy Program is the responsibility of the Assistant Secretary for Solar Energy in the Office of Conservation and Renewable Energy.

The proposed organization of the Office of the Deputy Assistant Secretary for Solar Energy is shown in Figure 15-2. This structure groups solar technologies that have principal markets in common under a single office to facilitate market analysis and commercialization. Close coordination is maintained with programs under the Deputy Assistant Secretary for Conservation because of their synergistic effect on solar development, as well as outreach and information programs in CS. The Solar Energy Program is also coordinated closely with the Economic Regulatory Administration and with the Federal Energy Regulatory Commission for related solar energy matters, such as formulating procedures and retail electric rate standards for solar technologies.

Among the outreach programs managed by the Deputy Assistant Secretary for Conservation and Solar Energy that deal with solar

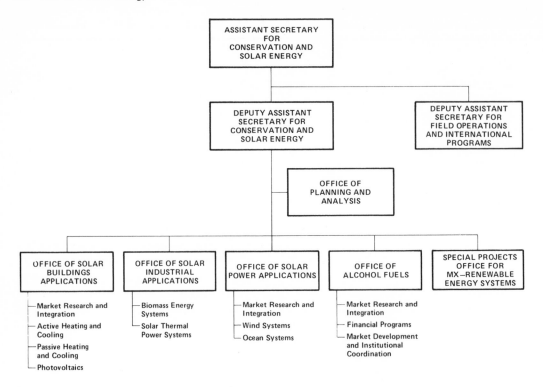

Figure 15-2. Proposed Organization of Solar Programs in Conservation and Solar Energy.

energy are the Energy Extension Service, which provides funding for state-level energy information programs for all types of users. It is modeled after the Department of Agriculture Extension Service. The Residential Conservation Service oversees a residential energy-audit program that is mandated by the National Energy Conservation Policy Act, P.L. 95–619.

# 16. Solar Energy Program Technology

The DOE Solar Energy Program encompasses the development of supply systems that derive energy directly or indirectly from the sun. The program is comprised of Biomass Energy Systems, Photovoltaic Energy Systems, Wind Energy Conversion Systems, Solar Thermal Power Systems, Ocean Systems, Agricultural and Industrial Process Heat, Active Heating and Cooling, and Passive and Hybrid Solar Heating and Cooling. A brief description of these eight major technology programs follows:

- *Biomass Energy Systems.* Biomass is renewable organic material such as terrestrial or aquatic vegetation or animal, agricultural, and forestry residues. This material contains stored chemical energy produced by plants from solar energy. Biomass can be collected and burned directly, or converted to liquid fuels (fuel oils, alcohols), or gaseous fuels (medium-Btu gas, synthetic natural gas), or other energy-intensive products (hydrogen, ammonia, petrochemical substitutes) that can supplement or replace similar products made from conventional fossil fuels.

- *Photovoltaic Energy Systems.* Sunlight is converted into electricity by "solar cells," which can be made from a number of different semiconductor materials. Intensive research is under way to create improved, high-efficiency, lower cost photovoltaic devices. In general, there are two generic types—*flat-plate arrays* that operates on direct sunlight at normal intensity and *concentrators* that increase the intensity of sunlight as much as 2000 times.

- *Wind Energy Conversion Systems.* Wind Energy Conversion Systems employ a wind-driven machine to turn an electrical generator or mechanical device. Wind derives its energy from the sun's heating of the earth's surface and atmosphere.

- *Solar Thermal Power Systems.* In a solar thermal power system the sun's heat is concentrated and used to heat water or some other fluid to provide industrial process heat or to drive a turbogenerator. The primary objective is to provide an alternative to fossil fuels for industrial and utility applications. Applications that provide both heat and electricity, called *total energy systems,* are also included in this program.

- *Ocean Systems.* Ocean-stored solar energy has potential as a renewable source of baseload electricity and energy-intensive products. Systems currently being studied and developed include ocean thermal energy conversion, salinity gradients, ocean currents, and wave energy. These concepts, at various stages of development, use resources that are available in different geographic areas.

- *Agricultural and Industrial Process Heat.* Agricultural and industrial process heat applications use a range of solar collection systems to produce hot air, hot water, and steam within three primary temperature ranges (low, less than 212° F; intermediate, 212° to 350° F; and high, greater than 350° F) to support farm and industrial operations. Depending on the system design and application, heat from these collection systems is injected into the process directly or through heat exchangers. The actual energy uses and range of temperatures are more diverse than in building space-heating applications and require specific process designs.

- *Active Solar Heating and Cooling.* Active solar heating and cooling systems employ predominately flat-plate collector technology. Modular or site built collection systems convert insolation into thermal energy by absorbing radiation. Mechanical subsystems transfer the heat into the building using air or liquids, where it goes directly to heat space, heat service water, or is stored for later use. Swimming-pool heating, domestic water heating, and to some extent space heating, are the leading commercial applications. Solar cooling technology, which provides for more economic year-round employment of solar collection systems, has not yet been demonstrated to be cost-effective.

- *Passive and Hybrid Solar Heating and Cooling.* Passive and hybrid solar buildings employ designs that maximize the benefits of natural energy flows and minimize dependence on conventional energy resources and mechanical equipment. Passive solar-heating systems use elements of the building to collect, store, and distribute energy. Passive cooling also uses elements of the building to store and distribute energy and, when prevailing conditions are favorable, to discharge heat to the cooler part of the environment (sky, atmosphere, ground). When other solar technologies (active systems or photovoltaics, for example) are integrated

in the design, the result is considered a hybrid solar application.

Each of these programs is discussed in detail in this chapter. It must be emphasized that the active, passive, and process heat programs are not budget-line items. Instead, the estimated funding for these programs is derived from functional budget lines of the DOE Office of Solar Applications.

A summary of the FY 1981 funding estimates for the DOE Office of Solar Technology is shown in Table 16-1.

The funding estimates of the DOE Office of Solar Applications are organized by function according to the current state of development for the commercial or nearly commercial solar technologies. Estimates of expenditures by technology are shown in Table 16-2.

The following sections describe in detail the Solar Energy Program by technology. Descriptions are structured according to the FY 1981 budget (i.e., each is divided into its subprograms, which are described in terms of relevant technical aspects and project status). A milestone chart for each subprogram presents major decision points through FY 1985 and funding levels for FY 1979 through FY 1981.

## BIOMASS ENERGY SYSTEMS

Biomass is renewable organic material such as terrestrial or aquatic vegetation, and animal, agricultural, and forestry residues. Biomass can be collected and burned directly, or it can be converted to other usable energy forms,

### Table 16-1. Office of Solar Technology FY 1981 Budget.

| TECHNOLOGY | BUDGET AUTHORITY* (DOLLARS IN THOUSANDS) | | |
| --- | --- | --- | --- |
| | APPROPRIATION FY 1979 | APPROPRIATION FY 1980 | REQUEST FY 1981 |
| OFFICE OF SOLAR TECHNOLOGY | | | |
| Biomass Energy Systems | 42,400 | 56,000 | 63,000 |
| Photovoltaic Energy Systems | 103,800 | 147,000 | 140,000 |
| Wind Energy Conversion System | 59,555 | 63,400 | 80,000 |
| Solar Thermal Power Systems | 98,300 | 121,000 | 117,500 |
| Ocean Systems | 41,145 | 40,000 | 39,000 |
| TOTAL | 345,200 | 427,400 | 439,500 |

*Includes Operating Expenses, Capital Equipment, and Construction.

## Table 16-2. Office of Solar Applications
## Estimated Funding by Technology.

| | BUDGET AUTHORITY (DOLLARS IN THOUSANDS) | | |
| --- | --- | --- | --- |
| | APPROPRIATION | APPROPRIATION | REQUEST |
| TECHNOLOGY | FY 1979 | FY 1980 | FY 1981 |
| OFFICE OF SOLAR APPLICATIONS | | | |
| Active Heating and Cooling | 90,413 | 67,300 | 51,700 |
| Passive Heating and Cooling | 23,900 | 33,900 | 43,850 |
| Agricultural & Industrial Process Heat | 21,768 | 35,000 | 43,000 |
| Photovoltaics | 16,800 | 13,200 | 35,600 |
| Wood | 300 | 2,600 | 8,650 |
| TOTAL | 153,181 | 152,000 | 182,800 |

such as gaseous or liquid fuels or energy-intensive chemical feedstocks. Principal applications for biomass products include industrial process heat generation; electrical generation; production of heat and power for onsite residential, industrial, or agricultural applications; and the production of liquid fuels for transportation. Figure 16-1 summarizes the major sources of biomass and the uses of energy from biomass to meet national energy needs.

## Program Strategy

The primary objective of the Biomass Energy Systems Program is to produce fuels and chemicals economically from biomass to reduce United States dependence on petroleum and natural gas. To achieve this displacement, the program supports the earliest possible transfer of promising biomass production and conversion technologies to private industry. The program has been reoriented to meet national and regional needs. Biomass resources will be matched with regionally identified markets. New sources of biomass and increased yields from energy crops will be vigorously pursued. Conversion technologies will be applied and developed to match the biomass resources with end-uses. Emphasis will be on near-term impact, increased liquid fuels, and new sources of biomass.

Biomass currently provides approximately

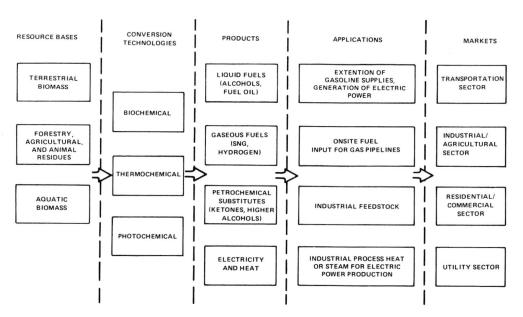

Figure 16-1. Biomass and applications.

1.5 quads (2 percent) of the United States energy requirements each year. The goal of the Biomass Energy Systems Program is to make an additional 0.5 to 1.5 quad per year available before 1985, through direct combustion of biomass as well as by converting it into gaseous and alcohol fuels; an additional 6.0 quads per year before 2000, through improved biochemical and thermochemical conversion technologies; and a total of 8.0 to 10.0 quads per year after 2000, as the result of possible additional contributions from innovative R&D on terrestrial and aquatic energy farms.

The program strategy is based on the grouping and phased development of biomass technologies and their applications according to the time frame in which they are expected to enter the commercial market. For technologies expected to enter the market in the near term (by 1985), efforts are directed toward process improvements and, where necessary, toward demonstrations in commercial-scale applications. Transfer of the technology to private industry will be assisted by dissemination of the technical and economic data obtained from the demonstration projects. Institutional barriers to commercialization will be identified and appropriate government actions will be recommended to achieve significant industrial production capacity.

Projects to support technologies that are expected to enter the market between 1985 and 2000 are directed primarily toward laboratory-scale investigations, process economics studies, and engineering design models using process development units (PDUC) and, if successful, large experimental facilities (LEFC).* The program direction will be reviewed continuously by evaluating promising technologies, assessing resources, determining market needs, and performing economic analyses. The economic feasibility of promising technologies will be demonstrated at energy farms and pilot conversion plants.

_____

*Process development units are small working models, larger than laboratory-scale models, that are built to establish the physical parameters of operations. Large experimental facilities are engineering models, larger than process development units and smaller than commercial units, that are built as an intermediate step in development of the technology.

Technologies expected to affect the commercial market after the year 2000 are now being developed through applied R&D.

## Program Structure and Budget

The Biomass Energy Systems Program is structured to reflect the categories of technology R&D to be pursued and the time period in which these technologies will begin to enter the commercial market. The program elements are:

- Technology support (before 1985)
- Production systems (after 1985)
- Conversion technology (after 1985)
- Research and exploratory development
- Support and other

In addition, related wood commercialization activities carried out through the Office of Solar Applications will be continued under the Wood Resources Manager in close coordination with the Biomass Program. The Biomass Program also cooperates with the Department of Agriculture and other government agencies pursuing related research.

Technologies expected to enter the commercial market in the near term include direct combustion and low-Btu gasification of wood and wood residues as sources of industrial process heat, electricity, and residential heat, fermentation of starches and sugars to grain alcohol that can be mixed with gasoline to extend gasoline supplies, anaerobic digestion of manures to produce a gaseous fuel, and direct combustion of agricultural residues for use by the agricultural sector.

Direct combustion of wood is expected to have the largest effect on the energy supply of the solar technologies that are ready now for commercialization. Gasohol (a mixture of 90 percent gasoline and 10 percent biomass-derived alcohol) also shows considerable promise as a supplement to the nation's transportation fuel supplies in the near term. President Carter's recent initiatives recognize this potential and promote the accelerated production and commercial use of alcohol as a fuel or fuel supplement.

Biomass supply and conversion technologies now being evaluated for entry into the commercial market after 1985 include energy farms to provide reliable resource supplies, and the production of medium-Btu gas and fuel oil by thermochemical conversion. Technologies expected to become commercial in the future include aquatic energy farms, development of exotic plant species, and biophotolysis and other advanced conversion technologies.

Requested FY 1981 funding for biomass systems was $73 million, an increase of $17 million from FY 1980, including $10 million for wood commercialization. Table 16-3 summarizes funding by subprogram for FY 1979 through FY 1981. This budget includes significant increases in the subprogram areas that will develop near-term applications and those that will support high-risk technologies with potentially high payoffs in the longer term.

## Program Detail

Technology Support. The Technology Support subprogram element will provide technical support, systems demonstrations, and environmental research in support of activities nearing commercialization. Technologies expected to have commercial impact by 1985 include direct combustion and low-

Btu gasification of wood residues, on-farm direct combustion of agricultural residues and anaerobic digestion of manures, and ethanol production from the fermentation of grains and sugar crops. Specific projects will emphasize improvements in process performance, economic competitiveness, and reliability; assurance of acceptable process operating characteristics; reduction of environmental impacts; assessment of resource availability and demand; and investigation of international development opportunities.

*Direct combustion and low-Btu gasification systems.* The most significant near-term conversion processes are the direct combustion and low-Btu gasification of wood or other biomass to heat buildings and produce industrial process heat and steam for driving turbines. The forest products industry is already a large user of fuels from wood and wood-process residues. A portion of this program element is aimed at ensuring economic availability of terrestrial residues, such as noncommercial standing forests and logging and mill residues at conversion sites. Direct combustion technology for using wood residues to produce process steam or electricity for use on or near the site is relatively well established. There is need for additional work,

## Table 16-3. Biomass Energy Systems Budget.

| BIOMASS PROGRAM | BUDGET AUTHORITY* (DOLLARS IN THOUSANDS) | | |
|---|---|---|---|
| | APPROPRIATION FY 1979 | APPROPRIATION FY 1980 | REQUEST FY 1981 |
| OFFICE OF SOLAR TECHNOLOGY | | | |
| Biomass Energy Systems | | | |
| Technology Support | 16,406 | 16,000 | 22,000 |
| Production Systems | 5,085 | 5,700 | 11,600 |
| Conversion Technology | 14,695 | 24,500 | 11,500 |
| Research and Exploratory Development | 5,213 | 8,300 | 14,400 |
| Support and Other | 1,001 | 1,500 | 3,500 |
| TOTAL OFFICE OF SOLAR TECHNOLOGY | 42,400 | 56,000 | 63,000 |
| OFFICE OF SOLAR APPLICATIONS | | | |
| Wood Energy Commercialization | | | |
| Market Test and Applications | — | 1,000 | 6,000 |
| Market Analysis | 300 | 500 | 850 |
| Market Development and Training | — | 1,100 | 1,800 |
| TOTAL OFFICE OF SOLAR APPLICATIONS | 300 | 2,600 | 8,650 |

*Includes Operating Expenses, Capital Equipment, and Construction

however, in the design and feasibility studies on applying this technology in special situations; e.g., the retrofit of industrial or utility boilers now using oil or natural gas for wood combustion. Similarly, demonstrations are needed for low-Btu gasifiers of wood as add-on systems for conventional oil- or gas-fired boilers.

In direct combustion of wood and wood residues, commercialization efforts are already under way; they are discussed later in this section. The direct retrofitting, without derating, of oil- and gas-fired boilers represents an important research need for potential applications of direct combustion technology. The Aerospace Research Corporation is developing a wood-fueled combustor that, by burning green woodchips, can achieve heat-release rates as high as 40,500 Btu per cubic foot per hour. The construction of a large combustion furnace phototype, with a firing capacity of 15 tons per hour of green wood, is nearly complete. The actual retrofitting of a boiler system is planned for FY 1981. Also planned is an evaluation of the potential of this combustion system to generate electricity for small-scale applications using gas turbine technology.

These Technology Support activities will ensure that the necessary bases are laid for market development and demonstrations.

During the early 1980s, low-Btu gasification units will be evaluated for process heat requirements. Plans are under way to install an improved, traveling-grate gasification system to replace a commercial-size oil-fired boiler. A number of small pilot projects in residue harvesting, preparation, and combustion may be supported to prove the technical and economic viability of specific systems.

*On-farm systems.* Anaerobic digestion is a biochemical process for the production of medium-Btu gas from organic wastes such as manure. The process involves several steps producing bacterial action at essentially ambient conditions (to 65° C) and requires a moisture content of 80 percent or greater in the process. Technology development efforts are being directed toward increasing yields and modifying reaction to decrease costs and

toward increasing the potential for using by-products as animal feed. Efforts will be directed toward developing simple and relatively inexpensive facilities for use on cattle feedlots and individual dairy farms. The primary resource is cattle waste, which is readily collected from feedlots. The technology is sufficiently well developed to be marginally competitive for onsite power generation in selected applications, such as farms.

Several process development units (handling 0.5 to 3.0 tons per day of oven-dried feedstock) for digesting animal manure are in operation now; and data will be collected and analyzed by FY 1981. One current project is studying anaerobic digestion of animal manures to produce methane at a Bartow, Florida, feedlot/anaerobic digestion facility designed to process 25 oven-dried tons per day of feedstock. The protein-rich solid residues from the process will be concentrated and fed to cattle as a dietary supplement. At Cornell University, researchers are attempting to improve the economic and technical aspects of anaerobic digestion systems for use in small-scale agricultural operations. Initial results show that small dairy farms (100 cows) can produce gas for $3.50 per million Btu, at a capital cost of $200 per head with a 3.6-year payback period. The Department of Agriculture supports DOE in these efforts.

Ethanol production for transportation fuel (10 percent blend of ethanol with gasoline to yield gasohol) has been increasing significantly in the past several years, aided by federal and state excise-tax incentives. Current production capacity is approximately 400 million gallons per year.

In order to achieve these national targets, some of the production, particularly in the early 1980s, will come from farms, communities, and small businesses. To assist this sector, DOE sponsored the development of *Fuel from Farms: A Guide to Small-Scale Ethanol Production*, which has been widely disseminated. SERI has published *A Guide to Commercial Scale Ethanol Production and Financing*. A representative 25-gallon-per-hour facility became operational in 1981. The major objective is the development of a basic design for a rural-based ethanol production process that meets the requirements of safety, product quality, ease of operation, a minimal work force, and energy efficiency. Assessments of the economics of on-farm ethanol production facilities will also be completed as soon as possible.

*Alcohol production systems.* The near-term focus in fermentation is R&D for improving the net energy balance and process economics of producing alcohol from sugar and starch crops. Figure 16-2 shows the process and input.

Most activity will focus on equipment design, testing, and demonstrations as well as on feedstock development. The systems developed in this subprogram will be for larger applications than the rural fermentation facilities. A process development unit to test fermentation of cellulosic feedstocks was scheduled to begin tests late in FY 1981. The Solar Energy Research Institute (SERI) has recently assumed management of a flexible biomass-to-ethanol conversion facility for testing various combinations of feedstock pretreatment, enzymatic hydrolysis, fermentations, and product recovery using cellulosic feedstocks.

Milestones and funding for the subprogram are shown in Figure 16-3.

**Production Systems.** The Production Systems subprogram encompasses expansion of the resource base for biomass energy production by developing specific plant species and intensive management techniques. Biomass resources include energy crops grown for their energy content, as well as agricultural or forest residues suitable for conversion processes. These resources include both terrestrial and aquatic biomass. Terrestrial biomass resources considered are silviculture resources, residues (field crop residues, animal manures, and forest residues), and herbaceous resources (grasses, grains, and other nonwoody plants). Aquatic biomass resources include algae and other aquatic plants grown in fresh, coastal brackish, or marine water.

The silvicultural techniques under investigation include site preparation, fertilization, irrigation, mechanical and chemical control of herbaceous competition, and harvesting schemes. Results of the research projects will also allow recommendations to be made on planting densities, lengths of rotation, and the number of crops between plantings for par-

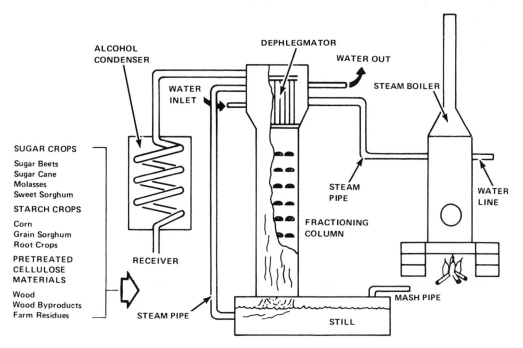

Figure 16.2. Ethanol collection from biomass after fermentation.

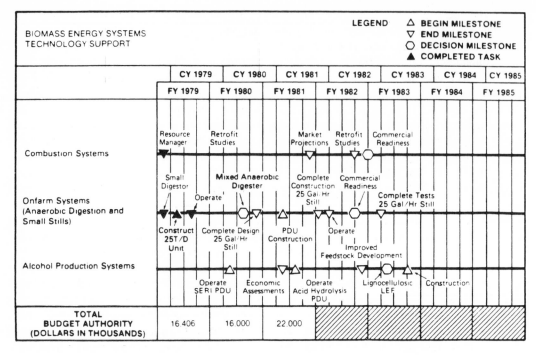

Figure 16-3. Biomass energy systems technology support milestones and funding.

ticular species or mixtures of species. The projects range in size from greenhouse and field plots to large test farms. After 1985, the development of herbaceous and silvicultural energy farms will supplement existing supplies of biomass for conversion to fuels and chemical feedstocks. The objectives for these energy farms include selecting the most promising species and developing site preparation, planting, and management techniques to obtain economical energy production.

Research on species and production management for herbaceous and silvicultural farms was initiated in FY 1978. Three regional silvicultural farms are scheduled for development. By scaling up the test plot and test farm studies, these farms will provide data to assess the technical and economic feasibility as well as the environmental effects of producing an assured supply of woody biomass on large tracts of land in different parts of the country. Start-up operations are scheduled for FY 1981. Under a proposed program memorandum of understanding with the Department of Agriculture (USDA), the USDA will support the Biomass Energy Systems Program in developing farm and forest crops for

high energy value and energy crop management practices and systems. Completed research includes a prototype machine developed by the USDA Forest Service, Southern Forest Experiment Station, for harvesting forest residues for fuel. Current investigations into herbaceous crops include species screening and plant breeding studies and evaluations of the effects of row spacing, fertilization, irrigation, and the frequency of harvest on sugarcane and tropical grass yields. The largest aquatic research project under way is the 0.25-acre test plot for monitoring the growth and nutrient requirements of giant kelp off the southern California coast. Although kelp has been grown as quickly as 2 feet per day, it is hoped that nutrient upwelling will accelerate this growth. In the long term, aquatic energy farms may provide an additional biomass resource. Figure 16-4 shows the major milestones and budget for the Production Systems subprogram.

Conversion Technology. Biomass material can be converted to liquid fuels (fuel oils, alcohols) or gaseous fuels (medium-Btu gas, synthetic natural gas) or other energy-inten-

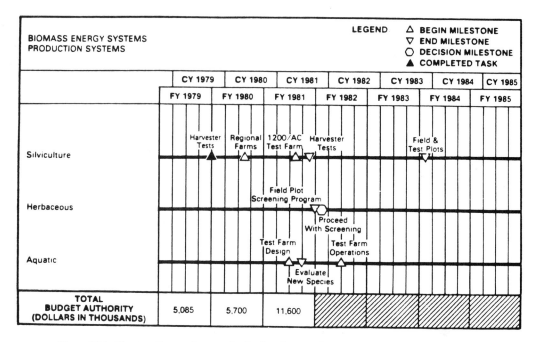

Figure 16-4. Biomass Energy Systems Production Systems Subprogram milestones and funding.

sive products (hydrogen, ammonia, petrochemical substitutes) to supplement or replace similar products made from conventional petrochemicals. Technologies for converting biomass into liquid and gaseous fuels fall into the general categories of thermochemical and biochemical conversion.

The goals for developing Conversion Technology are to improve system efficiencies, decrease process costs, and eventually increase the size of biomass converters. The goals will be met by conducting market assessments to identify optimal product strategies and cost goals, performing laboratory-scale experiments, developing process development units for promising technologies, and by scaling up to large experimental facilities where the concept appears feasible.

*Thermochemical conversion.* In thermochemical conversion processes, biomass is decomposed by using various combinations of elevated temperatures and pressures and by chemical reactants. Thermochemical conversion technologies include gasification, direct liquefaction, and pyrolysis.

Gasification and liquefaction can be used to produce gaseous and liquid fuels from organic matter. For example, gasification produces medium-Btu gas that can be used directly or upgraded into a wide range of derived fuels, such as synthetic natural gas, methanol, ammonia, and petrochemical subsitutes. The program seeks to develop processes that yield a large volume of products in a quicker reaction time and that can compete with products of larger scale facilities built for other feedstocks such as oil, natural gas, and coal.

Major objectives are to simplify the conversion process, improve conversion efficiency and productivity, and reduce production costs of gasification to medium-Btu gas. Medium-Btu gas, predominantly carbon monoxide and hydrogen, is expected to be economically attractive for use near the production facility either as gas or as an intermediate to liquid fuels or chemical feedstocks. Another major area of thermochemical biomass conversion for the midterm is liquefaction. In the liquefaction process, biomass is converted directly to fuel oils by reaction with carbon monoxide and hydrogen under high temperature and pressure in the presence of catalysts.

Seven of the most promising gasification processes have been chosen as the basis for constructing seven process development units.

Three units are in operation; two are under construction, and two more are planned. After an 18-month to 2-year period of process development unit operation, it will be decided whether to proceed to pilot plant construction or commercialization for the most promising processes. Various high-performance reactors are being tested to determine technical and economic feasibility. A major emphasis in gasification research is the simplification of the process train. For example, it is possible to produce medium-Btu gas from biomass without adding oxygen, a costly component of coal gasification technology. The use of catalysts is being tested also to selectively control the composition of the synthetic gas produced and to increase gas yields. Catalytic systems might eliminate shift converters and methanation reactors. The process development units being evaluated are attempting to exploit these process options. Several studies are under way to support the process development unit activities, including resource allocation modeling to evaluate the regional economic and resource implications of gasification facilities, studies of the potential for retrofitting coal gasifiers to operate with wood, technical and economic process assessments, and catalyst studies.

Increased emphasis is now being placed on systems studies and the development of integrated processes for upgrading gases to liquid fuels or chemicals such as methanol, other higher alcohols, and gasoline. Given a positive outcome of an independent review of the thermochemical conversion program during FY 1980, the Biomass Energy Systems Program plans to support the design and construction of a 100-oven-dry-ton-per-day pilot plant which will scale up the results of the most promising process development units and provide process information for the design of a commercial facility.

An experimental liquefaction facility for production of fuel oil from biomass has been operated since 1977 in Albany, Oregon. Based on a Bureau of Mines process developed at the Pittsburgh Energy Research Center, this plant operates at a scale that permits analysis of process economics. It is equipped to pretreat biomass and to react with carbon monoxide and hydrogen at elevated temperatures and pressures. The unit is sufficiently flexible to permit technoeconomic evaluation of a variety of biomass-to-oil processes with minimal equipment modification. A decision will be made by 1982 whether to proceed with pilot plant construction. Recent bench-scale work at the Lawrence Berkeley Laboratory has identified a new process for converting wood to a heating oil. A test trial of the process development units was moderately successful, and further trials of continuous aqueous processes are planned.

*Biochemical conversion.* Anaerobic digestion involves controlled decomposition of organic matter in the absence of air to produce methane and carbon dioxide. This medium-Btu gas can be burned directly, upgraded to pipeline-quality gas such as synthetic natural gas, or converted to other materials. In addition, the chemical industry uses methane as a starting material for many other chemicals. Anaerobic digestion has great potential because it yields a wide range of products and by-products of many uses. Figure 16-5 shows a schematic diagram of an anaerobic digestion unit.

For applications after 1985, the technology is being expanded beyond the use of manures toward anaerobic digestion of more abundant and widespread feedstocks, such as agricultural and forestry residues. A major objective is to develop pretreatment technology that will enable efficient anaerobic digestion of lignocellulosic (wood) feedstocks. Other goals include engineering improvements and cost reductions in anaerobic conversion systems and their components as well as demonstrations of equipment.

Anaerobic digestion of biomass residues other than manures is expected to reach the commercial market stage after 1985. In FY 1977, bench-scale experiments were initiated. In FY 1979, anaerobic digestion research was directed toward using agricultural and forestry residues as digestion feedstocks. The goal was to broaden the resource base for anaerobic digestion as a source of synthetic natural gas. A contract for a process development unit to scale up laboratory efforts on digestion of

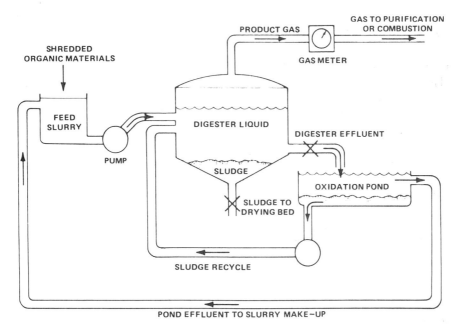

Figure 16-5. Unit for continuous production of methane by anaerobic digestion.

nonmanure feedstocks was awarded in FY 1979, and expansion of this effort in FY 1980 is expected to accelerate technology development and enable earlier technology transfer. Operational experiments are scheduled for FY 1982.

Approximately 10 percent of all ethanol used today is produced by the fermentation of grains and molasses. The use of ethanol as a versatile alcohol component of gasohol may expand rapidly because of the federal highway excise-tax exemption of $0.04 per gallon for gasohol. Other possible fermentation products include butanol, acetone, acetic acid, and other organic chemicals that can replace compounds now obtained from petroleum.

The main opportunity for cost reduction in the ethanol fermentation process involves development of less expensive sources of sugars to ferment. A major goal of fermentation conversion technology research has been to broaden the feedstock base beyond those feedstocks available for near-term fermentation systems. Wood, corn stover, and wheat straw may become fermentable feedstocks. Efficient processing requires separation of the material into its three constituents (cellulose, hemicellulose, and lignin) by pretreating the feedstock. The cellulose is converted to glucose by hydrolysis, and the glucose is fermented. The fermentation product is then distilled to yield ethanol suitable for automobile fuel. To achieve this objective and reduce fermentation costs, current research activities are studying acid or enzymatic hydrolysis of biomass feedstocks to produce fermentable sugars. Discovery of an effective yet inexpensive pretreatment for disrupting cellulose crystallinity and the close physical and chemical association between cellulose and lignin would constitute a major breakthrough toward the use of cellulosic materials for ethanol production.

Basic research on cellulosic feedstocks for fermentation began in FY 1977, with evaluation of commercial feasibility scheduled for FY 1983. Chemical and physical pretreatment methods are both being investigated to develop more efficient separation. Current Biomass-Energy-Systems–sponsored research is directed toward improving the reaction time and efficiency of the enzymatic process, and finding more effective methods for handling the acid solutions to reduce corrosion and pollution.

Several approaches to each process step are

being tested in laboratory experiments. A process development unit is being designed to evaluate the alternatives and to develop cost estimates and a conceptual design for a pilot plant. The process development unit, which will handle about 3 dry tons of biomass per day, is expected to begin operation in FY 1981. Major milestones and funding for the thermochemical and biochemical subprograms are given in Figure 16-6.

**Research and Exploratory Development.** Activities currently supported under this program element will identify and develop those biomass feedstocks having significant production potential but that have not been investigated extensively. Research is being conducted on land-based marine and freshwater plant production, the production of hydrocarbon-bearing plants, and the production of hydrogen through biophotolysis.

*Land-based aquatic biomass production.* The University of California, Berkeley, is evaluating the production and harvest of microalgae in a 1000-square-mile pond. The objective is to develop a prototype integrated wastewater treatment–biofuel production system that includes anaerobic digestion of the microalgae produced. Woods Hole Oceanographic Institute is determining the maximum rates of biomass production of selected species of seaweeds and freshwater plants in 0.25-acre ponds at Fort Pierce, Florida. Sustainable yields for freshwater hyacinth of 25 tons per acre per year have been demonstrated in smaller test facilities.

*Advanced feedstock production.* Projects supported in advanced feedstock development include the breeding and evaluation of "exotic" crops (e.g., Lawrence Berkeley Laboratory work on latex-bearing plants), as well as basic research on plant improvement.

*Euphorbia tirucalli* produces a milky juice called latex, one-third of which is composed of hydrocarbon-like materials that may be suitable as refinery feedstocks. Experiments

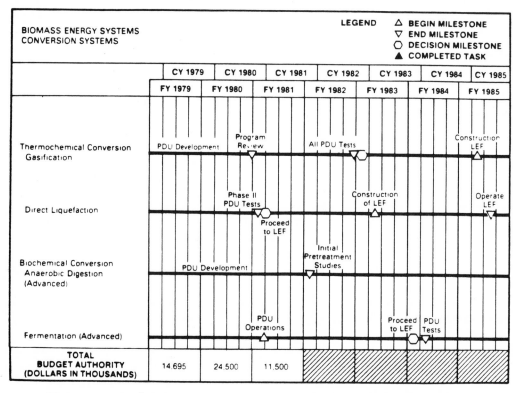

Figure 16-6. Biomass Energy Systems Conversion Systems Subprogram milestones and funding.

are under way to analyze these hydrocarbons and to optimize extraction techniques. Included in the projects are genetic selection studies to improve the yield of hydrocarbons.

Completed aquatic production studies include those on the growth of filamentous algae in ponds, growing and harvesting algae biomass in a 0.1 square-mile system, and the cultivation of macroscopic marine algae and freshwater aquatic weeds. Aquatic system projects include some large-scale marine biomass farming technology, as well as freshwater and marine production, harvesting, and conversion technologies.

*Biophotolysis.* Investigations at the University of California, Berkeley, have demonstrated the continuous production of hydrogen by blue-green algae through biophotolysis. Current studies are focused on improving the present rate of hydrogen production, outdoor cultures, and comparisons of interacting hydrogenases in vivo and in vitro. Biophotolysis research is also under way at the Solar Energy Research Institute.

*Innovative conversion processes.* Projects in new conversion technologies include advanced biochemical conversion research conducted primarily by universities. An example is the development of single-step fermentation of lignocellulosic material at Massachusetts Institute of Technology.

Milestones and funding for this subprogram appear in Figure 16-7. Additional milestones will be specified after the scope and nature of research plans have been further defined.

Support and Other. Various systems studies and environmental assessments, as well as general program support, are included in Support and Other. Mission analysis studies have been performed for several biomass technology development programs. Market analyses have indicated that 19 of 50 missions (combinations of resource conversion technologies and applications) could penetrate the market during the period from 1985 through 2020. Future mission analysis will evaluate the technical paths available to the Biomass Program and assess their potential commercial impact.

Environmental and resource assessments have formed the other major portion of this program element. Projects have been implemented that deal with resource availability at both the national and regional levels. Environmental assessments will be scheduled in close conjunction with technology development to ensure that environmental costs and effects will be considered before widespread deployment of near-term technologies.

Additional activities funded under this program element include program management support and consultation for program planning, preparation, and evaluation of program solicitations and participation in overall program technical evaluations and in coordination monitoring.

In FY 1980, this element supported market, cost, and net energy balance analyses of alternative biomass systems. The program will maintain close interactions with target markets (e.g., farmers and industry) to make technologies available to potential end-users. A data bank will be maintained with available cost data on production and conversion processes. Regional assessments of environmental impacts of technologies will be pursued in areas with a high potential for near-term impact.

*Wood commercialization.* Wood resources are applicable in both the industrial and residential markets. The formulation of reliable, economic fuel supplies in the desired form and development of distribution networks are required to increase penetration into each sector. Because the supplies are site-specific, this issue must be addressed on local or regional bases for specific sectors.

Direct-combustion hardware systems are available for new installations for all applications in the industrial and utility markets. Systems are available for cogeneration, coal cofiring, and multifuel operation (oil or gas). Burners, stoves, and furnaces are available for the residential market. However, technology for performance improvement, multifuel capability, product certification, installation, and operational safety need attention. Projects in these areas will be conducted; for example, fireplace installations are extremely inefficient

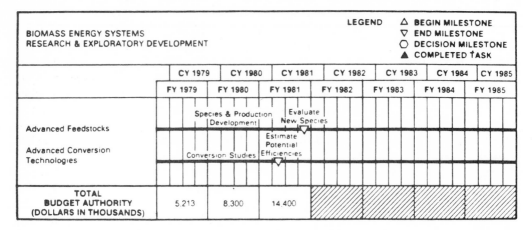

Figure 16-7. Biomass Energy Systems, Research and Exploratory Development Subprogram, milestones and funding.

space-heating devices, and techniques will be investigated for conversion to stove-type operations.

Wood Commercialization includes a variety of national and regional market analyses and resource supply assessments. Wood availability information currently compiled by the U.S Forest Service will be synthesized and provided to potential industrial uses and wood suppliers to accelerate development of a fuel wood infrastructure.

Retrofit of gas- or oil-fueled industrial or utility steam systems is possible, particularly in those cases where the original burners were grate-fired coal systems. Retrofit in other cases requires technology improvements in suspension burning and the development of gasifiers that produce low-Btu gases or both.

Low-Btu gas systems may replace oil and gas in several industrial applications, including process heat production and close-coupled operation of diesels, spark ignition engines, and turbines for production of electricity or shaft power. Such systems are unavailable presently and will be developed.

Mobile systems, which can convert wet or remotely located forest (and agricultural) residues into solids and petroleum-like liquids, are being developed in the United States. The combustion characteristics of these products will be evaluated to provide suitable energy-conversion systems.

Air emissions are a significant concern for Wood Commercialization and are complicated by the lack of federal performance standards.

Many industrial wood-burning applications and the residential space-heating market use systems that consume less than 100-million-Btu-per-hour input. These sources are presently unregulated under the Clean Air Act and the Powerplant and Industrial Fuel Use Act. Wood combustion has significantly fewer problems than coal; however, conventional utility and industrial direct combustion systems will probably produce unacceptable quantities of particulates prior to flue gas cleanup. Control technology is available (i.e., cyclones, electrostatic precipitators, baghouses), but equipment and operating costs may be high. The emissions from cumulative residential point sources may be even more objectionable, particularly in nonrural areas.

During FY 1980, the program is supporting a series of four regional efforts in Wood Commercialization planning. These plans will include activities to identify and use regional wood resources, such as identification of boiler systems that can be converted to wood use, supply studies or inventories, information needs, educational needs, and technical assistance programs.

In FY 1981, efforts at the federal level may expand with emphasis on three areas of activity. The first is Market Analysis, including barrier analysis and economic assessments. Preliminary studies were initiated at the regional level in FY 1981. The second area is the demonstration of wood combustion in a number of industrial applications. These "project-oriented" Agricultural and Industrial

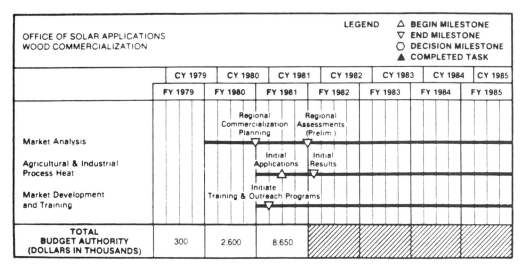

Figure 16-8. Office of Solar Applications, Wood Commercialization Subprogram, milestones and funding.

Process Heat activities will focus on assessing the potential process heat markets, determining the availability of wood residues, and providing working demonstrations for a variety of process needs. The first projects will be funded in FY 1981 with early results from small demonstrations projected for early FY 1982. The final area of effort is Market Development and Training, in which regional outreach activities will provide specialized information services and technical assistance to individuals or groups on all aspects of wood combustion. The programs were developed and initial activities begun by the first quarter of FY 1981.

Milestones for residential, industrial, and utility markets analysis, and estimated funding for industry and utility research and development are presented in Figure 16-8.

## PHOTOVOLTAIC ENERGY SYSTEMS

Photovoltaic systems offer the promise of a clean, reliable, environmentally acceptable source of electrical power. By achieving President Carter's DPR goal in the year 2000, photovoltaic systems could provide electricity to 22 million homes or displace 1 quadrillion Btu of primary fuel per year (78.5 million barrels of oil). Equally important, photovoltaic systems offer developing countries a chance to generate their own energy resources

without resorting entirely to overseas energy suppliers. Already, the cost of photovoltaic modules has decreased significantly—from $22 Wp* in 1976 to as low as $7/Wp in 1979. Installed system costs have dropped from $50 to $60/Wp in 1976 to $16 to $20D/Wp in 1979. Costs are expected to drop $1/Wp in 1983.

Photovoltaic systems have a wide variety of applications. They have been used for water pumps in Nebraska, dust-storm warning signs in Arizona, a village power system on an American Papago Indian reservation, and a variety of other uses in Asia and Latin America. The markets seem quite large if costs can be lowered.

Consistent with addressing these potential domestic and international markets, the objective of the DOE Photovoltaic Program is to reduce photovoltaic systems costs to a competitive level in both distributed and centralized grid-connected applications and to resolve technical, institutional, environmental, and social issues related to widespread adoption of photovoltaic energy systems.

Photovoltaic systems such as those for pri-

---

*Peak watt (Wp) is a measurement unit used for performance rating of solar panels and arrays. A panel rated at 1 Wp will deliver 1 watt under specified standard operating conditions with a solar insolation of $10^3$ watts per square meter. January 1980 dollars ($1980) are used in this section except where noted.

vate residences include not only basic hardware components but everything in the supply chain, from raw materials to construction, including costs of marketing and distribution. The program strategy is directed toward the development of technologies and infrastructures that will yield technically, economically, and socially feasible energy systems in primarily grid-connected applications.

Photovotaic Energy Systems provide a clean, simple method for direct conversion of sunlight to electrical energy. These require no complex machinery, heat engine cycle, or intermediate conversion to heat; in addition, photovoltaic systems are silent. Since they are intrinsically modular, a wide range of system sizes and models can be designed to fit almost any need.

First developed for use in the space program, photovoltaic solar cells absorb sunlight and convert it directly into electricity. Most solar cells are made by combining two very thin layers of semiconductor material. Each layer is prepared by adding different materials, such as silicon, to the major constituent, so that one layer (N) demonstrates negative electrical properties and the other layer (P) has positive properties. Terminals of an external electrical circuit are attached to the front and back of the cell. Sunlight knocks electrons loose from some of the silicon atoms, leaving "holes" in the atomic structure. Because of the electrical properties of the cell, the increment of free electrons creates voltage in the cell, and a current can be drawn through the external circuit. If a load, such as a direct current electrical motor, is placed in the circuit, it will be operated by the current. Thus, sunlight (specifically, photons of energy) is directly converted into electricity. Figure 16-9 illustrates the basic operation of a solar cell. Sunlight penetrates the 0.5-micron-thick N-layer and liberates electrons near the junction of the two layers. These electrons are collected by the wire attached to the front of the cell and can be used as ordinary direct current.

The major component in all photovoltaic (PV) systems is the array, which is composed of a number of electrically interconnected sealed panels (modules), each containing many solar cells. (See Figure 16-10.) Today, arrays account for approximately one-half of the total cost of photovoltaic systems. Other system components—called the balance of system (BOS)—include power conditioners (to convert direct current to conventional alternating current), storage mechanisms (bat-

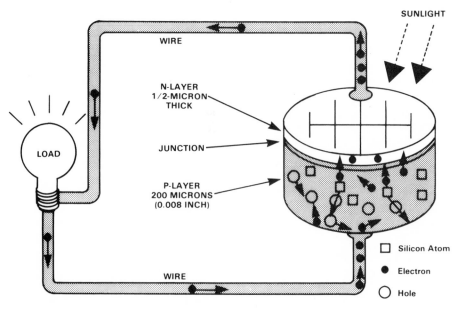

Figure 16-9. Photovoltaic principle.

Figure 16-10. Mead, Nebraska, irrigation project uses 120,000 individual solar cells to produce 25 kW of electricity.

teries), controls, structural parts, and installation and are used to provide electrical power compatiable with user needs.

Photovoltaic systems may be used to generate power in large central stations for sale to customers, or in distributed systems in which the consumers produce the power themselves. Major markets for photovoltaic power systems can be subdivided into stand-alone, or off-grid, applications (such as remote telephone relay stations) and grid-connected applications. Principal grid-connected applications are considered to be residential (5 to 20 kWp), community and intermediate load center (50 to 2000 kWp), and central station (more than 20 MWp). Figure 16-11 illustrates the major components of a residential photovoltaic system.

## Program Strategy

Key elements of the Photovoltaic Program strategy include:

- Substantial reductions in the price of components and subsytems via
  1. Aggressive, advanced research and development to bring advanced concepts to the point of technical feasibility
  2. Intensive technology development to identify, develop, and suitably demonstrate cost-effective designs and production processes for components for which technical feasibility has been proven, thereby establishing their technology readiness
- Definition, development, design and real-world testing of complete photovoltaic systems in a series of tests and applications to demonstrate that (1) such systems are feasible by developing actual cost and performance data and (2) they are ready for commercialization
- Development of a substantial body of experience, confidence, and expertise within the private sector by both users and suppliers of photovoltaic systems.
- Careful study and implementation of commercialization strategies that will encourage market penetration and capacity

International markets potentially could fill the intermediate market "gap" between remote and grid-connected applications for photovoltaic systems in the United States. To help realize this potential and help Third World countries with their energy needs, an International Photovoltaic Plan was submitted to Congress early in FY 1980.

Overall responsibility for carrying out the above strategy rests with the DOE Assistant Secretary for Conservation and Solar Energy. Within this office, the Office of Solar Technology and the Office of Solar Applications see that the activities to support the strategy are executed. Basic research in the solar energy area is the responsibility of the Office of Energy Research.

Other participants in the program include the Solar Energy Research Institute, which is designated Lead Center for Advanced Re-

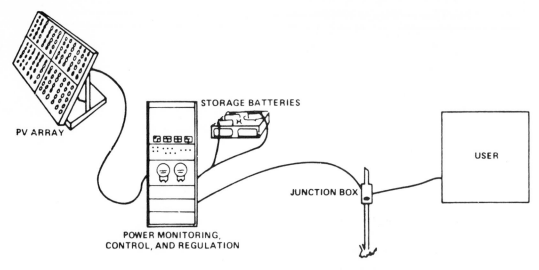

Figure 16-11. Major components of a photovoltaic system.

search and Development, and the Jet Propulsion Laboratory, lead center for Technology Development and Applications. National laboratories and principal organizations doing photovoltaic work include Sandia Laboratories, NASA-Lewis Research Center, the MIT-Lincoln Laboratory, and the Aerospace Corporation.

The photovoltaic program is directed toward the development of economically competitive, commercially available, photovoltaic power systems that provide safe and reliable energy for a wide range of applications. Distributed, grid-connected, residential systems providing both electricity and heat should be able to displace significant amounts of centrally generated electricity, first in the Southwest and, subsequently, throughout most of the United States. Intermediate-size commercial, institutional, and industrial onsite systems should provide a similar option. Finally, utilities should ultimately be able to augment their generating capacity with larger scale systems.

Initial competitive or break-even pho-tovoltaic system price goals have been established in each of three major applications areas: remote/international, residential, and utility. Achievement of the goals will mean that the photovoltaic systems have reached the point of commercial readiness in the application class. Commercial readiness for residential applications and for intermediate load center onsite applications is to be reached in 1986 and for central station applications in 1990. At these times, the federal government will choose among a number of final incentives to accelerate the market penetration of photovoltaic systems. Specific price goals are presented below in Table 16-4.

There are a number of DOE-sponsored application tests of photovoltaics. They include a remote radar unit at the Naval Weapons Center, China Lake, California; power units for six National Weather Service sites; and a system that augments an existing diesel grid to provide electricity at the Mount Laguna Air Force Station, California. Photographs of the Mount Laguna system, the largest systems installed to date, and of a system that provides

### Table 16-4. Photovoltaic Program Cost Goals (1980 Dollars).

| TIME | MODULE PRICE (FOB) | SYSTEM PRICES | PRIME APPLICATIONS |
|------|--------------------|--------------------|--------------------|
| 1982 | $2.80/Peak Watt | $6-13/Watt | Remote/International |
| 1986 | $0.70/Peak Watt | $1.60-2.20/Watt | Residences |
| 1990-2000 | $0.15-0.50 Peak Watt | $1.10-1.30/Watt | Utilities |

power to the office building at an Ohio radio station, are given in Figures 16-12 and 16-13.

Recently, DOE announced the award of nine contracts for test applications of photovoltaic projects. Included in the awards are electrical supply systems for commercial and public buildings. Typical systems are a power system for pumping water at Sea World in Orlando, Florida; an array to provide electricity to a high school in Beverly, Massachusetts; and a system to provide electricity to the Phoenix Sky Harbor International Airport.

In addition to the photovoltaic application experiments discussed earlier, the Federal Photovoltaic Utilization Program (FPUP) relies on the federal government as a market in developing the low-cost private photovoltaic production capability necessary to reduce the consumption of fossil fuels in the United States. Cycle I and Cycle II of FPUP focused on small remote applications, generally less than 1 kWp. These included forest-ranger stations and other remote sites where small isolated power systems are needed. These two cycles were initiated in FY 1978 and FY 1979 and are now in the construction phase. These photovoltaic application systems illustrate the variety of applications under consideration. Obviously, not all of them will be competitive

with other energy systems, but they do furnish information on a variety of system designs and system applications.

## Program Structure and Budget

The Photovoltaic Program has the following budget structure:

- Advanced Research and Development
- Technology Development
- Systems, Engineering, and Standards
- Tests and Applications
- Market Analysis
- Market Test and Appplications

The Office of Solar Technology is responsible for the first for subprograms and the Office of Solar Applications is responsible for the latter two.

In the Office of Solar Technology, the Advanced Research and Development subprogram brings advanced photovoltaic concepts to the point of technical feasibility. The Technology Development subprogram establishes technology readiness for promising concepts and for other system elements. The Systems, Engineering, and Standards activity provides the link between technology and systems by

Figure 16-12. Mt. Laguna photovoltaic system.

Figure 16-13. Photovoltaic-powered radio station WBNO, Bryan, Ohio.

defining system requirements, carrying out design studies, establishing standards for system evaluation, and conducting breadboard tests of systems. The Test and Applications activity carries out application experiments to explore system feasibility and system readiness experiments to prove that components that have reached technology readiness can be integrated into workable systems.

In the Office of Solar Applications, the Market Analysis activity identifies potential markets for photovoltaic systems, establishes cost goals for system performance, and analyzes the scope of incentives needed to accelerate the introduction of photovoltaics into the marketplace. The Market Tests and Applications activity furnishes policymakers and others with information on how systems are received in real-world conditions.

Table 16-5 presents funding estimates for each program element. The subsequent sections describe these program elements in detail.

## Program Detail

**Advanced Research and Development.** The Advanced Research and Development subprogram investigates concepts, materials, and structures leading to low-cost solar cells ($0.15 to 0.40/Wp module price), which

should allow a photovoltaic systems cost of $1.30/Wp or less in 1990. This will enable photovoltaics to expand emerging markets and make central utility applications cost-effective.

Proving the technical feasibility of advanced concepts, materials, and cells is a major objective of the Advanced Research and Development subprogram. To do this, it must be demonstrated that these structures have potential for high conversion efficiencies while retaining attractively low costs. Higher conversion efficiencies are also being sought with advanced cells used with concentrated sunlight.

The Advanced Research and Development subprogram includes three major elements:

- Advanced materials/cells research
- High-risk research
- Research support and fundamental studies

In the advanced materials/cells research area polycrystalline silicon, cadmium sulfide, gallium arsenide, amorphous silicon, and other materials are being investigated in order to determine which of these should be carried through to the Technology Development phase on the basis of proven technical feasibility.

High-risk research efforts are directed

## Table 16-5. Photovoltaic Energy Systems Budget.

| | BUDGET AUTHORITY (DOLLARS IN THOUSANDS) | | |
|---|---|---|---|
| PHOTOVOLTAIC PROGRAM | APPROPRIATION FY 1979 | APPROPRIATION FY 1980 | REQUEST FY 1981 |
| OFFICE OF SOLAR TECHNOLOGY | | | |
| Advanced Research and Development (AR&D) | 33,087 | 47,000 | 50,000 |
| Technology Development (TD) | 40,546 | 61,000 | 60,000 |
| Systems, Engineering, and Standards (SES) | 6,492 | 14,500 | 15,000 |
| Tests and Applications (T&A) | 23,675 | 24,500 | 15,000 |
| TOTAL OFFICE OF SOLAR TECHNOLOGY | 103,800 | 147,000 | 140,000 |
| OFFICE OF SOLAR APPLICATIONS | | | |
| Systems Development (SD) | 1,000 | 0 | 0 |
| Market Analysis (MA) | 300 | 1,000 | 400 |
| Market Tests and Applications (MTA) | 15,000 | 10,200 | 32,200 |
| Market Development and Training (MDT) | 200 | 1,200 | 0 |
| Solar International (SI) | 0 | 0 | 0 |
| TOTAL OFFICE OF SOLAR APPLICATIONS | 16,500 | 12,400 | 35,600 |

toward those materials and concepts that offer promise for eventual, efficient, low-cost photovoltaic conversion. Emerging materials (i.e., materials whose intrinsic properties indicate a potential for low-cost, greater-than-10-percent efficiency cells in thin-film form), advanced concentrator concepts, and electrochemical photovoltaic cells are investigated in this category. Also included in the high-risk research effort is the innovative-concept program designed to support exploratory research to foster new, more efficient, cost-effective approaches to photovoltaic energy conversion.

The research support effort seeks to furnish research tools and to anticipate potential problems in the advanced materials/cells research and high-risk research areas. Measurements and evaluation functions, devaluation functions, studies of technical issues, and program planning constitute this area of activity.

The primary objective of the Advanced Research and Development subprogram is to develop a workable, low-cost solar cell. Ways of raising the efficiency of solar cells are being researched; theoretically, this should lower costs and conserve materials and space. The Advanced Research and Development goals are to:

- Demonstrate 10 percent or greater thin-film solar cell efficiency in FY 1980
- Demonstrate 12 percent or greater advanced technology solar cell efficiency by the end of FY 1982
- Demonstrate technical feasibility for one advanced material by the end of FY 1982, and for a total of four by the end of FY 1986
- Demonstrate 14 percent or greater advanced technology solar cell efficiency by the end of FY 1985
- Demonstrate 30 percent or greater monolithic multijunction concentrator efficiency by the end of FY 1984

*Advanced materials/cells research.* Polycrystalline silicon, cadmium sulfide, amorphous silicon, and gallium arsenide are the thin-film cell technologies that are being stud-

ied in the advanced materials/cells research effort. Cadmium sulfide and polycrystalline silicon are the first candidates for rapid advancement to Technology Development.

The goals of cadmium sulfide thin-film studies are to achieve at least 10 percent efficiency in FY 1980. Additional goals are to demonstrate cell stability for large-area encapsulated cells and 8 percent conversion efficiency for small-scale arrays by the end of FY 1982. Concurrently, analyses evaluating the feasibility of low-cost production (less than \$0.70/Wp) are planned to further evaluate the performance and price characteristics of this technology.

Research on thin-film polycrystalline silicon is concerned with obtaining large-grain films and cell efficiencies of greater than 10 percent. Thin-film polycrystalline silicon cells hold the promise of significantly reducing the cost of photovoltaic energy conversion while maintaining the positive attributes generally associated with silicon; e.g., abundance, relatively high efficiency and stability, and the existence of a broad technology base.

Amorphous silicon is being investigated to understand the fundamentals of the defect-state passivation process, which has led to solar cells with 6 percent efficiency for hydrogenated amorphous silicon. This effort should result in the development of amorphous cells with an efficiency of 10 percent.

Gallium arsenide research concentrates on the deposition of films with large-grain sizes, obtaining an understanding of the effects of grain boundaries on cell performance and an investigation of several junction formation techniques. Because of its steep absorption edge and high single-crystal efficiency, gallium arsenide offers the potential of a 12 percent efficient thin-film cell. The objective of efforts in this area is to achieve a thin-film cell conversion efficiency of 10 percent in FY 1980 and of 12 percent by FY 1983.

*High-risk research.* Research in this area covers emerging materials, advanced concentrator concepts, electrochemical photovoltaic cells, and innovative concepts. The emerging materials are promising new materials and device concepts for photovoltaic conversion that have a potential in the long-term future. Among the materials presently being investigated are $Zn_3P_2$, CdTe, ZnSiAs, $Cu_2O$, CdS/InP, $Cu_{2=x}Se$, BaS, polyacetylene, and indium tin oxide.

Research on advanced concentrator concepts has also been initiated. Multijunction concentrator cells are being investigated for a possible efficiency of at least 30 percent. Other concentrating concepts, such as the luminescence converter, are also being studied.

The innovative-concept program issues periodic solicitations to explore photovoltaic conversion concepts that may be high risk but also offer the promise of low-cost photovoltaic systems. Projects awarded in the innovative-concept program are such that, at the end of the contract period, it is possible to project the utility of further research and to evaluate the feasibility of incorporating the new concept into other areas of the Advanced Research and Development activity.

*Research support.* Research support activities include measurements and evaluation functions, studies of technical issues, and program planning. A Photovoltaic Material/Device Evaluation Laboratory is the basis for the measurements and evaluation task. The laboratory, which will be fully operational by the end of 1980, already has begun to evaluate the technical progress of photovoltaic research subcontractors, provide measurements procedures to them, and conduct routine reviews of existing techniques, instrumentation, and developments. The technical issues area addresses problems of environmental effects, material availability, cell reliability and stability, and projected cell costs. Each of these problems is addressed individually; plans are formulated for initiating research studies where necessary.

Anticipated accomplishments for FY 1981 include:

- Demonstrate 10 percent thin-film cell efficiency for at least three cell materials or configurations

- Continue exploratory development for two advanced material technologies
- Complete the evaluation of two experimental materials for photovoltaic cell construction
- Demonstrate 10 percent electrochemical cell efficiency
- Complete environmental assessment of cadmium as a photovoltaic material

Milestones and funding for this subprogram are presented in Figure 16-14.

**Technology Development.** The Technology Development program consists of two principal activities: collector and balance-of-system technology development. Collector technology development attempts to reduce the cost of manufacturing photovoltaic cells and modules, both for flat-plate and concentrating systems. The balance-of-system activity seeks to reduce the noncollector systems costs that affect the life-cycle cost of a photovoltaic system. The strategy of the Technology Development activity is to pursue simultaneously those technology options that continue to show promise of meeting the program goals, thereby increasing the probability that at least one technology will be successful. Areas investigated in the Technology Development subprogram include:

- Flat-plate collectors and concentrator collectors
- Advanced collector technologies
- Balance-of-system components
- Total energy (hybrid) systems

Technologies for both flat-plate silicon array collectors and concentrating array collectors have demonstrated feasibility. Flat-plate silicon arrays are being developed under the Low-Cost Solar Array (LSA) Project, which addresses all steps in the array production process. These steps are the production of raw polysilicon growth of silicon sheets, creation of an individual solar cell, encapsulation, and high-volume automated array assembly. Emphasis is placed on improving quality while reducing cost during each phase. Develop-

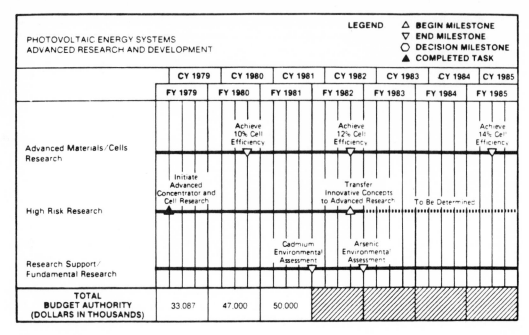

Figure 16-14. Photovoltaic Energy Systems Advanced Research and Development Subprogram milestones and funding.

mental work in concentrating systems is two-fold. The first efforts are aimed at developing more economical concentrators; the second, at producing solar cells capable of operating at high efficiencies in concentrated sunlight.

Advanced technology concepts that have demonstrated technical feasibility in the Advanced Research and Development subprogram will be carried through the developmental phase in Technology Development.

The Technology Development subprogram also is involved in the development and coordination of balance-of-system components, which include array structures and installation, power conditioners, and storage subsystems. Effort is directed toward innovative ideas, improving production techniques, and cost reduction.

Photovoltaic total energy systems, which produce both electrical and thermal energy, are expected to achieve early cost effectiveness. Total energy systems include flat-plate and concentrator arrays and can be used in residential, agricultural, industrial, and commercial applications.

Specific goals for the Technology Development subprogram include achievement of:

- $2.80/Wp collector technology readiness in FY 1980 for remote and international applications
- $2.80/Wp collector commercial readiness in FY 1982
- $0.70/Wp collector technology readiness in FY 1982
- $0.70/Wp residential and intermediate load center balance-of-system commercial readiness in FY 1986
- $0.70/Wp collector commercial readiness in FY 1986
- $0.15 to 0.40/Wp 12 to 14 percent thin-film collector technology readiness by FY 1987 for central station markets
- $.015 to 0.40/Wp thin-film collector commercial readiness established in FY 1990

Technology readiness is considered to have been achieved when all subsystems and components of a photovoltaic system, including manufacturing, have been individually demonstrated in a system that would be competitive if components were to be mass produced and when resulting prototype systems are available for testing. Commercial readiness is achieved when actual systems are offered for sale at a competitive price for a particular application.

*Flat-plate collectors.* The feasibility of low-cost silicon flat-plate collectors has been established. The goal is to demonstrate technology readiness costs below $.070/Wp, with 11 percent efficiency, by the end of FY 1982. This cost can be achieved with current technology if significant cost reductions are realized in the silicon raw material and development of automated production facilities. This cost reduction will be accomplished through Experimental Process System Development Unit design, test, and evaluation for silicon material refinement to achieve production costs of less than $14 per kilogram. A similar experimental process unit will be established for ribbon pullers. Analyses suggest that significantly lower costs than $0.70/Wp are possible with silicon technology.

*Concentrator collectors.* Progress in concentrator collector technology will be aided by advances made in silicon and single-crystal gallium arsenide devices. It is anticipated that concentrator designs will result in overall collector prices equal to or less than flat-plate prices, including the added cost of tracking. The objective is to establish a $2.80/Wp concentrator cost by FY 1982. More advanced tracking concentrator designs incorporating higher efficiency solar cells could provide further price reductions. A $0.70/Wp concentrator collector cost is targeted for FY 1982, with commercial readiness in FY 1986. A goal of $0.50/Wp for concentrator collectors is planned for FY 1986 with commercial readiness in FY 1990.

*Advanced collector technologies.* After technical feasibility is established in the Advanced Research and Development subprogram, two classes of advanced collectors will be pursued. The goal will be to achieve technical readiness by FY 1988 with a cost target from $0.15/Wp to 0.40/Wp. The classes include (1) advanced thin-film materials such as calcium sulfide and polycrystalline, and/or amorphous silicon, and (2) advanced concentrator concepts using multijunctions and spectrum splitting.

*Balance-of-system components.* There are numerous components other than the collector within a photovoltaic system. These include land, structure, direct and alternate current wiring, power conditioning and control, and energy storage. As the cost of collectors decreases, the cost of the balance-of-system components becomes more important. Numerous projects are under way to ensure that these balance-of-system costs are low enough so that photovoltaic systems will penetrate the marketplace when the collectors and modules reach commercial readiness. A comprehensive Technology Development plan will be completed in late FY 1980.

*Total energy (hybrid) systems.* Total energy or hybrid systems, which produce both electrical and thermal energy, have a potential for lower initial cost than separate thermal and photovoltaic systems because of savings in collector materials, support structures, land use, installation, and maintenance costs. Technology readiness for these systems is targeted for FY 1982 and commercial readiness in FY 1986. Both flat-plate and concentrating total energy systems are being pursued.

Anticipated accomplishments for FY 1981 include completion of:

- Automated module assembly prototype equipment design
- Prototype silicon sheet equipment design
- Module encapsulation process
- Conceptual designs for advanced power conditioners

Milestones and funding for this subprogram are given in Figure 16-15.

**Systems, Engineering, and Standards.** The Systems, Engineering, and Standards subprogram identifies and studies the requirements for photovoltaic systems in residential, community, and intermediate load center and central station applications. This includes sys-

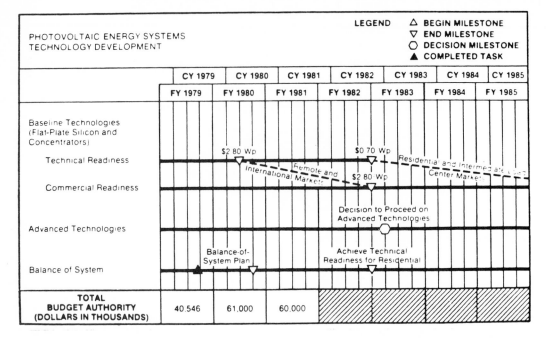

Figure 16-15. Photovoltaic Energy System, Technology Development Subprogram, milestones and funding.

tem and design analyses, breadboard and prototype system fabrication and testing in controlled environments, and the establishment of performance and reliability standards. This subprogram involves work in three areas: (1) system definition, (2) system development, and (3) performance criteria and test standards. The system definition element determines performance and interface requirements of future photovoltaic applications. This requires analysis of institutional and legal constraints and the characteristics of the competing technologies. System definition also furnishes system design guidelines, technical performance requirements, and cost goals.

System development includes photovoltaic system prototype development, testing, and evaluation of installation and maintenance techniques. Proven photovoltaic system designs will be made available for additional evaluation in application experiments and commercial systems.

The performance criteria and test standards area develops necessary criteria and standards to ensure nationwide uniformity in the specification of photovoltaic products and design

processes. Representatives from professional, trade, and consumer organizations assist in the standardization process. Other activities include recommendations for accreditation of test laboratories and product certification.

Specific goals for the Systems, Engineering, and Standards subprogram include:

- Establishment of component, subsystem, and system technology standards by coordinating photovoltaic system technology development with application requirements and objectives
- Development and implementation of a uniform system definition methodology to provide specific engineering component, subsystem, and system definitions, as well as to establish system application requirements and interfaces
- Provision of necessary breadboard and prototype system testing capability for the program as a whole and for the Tests and Applications subprogram in particular
- Establishment and advocacy of industrywide photovoltaic performance criteria and standards, including system

reliability guidelines and quality assurance in four applications areas: remote, residential, intermediate, load centers, and central station

For FY 1981, anticipated activities include continued work on interfacing photovoltaic systems with other energy systems and determining final performance criteria for selected photovoltaic systems. Milestones and funding

for this subprogram are presented in Figure 16-16.

Tests and Applications. The Tests and Applications subprogram obtains operational experience with complete photovoltaic systems in a wide range of applications, including experiments in remote, residential, intermediate load center, and central station applications. In the last three experimental areas, interac-

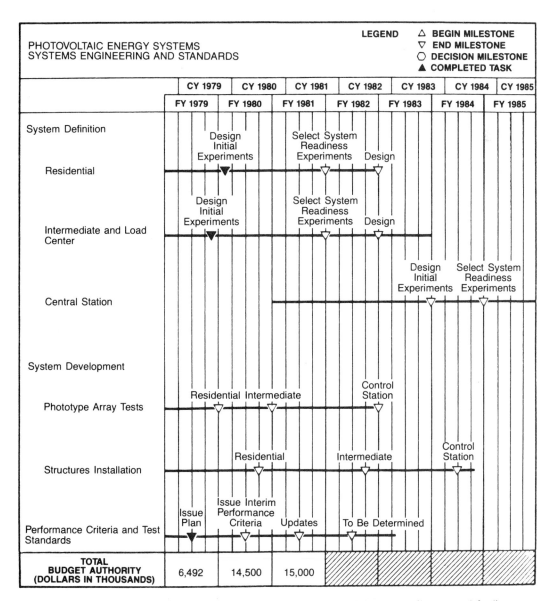

Figure 16-16. Photovoltaic Energy Systems Engineering and Standards Subprogram milestones and funding.

tion with electrical utility grids will be emphasized. The Tests and Applications activities have provided early evidence of reliable photovoltaic system operation under a variety of adverse climatic conditions. In addition, photovoltaic application experiments conducted in the Tests and Applications subprogram provide valuable information in planning and designing international application experiments in developing countries. There will be several such international application experiments jointly conducted by the United States and foreign governments as a result of the United States recent agreement with Saudi Arabia–Italy.

The Tests and Applications subprogram experiments are divided into two activities:

- Initial system evaluation experiments
- System readiness experiments

The initial system evaluation experiments activity is concerned with evaluating the performance and reliability of photovoltaic components and systems in specific applications. For each experiment, data are collected in areas such as environmental acceptability, maintainability, durability, operability, operational safety, and cost.

The same objectives are also pursued in the system readiness experiments. However, the photovoltaic systems evaluated in this area are composed of components and subsystems considered to be prototypes of commercially producible systems based on design data acquired in the initial system evaluation experiments. Successful completion of a system readiness experiment will indicate that the system can be manufactured successfully to compete in the marketplace, and thus the application will be judged to have achieved commercial readiness. On the other hand, problems would indicate a need for more development work.

Specific goals for the Tests and Applications subprogram include:

- Demonstrate system feasibility for residential and intermediate load center applications in FY 1982
- Demonstrate system readiness for resi-

dential and intermediate load center applications beginning in FY 1984 with a $1.60/Wp photovoltaic-system price
- Demonstrate system feasibility for central stations by FY 1985
- Demonstrate system readiness for central stations beginning in FY 1989 with $1.10 to 1.30/Wp photovoltaic-system price range

*Initial system-evaluation experiments.* Initial system-evaluation experiments develop basic information about the suitability and acceptability of photovoltaic power systems in various application areas, operation and maintenance of photovoltaic systems in a realistic environment, and actual costs of installation and operation. Initial system evaluation experiments currently under way are listed in Table 16-6.

These projects were scheduled to be operational in FY 1980. Initial design efforts for residential projects began in FY 1980, and construction was scheduled to be completed in FY 1981. The initial effort for central stations will be devoted to component and subsystem testing in a central station test facility, and a preliminary design of the first experiment will begin in FY 1981. Construction will start in FY 1983; projects will be operable the following year.

*System readiness experiments.* The main purpose of the system readiness experiments is to prove that technology-ready components and subsystems from the Technology Development subprogram can be assembled into workable and cost-effective complete photovoltaic systems. These experiments will be designed using performance data from the initial systems evaluation experiments.

For both residential and intermediate load center applications, technology-ready modules and other components are expected to be available in FY 1982. Accordingly, the system readiness experiments projects in these two applications will be in the design phase by FY 1982, with installation scheduled to take place in FY 1982–FY 1984. By the end of FY 1984, system readiness for $1.60/Wp

Table 16-6. Initial Systems Engineering Experiments.

| SYSTEM | LOCATION | TYPE | SIZE (KWp) | FUNCTION | ESTIMATED OPERATION |
|---|---|---|---|---|---|
| Mississippi County Community College | Blytheville, AR | Concentrator | 250 | Total energy system for college building | May 1980 |
| Northwest Mississippi Junior College | Senatobia, MS | Flat Plate | 200 | Electricity for college | May 1980 |
| Newman Power Station | El Paso, TX | Flat Plate | 18 | Electricity | Nov 1980 |
| Okahoma Center for Science and Arts | Oklahoma City, OK | Flat Plate | 150 | Electricity | Nov 1980 |
| Shopping Center | Lovington, NM | Flat Plate | 150 | Electricity | Dec 1980 |
| Beverly High School | Beverly, MA | Flat Plate | 150 | Electricity | Dec 1980 |
| Dallas-Fort Worth Airport | TX | Concentrator | 27 | Total energy for utility system | Nov 1980 |
| BDM Building | Albuquerque, NM | Concentrator | 47 | Electricity | Nov 1980 |
| G.N. Wilcox Hospital | Kauai, HI | Concentrator | 85 | Electricity | Nov 1980 |
| Sea World | Orlando, FL | Concentrator | 110 | Total energy | Mar 1981 |
| Sky Harbor International Airport | Phoenix, AZ | Concentrator | 225 | Electricity | Mar 1981 |

residential systems and intermediate load center systems will be achieved. Subsequent system readiness experiments will incorporate lower cost systems.

For central-station applications, it is expected that $0.15 to $0.40/Wp collectors will be required for cost effectiveness in most parts of the country. However, technological readiness for the $0.15 to $0.40/Wp collector is not expected until FY 1987. Central-station projects are scheduled to be in the final design phase in that year, with partial operation in FY 1988. The full plant will be complete and fully operational in FY 1989.

Anticipated FY 1981 activities include completing construction of committed projects as shown in Table 16-6. Milestones and funding for this subprogram are shown in Figure 16-17.

Market Analysis. In general, the marketing strategy for photovoltaic systems is to define the characteristics (price, performance, system configuration) of a photovoltaic system necessary for a particular market sector, such as developing countries or United States residences. The Photovoltaic Program will identify early or "best" markets for the product and begin market development activities, such as market tests and information dissemination, in those best market areas.

Market research establishes specific cost and performance goals that products must meet in order to compete successfully in the marketplace. This activity is supported best on the federal level by carefully selected and well-focused market studies that result in a realistic appraisal of the marketplace. This is a continuing process, with market knowledge and cost and performance goals sharpening throughout the product development process.

Major markets for photovoltaic power systems can be subdivided into stand-alone, or off-grid, applications and grid-connected applications. Stand-alone applications consist of two segments: the present market for photovoltaics and the intermediate market.

The present-market segment includes only those products that are cost-effective or so close to being economic that commercial sales are under way and expected to grow in the future. This market is for remote applications, which have no access to grid power and which have power requirements of a few kilowatts or less—for example, cathodic protection of bridges and telephone-relay stations.

The intermediate-market segment consists of larger off-the-grid applications (i.e., 5 to 25 kW such as water pumping or village power, which do not have a substantial, recognized market but have the potential to support large photovoltaic sales prior to the

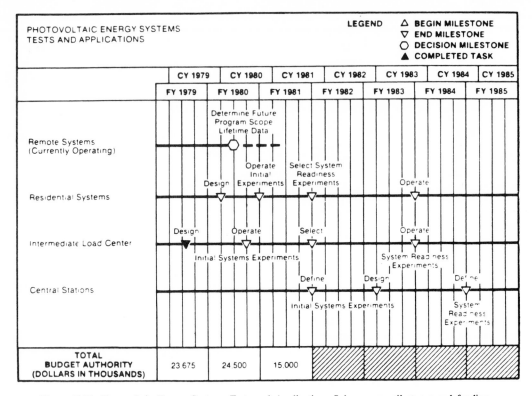

Figure 16-17. Photovoltaic Energy Systems Tests and Applications Subprogram milestones and funding.

development of grid-connected markets. Primary market sectors for off-grid applications are believed to be the federal government and developing countries that have not established extensive central-station power grids.

Specific thrusts of the Market Analysis program in the development of stand-alone photovoltaic systems will include:

- Complete market assessments of the federal, private and international markets
- Complete international program plan for photovoltaic systems
- Establishment of near-term price goals
- Complete multiyear system procurement strategy

The grid-connected market consists of two distinct application segments: (1) distributed applications connected to a utility grid, and (2) central-station electricity-generating facilities operated by utility systems. The distributed applications include private residences

(primarily single-unit), commercial and industrial establishments, and institutions such as schools and hospitals. It has been assumed that residential structures would operate without onsite storage, using the grid as a backup resource, although it is recognized that ongoing technical development as well as user requirements and preferences would make onsite storage a feasible option.

The market will comprise residential, commercial/industrial, and utility sectors in the United States and other developed and developing countries. Specific thrusts of the market analysis activities in the development of photovoltaic systems will include:

- Assess market requirements for product/process development—i.e., verification of price goals for residential, intermediate, and central-station applications; verification of design specifications based on user institutional requirements
- Identify distribution/marketing in-

frastructure requirements needing DOE price goals—i.e., installation volume required per region for application

- Verify "best and next best" regions for market penetration, both residential and those leading to utility markets
- Develop market-stimulation strategies on regional basis for commercial/industrial and utility sectors
- Complete barriers and incentive recommendation study for grid applications
- Develop procedures for cost-shared market field tests

**Market Tests and Applications.** The major activity of the Market Tests and Applications subprogram has been the Federal Photovoltaic Utilization Program (FPUP). Other Market Tests and Applications activities include the development of cost-sharing procedures for financial assistance to private and public photovoltaic users and the gathering of marketing information on early residential applications. Beginning in FY 1981, a substantial multiyear photovoltaic systems purchase program will be initiated.

*Federal photovoltaic utilization program.* The Department of Energy Act of 1978 Civilian Applications, Public Law 95-238, Sec. 208, authorized $13 million in FY 1978 to acquire life-cycle, cost-effective photovoltaic systems for federal facilities. That program was greatly augmented by the establishment of the Federal Photovoltaic Utilization Program (FPUP) under title V, part 4, of the National Energy Conservation Policy Act (Public Law 95-619), which authorized the appropriation of an additional $98 million for FY 1979 through FY 1981.

As indicated in the Solar Photovoltaic Energy Research, Development, and Demonstration Act of 1978 (Public Law 95-590), the key objectives of FPUP are to:

- Accelerate the growth of a commercially viable and cooperative industry to make photovoltaic solar electrical systems available to the general public as an option in order to reduce national consumption of fossil fuel
- Reduce fossil fuel costs to the federal government
- Stimulate the general use within the federal government of methods for the minimization of life-cycle costs
- Develop performance data on the program

To achieve the objectives established by Congress, the FPUP must stimulate industry and market development. It must support the applications that have been identified as having significant market potential in the private domestic arena and in foreign countries as well as in the federal agencies.

The FPUP is being conducted in five cycles, depending upon availability of funds, as summarized in Table 16-7.

The FPUP has funded 3,150 cost-effective photovoltaic systems in 11 federal departments at a cost of $19,559,190 for Cycles I, II, and III. Every state is represented by applications except North Dakota. There are also applications in the West Indies, the Pacific Islands, and Europe. Cycle IV will begin early in 1980, and Cycle V will be implemented in FY 1981, depending on appropriation of funds.

The Advisory Committee established by Public Law 95-590 has been chartered, and the members named. A rule for the monitoring and assessment of systems installed under the program was published in final form on November 7, 1979.

The program will be closely coordinated with the activities and experiments under way under Public Law 95-590. Technical support will be provided to the agencies that will draw on the technical expertise gained by both industry and DOE field centers in prior applications.

*Photovoltaic systems multiyear purchase strategy.* In FY 1981, DOE will initiate the first segment of a system procurement program that is planned to continue through 1985. This program, which will constitute the core of the DOE market development plan for photovoltaics, will be structured to encourage

Table 16-7. FPUP Application Cycles.

| FISCAL YEAR | CYCLE | MAJOR APPLICATIONS | COST EFFECTIVE |
|---|---|---|---|
| 1978 | I | Small Remote | Now |
| 1979 | II | Small Remote | Early 1980's |
| | III | Intermediate Remote | |
| 1980 | IV | Intermediate Remote | Mid 1980's on |
| | | Residential | |
| 1981 | V | Selected Intermediate Grid Connected | |

development of a supply and delivery infrastructure, as well as to provide support to the photovoltaic industry. Competitive procurement requests will be issued to fund several companies or teams to provide a few experimental systems that will undergo engineering field tests. A second phase of competition will then be announced to fund a larger number of the systems that were successful in the first phase. The cost of Phase 2 systems must be shared by the supplier and will be used in market tests in the most advantageous geographic areas. Bidders in both phases will be required to offer a complete package, including a marketing plan, prospective private- and public-sector users for the systems, and arrangements with local utility companies for a grid tie-in if applicable. The number of funded suppliers will depend upon the level of appropriations for the program in each fiscal year.

Funded systems will be used in a variety of applications for international, residential, industrial/commercial, and selected central-station markets. The purchase program will be accompanied by other market-development activities such as Market Analysis and education.

## WIND ENERGY CONVERSION SYSTEMS

One promising way to tap the sun's energy is to use wind systems. A small portion of solar energy received by the earth is converted into surface winds as a result of uneven heating of the atmosphere. Natural forces tend to concentrate this resource so that a reasonably windy site has roughly the same annual energy available per square foot of collector as a good insolation site. This energy source is continually renewable and wind power systems are not expected to have major adverse effects on the environment.

Wind energy systems have been in use for many centuries, traditionally for irrigation and milling operations. Since the turn of the century, wind energy systems also have been used for generating electrical power. Generally, these systems have been small, ranging in output from a few watts to a few kilowatts. Their commercial availability has declined since the advent of inexpensive and reliable power from the Rural Electrification Administration in the 1930s. The notable exception to the smaller machines was the 1.25-MW Smith-Putnam wind turbine located at Grandpa's Knob, Vermont, which fed power into the local grid from 1943 until April 1945.

A number of intermediate-sized units were tested in Europe from the 1930s through the 1950s. The low cost of fossil fuels and the promise of nuclear power in the post–World War II era caused the worldwide demise of interest in wind power until the mid-1970s.

Wind systems convert the kinetic energy of wind into mechanical power. In addition to widespread use for mechanical water pumping, electrical power generation is generally considered to be the principal application for wind systems, although heating, cooling, and other applications are also practical. Small-scale systems (1 to 100 kW) are expected to be employed onsite for residential and farm uses; intermediate-scale systems (100 to 1000 kW) for larger farms, irrigation, small utilities, and remote communities; and large-scale systems (greater than 1 MW) for electrical utility and industrial uses. The federal Wind Program is developing machines in these general classes for residential, agricultural, industrial, and utility sectors.

An energy cost goal of $0.03/kWh to $0.04/kWh ($1980) has been established for both small and large wind systems. This goal is levelized, life-cycle energy cost. It is anticipated that an initial market will begin to form at about $0.04/kWh to $0.07/kWh, a level sufficient to support production of early systems in moderate quantities.

Large-scale systems are expected to compete first in utilities that are heavily dependent on oil and also in areas with a large hydroelectricity capability, where the conventional system can serve as a backup. Later penetration into coal and nuclear systems is expected as costs of wind turbines are reduced through mass production and advanced designs and as capital and fuel costs escalate. Intermediate-scale systems will be competitive for onsite use by industry or small utilities at roughly the same costs. At oil costs of $30 per barrel, fuel costs of $0.05/kWh are now common for utilities heavily dependent on oil, such as in Puerto Rico and Hawaii. Principal competition for small-scale, onsite wind systems is utility-generated electricity where initial competitiveness is dependent on local cost of power and the rate structure, including utility fuel costs, block rate structures, and any demand charges to cover the cost of providing power when the wind system cannot meet instantaneous user demands. A market for small wind systems is projected for remote applications at an energy cost of $0.06/kWh to $0.10/kWh for large-scale use.

The backup power and demand charge uncertainties will be resolved as data are acquired from wind systems in actual use, and as national energy policy is implemented. Under the Public Utilities Regulatory and Policy Act of 1978, the DOE has proposed voluntary rate standards for solar, wind and other renewable systems that would prohibit unfair demand charges and other discriminatory practices. In addition, the Federal Energy Regulatory Commission has proposed standards for determining backup power, sellback, and interconnection costs for small power producers, including wind systems. If utilities do not impose demand charges on small-systems users to cover the cost of providing backup power, the cost goal for small systems may be relaxed somewhat. This would represent a much less severe technical challenge in the development of commercially feasible systems.

The solar energy DPR projected a wind energy contribution of 0.9 quad of conventional energy displaced in a Base Case in the year 2000, and a 1.7-quad impact in a Maximum Practical Case with an increased level of federal and state activities. The DPR Maximum Practical Case would use a mix of 30,000 to 40,000 large machines (1 to 2 megawatts-electric (MWe) and 1- to 2-million 1-to-10 kW small machines. This contribution is predicated on a successful R&D program to reduce energy costs and provide reliable machines, resolution of institutional barriers, and availability of appropriate incentives.

Wind systems are generally classified as horizontal or vertical axis machines. A typical horizontal axis machine is the 200-kW Mod-OA experimental machine shown in Figure 16-18. This machine swings (yaws) to follow changes in wind direction.

Under development is the Darrieus vertical axis machine—or "eggbeater"—that operates with the wind from any direction (see Figure 16-19). The main components of either type of machine are the rotor blades, gear train, electricity generator, structure, and tower. The principal engineering challenges are to design wind systems that can be manufactured cheaply, perform reliably, and capture a significant fraction of available energy.

The focus of the Wind Program is to achieve cost reductions in wind machines through:

- Increased capture of available energy in the wind
- Improved aerodynamic properties
- Improved control systems

Figure 16-18. MOD-OA wind turbine operating as part of the Clayton, New Mexico, municipal electric systems.

- Cost and weight reductions due to (1) advanced blade materials and fabrication techniques, (2) better understanding of wind loads and fatigue (crack) resistance, (3) better systems designs, and (4) manufacturing improvements

The strong dependence of these systems on wind velocity requires a thorough understanding of wind behavior in order to select appropriate wind machine sites. The wind in the atmospheric boundary layer closest to the ground is a composite of many different components or scales of air motion, which are both space- and time-dependent. Typical examples of such scales of winds for the United States are the trade winds, the prevailing westerlies, traveling synoptic systems, and breezes of various kinds. Further, the surface of the earth is not homogeneous as far as roughness and heat exchange are concerned.

All these factors contribute to making the wind a variable, complex kinetic energy source. Generally, the greatest wind energy potential is found over extensive flat land (Great Plains), along coasts (Northwest and Northeast), around islands (Caribbean and Hawaiian), where roughness is low and the surface friction is less, or where terrain features accelerate airflow. Great energy potential is also available at exposed higher elevations such as in the Rocky and Appalachian Mountains. Considerable deviations from these general situations occur on local time and space scales, which frequently are difficult to detect without direct measurements.

## Program Strategy

The objective of the Wind Energy Conversion Systems (WECS) Program is to accelerate the

Figure 16-19. Darrieus machine being tested at Sandia Laboratory, New Mexico.

development of reliable and economically competitive wind energy systems to enable the earliest possible commercialization of wind power. To achieve this objective, it is necessary to advance the technology, develop a sound and competitive industrial base, expand user awareness and acceptance, and address those nontechnical issues that could delay the introduction of new wind machines. Thus, the program's primary challenge is the development of economical wind systems capable of providing as many as 30 years of reliable and safe service.

Because of the relatively advanced state of aerodynamic technology, basic research is not stressed in the WECS Program. Instead, the program centers on mission and applications analyses, the development of more effective machines, testing prototype machines in util-

ity and user markets, and the accumulation and analysis of site-specific wind data. The basic wind technology features horizontal axis, propeller-type systems, although the DOE program is investigating the competitiveness of the Darrieus-type vertical axis machine. Also, many innovative concepts are now being investigated, and several of them have successfully entered the competitive small-systems development cycle.

The Wind Program strategy includes several key efforts:

- Undertake research and technology development leading to a series of progressively advanced systems with lower cost
- As machine and operating costs reach competitive levels for initial markets,

phase in demonstrations to foster user awareness and to stimulate an initial industrial production capability

- As costs approach the competitive level in the general market, replace the technology development and demonstrations with user incentives to stimulate adoption
- Resolve the institutional uncertainties of onsite wind machines, including backup and buyback rates, land use, zoning, financing, and reliability
- Provide planning and siting tools to permit users to assess the merits of wind systems in their applications

The progress of the technology and system development subprograms for large wind systems is illustrated in Figure 16-20. The MOD-1 is first-generation, 2000 kW experimental machine now operating near Boone, North Carolina. Even if produced in commer-

cial quantities (hundreds of units) and operated in areas with average wind speeds of 14 to 16 mph, a machine of the MOD-1 design would be competitive only in limited circumstances. Experience in design of that machine was used to design the MOD-2, a potentially commercial machine that is projected to be competitive in a significant fraction of the market. Similarly, MOD-2 design and operating experience will be used to design advanced machines, scheduled for initial operation in the mid-1980s, which should be economically preferable over larger markets if produced in quantity.

The DOE Wind Systems Program is managed at the program level from DOE headquarters, while project-level management is decentralized through several DOE field offices, and national, NASA, and Department of Agriculture laboratories. In turn, these agencies manage contracts that are issued to universities and industry where the majority

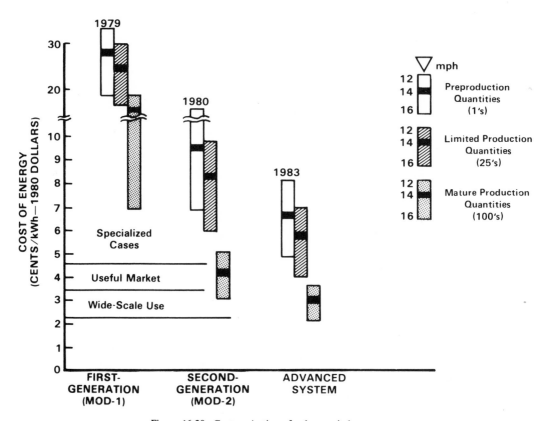

Figure 16-20. Cost projections for large wind systems.

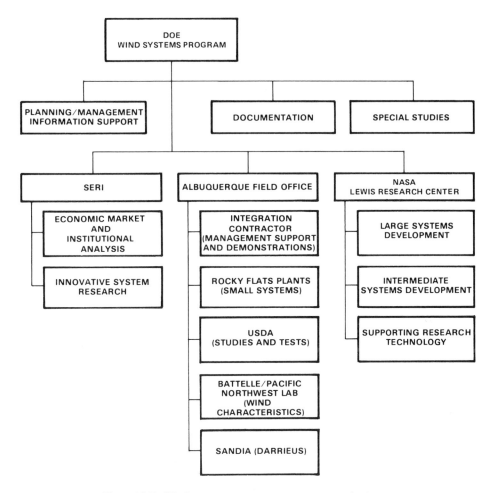

Figure 16-21. Wind systems program management organization.

## Program Structure and Budget

The Wind Systems Program budget has been restructured to reflect the functional approach to wind energy systems development. Table 16-8 shows the major subprogram activities and major products.

The estimated funding for the Wind Program is shown in Table 16-9. The FY 1981 estimate of $80.0 million is an increase of $16.6 million over the FY 1980 appropriation including FY 1980 construction funds. It includes funds to complete development of first-generation small wind machines and to continue development of advanced intermediate- and large-scale wind machines.

## Program Detail

The DOE Wind Systems Program is managed by the Wind Systems Branch in the Office of Solar Technology. In addition, a DOE resource manager for wind systems has been appointed to stimulate adoption of this technology.

**Research and Analysis.** The objectives of the Research and Analysis subprogram are to determine wind energy economics and markets on national, regional, and sector bases and to disseminate information to potential users. This subprogram also explores the

Table 16-8. Wind Energy Program Objectives, Activities, and Key Products.

| SUBPROGRAM | OBJECTIVES | ACTIVITIES | KEY PRODUCTS |
|---|---|---|---|
| Research and Analysis | Assess economics of wind energy<br>Assess barriers to acceptance<br>Determine potential of innovative concepts<br>Resolve environmental issues<br>Assess incentives<br>Study market needs | Marketing assessments<br>Incentives studies<br>Economic analyses<br>Evaluation of innovative concepts<br>Environmental monitoring and analysis | Information for decision-makers and users<br>Estimates of market size vs. cost<br>Concepts with potential for development<br>Environmental design handbooks |
| Wind Characteristics | Improve capability of evaluating good wind sites<br>Provide design requirements· for WECS<br>Develop wind survey techniques<br>Improve operational use of WECS | Collect wind characteristics data and develop wind models<br>Develop wind forecasting models | Large area wind survey techniques<br>National wind resource assessment<br>Performance and design data<br>Siting handbook |
| Technology Development | Reduce system costs<br>Improve performance<br>Provide analytical tools and component performance data to WECS producers and users | Components development research and tests<br>Systems designs | Higher performance, more reliable, lower cost systems and components<br>Improved and validated design models |
| Engineering Development | Determine operating and economic characteristics of user operated WECS<br>Reduce systems costs and improve performance of WECS | Design, fabricate, install, and operate WECS intertied with utilities<br>Improve systems design through use of technological advances | Utilities gain user experience<br>Improved systems performance and reduced costs<br>Validated WECS designs |
| Implementation and Market Development | Reduce cost through production efficiencies<br>Establishment of a commercialization infrastructure<br>Promote market demand growth<br>Reduce institutional barriers to WECS users | Initiate market buys for demonstration projects<br>Recommend incentives package for manufacturers and users<br>Set precedent for WECS assessments and standards development | High-efficiency competitive production facilities<br>User acceptance of WECS commercialized wind technology contributing to national energy |

potential of innovative wind system concepts such as augmentor and vortex systems. The anticipated activities emphasize establishing firm estimates of wind system markets and the relationship of these markets to required wind system costs. The funding also provides for detailed analyses of wind energy markets to determine technical, economic, and institutional problems encountered by the wind system owner. Research on environmental, social, and legal issues is included in this subprogram, as well as planning support for the overall program.

Current achievements include the initial Wind Systems Program mission analyses, completion of several market studies, and development of analytic tools to permit users to evaluate wind systems for specific applications. Because of the nature of the wind resource, these analytical tools required development. Currently underway are an analysis of the value of wind systems to rep-

## Table 16-9. Wind Energy Conversion Systems Budget.

| WIND ENERGY CONVERSION PROGRAM | BUDGET AUTHORITY (DOLLARS IN THOUSANDS) | | |
| --- | --- | --- | --- |
| | APPROPRIATION FY 1979 | APPROPRIATION FY 1980 | ESTIMATE FY 1981 |
| OFFICE OF SOLAR TECHNOLOGY | | | |
| Research and Analysis | 6,466 | 8,739 | 7,700 |
| Wind Characteristics | 4,447 | 5,401 | 6,200 |
| Technology Development | 9,925 | 13,181 | 17,100 |
| Engineering Development | | | |
| Small Systems | 8,095 | 7,700 | 10,000 |
| Intermediate Systems | 6,200 | 2,300 | 14,800 |
| Large Systems | 19,892 | 24,071 | 24,200 |
| Total Engineering Development | 34,187 | 34,071 | 49,000 |
| Implementation and Market Development | 4,530 | 2,008 | 0 |
| TOTAL | 59,555 | 63,400 | 80,000 |

resentative utilities, using wind data from 17 sites across the nation, and a more extensive characterization of the wind energy market. Proposed FY 1981 activities include:

- Market and impact analysis
  1. Complete estimates of wind market size as a function of wind energy cost
  2. Assess environmental and institutional issues
  3. Develop and improve economic and planning handbooks for small and large wind systems
  4. Analyze incentive options and alternative program strategies
- Innovative Concepts
  1. Evaluate diffuser-augmented wind turbine
  2. Explore other new concepts
  3. Refine cost-estimating techniques for innovative concepts
- Information dissemination
  1. Provide information to individuals, utilities, and state and local governments
  2. Develop information packages for targeted audiences

Milestones and funding for the research and analysis subprogram are presented in Figure 16-22.

Wind Characteristics. Wind energy is not uniformly distributed over the nation, nor is it available continuously. Moreover, it is strongly dependent on site topography, vegetation, and seasonal variation. The potential economics of wind systems depend strongly on the wind resource. This subprogram will provide wind characteristics information that directly affects the siting, reliability, operation, and economics of both large and small wind systems. The estimated funding of $6.2 million for FY 1981 reflects an increase of $799,000 over that of FY 1980. This increase is the result of added emphasis on developing and improving wind survey techniques; reassessment of identified but uncertain potentially high wind areas on a scale useful to farm or rural users and to utilities; and expanded evaluation and data collection at candidate utility test sites.

Major accomplishments of this subprogram include publication of a regional survey for the Pacific Northwest, and work on remaining regions is well under way to complete a national survey report by mid FY 1981. A siting handbook for small systems has been published, and a similar handbook for large machines will be published in FY 1980. Planned FY 1981 activities include:

- Wind energy prospecting
  1. Improve survey techniques to iden-

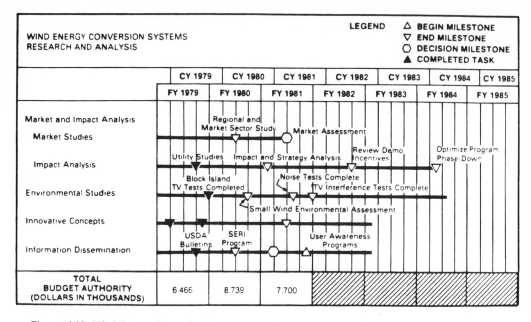

Figure 16-22. Wind Energy Conversion Systems research and analysis subprogram milestones and funding.

tify favorable sites for small and large machines

2. Refine siting techniques to incorporate experience with machines in the field
3. Conduct meteorological measurements at additional unidentified high-potential sites

- Support for design and operation of wind systems
  1. Continue evaluation of wind-gust models
  2. Improve analysis of wind shear, gust loading, and variations to improve machine operations
  3. Investigate short-term wind-forecasting techniques to improve electrical-utility load dispatching

- Site evaluation
  1. Evaluate wind data from the candidate utility system site
  2. Assist in selecting specific sites for field testing for both small- and large-scale systems

Milestones and funding for the Wind Characteristics subprogram are shown in Figure 16-23.

**Technology Development.** The Technology Development subprogram seeks to improve the performance and lower the cost of wind systems and their mechanical and electrical components. The estimated funding of $17.1 million for FY 1981 represents an increase of $3.9 million over that of FY 1980, a result of increased emphasis on subsystem and component R&D activities. A principal activity is the operation and maintenance of the small-systems test center at Rocky Flats, Colorado, for long-term testing of commercially available, privately and federally sponsored prototype systems. The Rocky Flats Test Center tested 18 small systems in FY 1979 and will be testing more than 50 by FY 1981. Some state programs for tax credits or other incentives are predicated on the activities at the Rocky Flats Test Center. Small machines must be tested there in order to receive certification as to ratings and reliability that qualify them for such benefits. Rocky Flats also plays a critical role in the development of standards for small machines. In addition to its long-term testing capability, Rocky Flats recently has developed a capability to conduct quick performance tests on small systems through the use of a controlled-velocity test

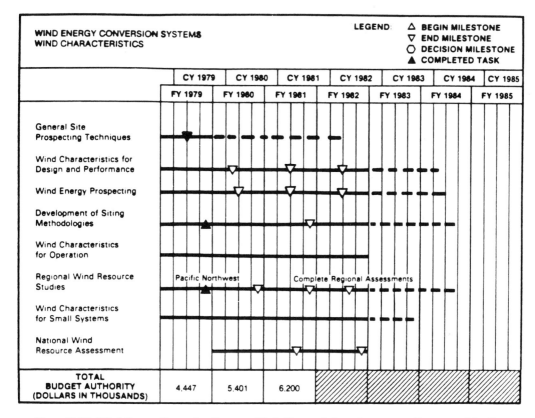

Figure 16-23. Wind Energy Conversion Systems, Wind Characteristics subprogram, milestones and funding.

track made available at the Department of Transportation, in Pueblo, Colorado.

This subprogram also provides for operation of the MOD-O test facility at Plum Brook, Ohio, whose focus is on developing and testing the advanced blade and control systems that are the principal cost elements in intermediate and large wind systems. Tests are scheduled for improved drive trains, generators, and wood, fiberglass, and steel blades made from advanced materials through improved fabricating techniques. Improved components and subsystems will be tested on the MOD-0A and MOD-1 machines.

Planned FY 1981 activities include:

- Small Systems Technology (Rocky Flats)
  1. Failure mode and effects analyses of small wind systems
  2. Trade-off and technology studies to identify design requirements

3. Testing of advanced components and commercially developed systems for different environmental and operating conditions

- Intermediate and Large Systems Technology
  1. Continued research on blades and subsystems
  2. Development and publication of engineering design handbooks covering performance prediction and structural, aerodynamic, and safety-environmental considerations to assist manufacturers and designers
  3. Development and publication of design handbooks for erection, installation, and operation-maintenance of intermediate and large systems
  4. Development and publication of computerized design model for intermediate-scale machines

5. Investigation of lightweight, high-efficiency drive train systems, including gearboxes

6. Incorporation of improved components and modifications for existing experimental systems based on technological developments and test results

9. Configuration studies to determine the system design and cost implications of advanced components and ideas

- Evolving Technologies
  1. Investigation of structural dynamics of Darrieus vertical axis wind turbine
  2. Continued analysis of 17-m Darrieus machine

Milestones and funding for this subprogram are shown in Figure 16-24.

Engineering Development. The Engineering Development activity provides for design and development of advanced small (1 to 100 kW), intermediate (100 to 1000 kW), and large (greater than 1 MW) wind systems. As a general principle, advanced machine designs are sought through competitive procurements, with more than one award, if possible, to allow competition through the design and fabrication process.

This activity also provides operational testing of several ongoing intermediate and large machine field tests of experimental machines in various utility applications:

- MOD-0A (200 kW rated): Clayton, New Mexico; Culebra, Puerto Rico; Block Island, Rhode Island; Oahu, Hawaii, to be dedicated in summer 1980.
- MOD-1 (2 MW rated), Boone, North Carolina.
- MOD-2 three unit cluster (each unit is 2.5 MW rated) Goldendale, Washington. The first unit began operation in the fall of 1980.

The FY 1981 funding estimate will provide

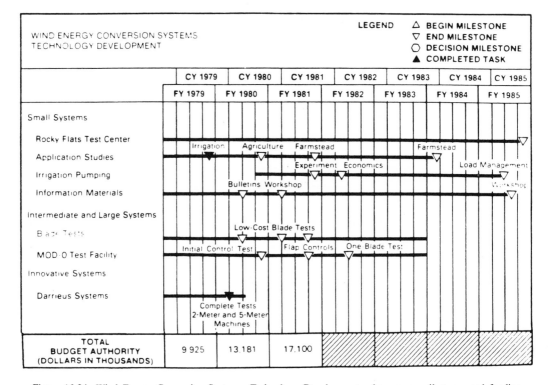

Figure 16-24. Wind Energy Conversion Systems, Technology Development subprogram, milestones and funding.

primarily for two competitive contractors for systems development of advanced intermediate-scale and advanced large-scale (greater than 1 MW) machines. This approach will increase the probability of achieving cost goals and ensure having at least two wind system manufacturers for intermediate- and large-scale machines and several manufacturers for small systems. At the intermediate size, one system will be a horizontal-axis machine and the second will be a vertical-axis Darrieus machine to allow a direct comparison of the two concepts. Milestones and funding for Engineering Development are given in Figure 16-25.

*Small systems development.* The first group of nine federally sponsored 1-, 8-, and 40-kW prototypes are being delivered to the Rocky Flats Test Center, and the remaining systems are in the final stages of fabrication. All nine prototypes will undergo initial testing at Rocky Flats. DOE has awarded two contracts each for the development of 4-kW and 15-kW prototypes. Also, DOE recently awarded a contract to Alcoa for the development of a

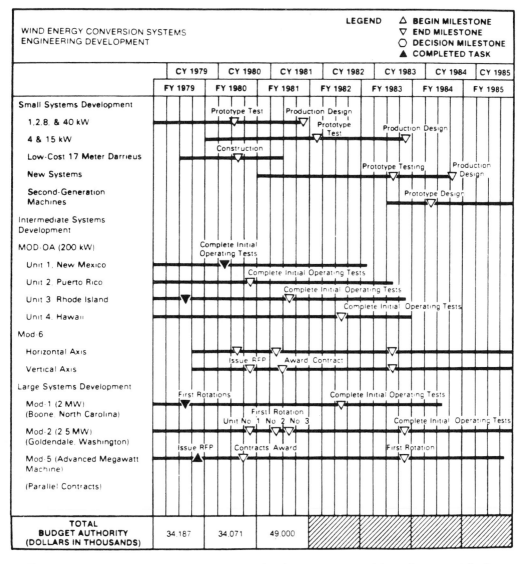

Figure 16-25. Wind Energy Conversion Systems, Engineering Development activity, milestones and funding.

low-cost, 17-meter Darrieus, with an option for three additional machines. The first machine was installed at Rocky Flats in FY 1980. Planned FY 1980 activities include:

- Completion of the development and initial testing of the 4-kW and 15-kW small-scale systems (two contractors each)
- Initiation of product improvement efforts for the more promising of the 1- to 2-kW, 8-kW, and 40-kW systems (about five contractors expected) as a result of current application and technical analysis
- Additional WECS development includes:
  1. Development of a limited number of second-generation systems for rural, residential, and farm applications (two to four contractors at 2 to 20 kW)
  2. Completion of applications testing using limited numbers of the 1- to 2-kW, 8-kW, and 40-kW systems

*Intermediate systems development.* The first Mod-0A (first generation, 200-kW) system at Clayton, New Mexico, has now operated for more than 5000 hours and delivered 500,000 kWh of electricity to the utility network. The Mod-0A at Culebra, Puerto Rico, has now been operating for more than 800 hours. The third Mod-0A machine, located at Block Island, Rhode Island, has been turned over to the utility and is entering operational testing. Kakuku Point, Oahu, Hawaii, has been selected as the fourth Mod-0A site, with machine installation completed in 1981.

The FY 1981 budget request will provide for parallel, competitive development of an advanced horizontal axis machine Mod-6 with a comparable, vertical axis Darrieus machine, and continued field testing and improvement of the four Mod-OA machines.

*Large systems development.* The world's largest windmill, the 2-MW Mod-1, is a first-generation experimental machine with a 200-ft-diameter blade. Installed at Boone, North Carolina, in April 1979, it is now undergoing extensive field testing by the Blue Ridge Electric Membership Cooperative. The Mod-1 has been successful in proving the technical feasibility of wind systems of the megawatt scale. An even larger Mod-2 (second generation, 2.5-MW, 300-ft-diameter) machine is currently being fabricated, with a 3-unit cluster scheduled for full operation in FY 1981. The site was recently selected near Goldendale, Washington, operated by the Bonneville Power Administration. The cluster will generate a total of 7.5 MW, enough electricity to serve approximately 2000 average homes at an energy cost of less than $0.08 to $0.10 per kilowatt-hour. This Mod-2 project will verify the numerous design improvements as compared to Mod-1, demonstrate that the electricity from multiple wind turbines can be successfully synchronized with an electrical utility grid, and provide needed operational data.

The estimated FY 1981 funding for large systems development will provide for an advanced MW-scale system (Mod-5) development with two competitive contractors. As the last planned large-scale, government-developed, horizontal axis system, the competitive design and development contractors for Mod-5 will ensure a choice of at least two manufacturers for potential users. The probability of reaching the cost goal of $0.06 to $0.08 per kilowatt-hour for preproduction machines will also be increased. The level of technology expected for Mod-5 should lead to an energy cost of $0.03 to $0.04 per kilowatt-hour if the units are produced in quantity. This effort will allow for the development of a competitive industry as large wind systems begin to move into the commercialization phase in the mid-1980s.

The FY 1981 funding estimate provides for:

- Continued development of advanced machines (Mod-5)
- Continued operation of the Mod-1 experimental machine
- Continued operation of the Mod-2 cluster

## SOLAR THERMAL POWER SYSTEMS

The Solar Thermal Program involves phased R&D activities to assist American industry in establishing the technical and cost readiness of mid- and high-temperature solar concentrating collector systems as a prerequisite to commercial implementation. The program builds on a base of component and subsystem R&D test facilities, and small-scale field experiments. It is about to enter a phase involving integration of these resulting designs into large-scale field experiments and pilot plants, in parallel with the development of second-generation concentrator designs and higher temperature system concepts.

Three classes of systems have been identified as having the potential to capture major shares of the United States primary energy market. The applicability of these systems to specific market sectors is based on their unique optical characteristics (hence, different geographic applicability and temperature capability) and modularity (hence, scale of application). The three major classes are linear-focusing distributed receivers (parabolic troughs and hemispherical bowls), point-focusing distributed receivers (parabolic dishes), and central receiver systems. As shown in Figure 16-26, the high temperature

heat from solar thermal systems can be used directly in industrial processes, in turbines to produce electricity, in cogeneration applications and ultimately to produce liquid and gaseous fuels.

The Solar Thermal Program is a key element in support of President Carter's 20 percent solar goal. Electricity generation and industrial heat applications of solar thermal concentrator systems are projected to contribute 3.0 quads per year of the roughly 19.0 quads implied by the goal. Futhermore, in low-temperature applications, it is expected that concentrating collectors will compete strongly with presently commercial nonconcentrating flat-plate collectors. Photovoltaic concentrator systems development will benefit directly from the technology base developed in the Solar Thermal Program.

The industrial and utility markets for solar thermal technology constitute more than half of the nation's present energy consumption. Promising applications for solar thermal systems include:

- Large central receiver electrical applications, including repowering of gas- and oil-fired peaking electrical generating plants and storage-coupled power plants of intermediate-capacity factor

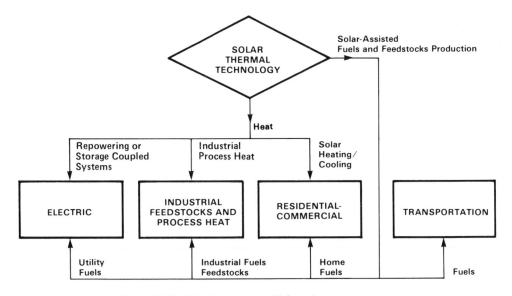

Figure 16-26. Solar thermal potential in major energy sectors.

- Industrial process heat systems that may use any of the collector types, depending on their temperature needs, geographic location, and process/plant requirements
- Remote power systems for irrigation, military, and other applications
- Small community power sytems, usually less than 10 MWe in size
- Total energy systems that provide electricity, process heat, and space-heating and space-cooling to industrial, commercial, and residential users
- Ultimately, storage-coupled process heat, for high-capacity-factor utility, and fuel and chemical production applications

The Solar Thermal Program is developing the four types of systems described below: central receivers, parabolic troughs, parabolic dishes, and hemispherical bowls.

A central receiver solar thermal power plant consists of three major components: (1) a field of movable mirrors called *heliostats*, which collect and reflect solar radiation; (2) a receiver to capture this concentrated radiation and create a controlled, high-temperature environment; and (3) a support tower upon which the receiver is mounted. At the receiver, solar radiation is used to heat a fluid (water, steam, molten salt, liquid metal, or gas) that is transported through pipes to a heat engine/generator. The generator converts heat energy into electricity, or the heat can be used directly for an industrial application. A storage subsystem can be added to the basic configuration to ensure constant input to the heat engine and to maintain plant operation during periods of low insolation. Figure 16-27 presents a schematic diagram of a central receiver system.

Central receivers for electrical power generation typically operate at 1000° F for steam- and liquid-cooled systems and at or above 1500° F for gas-cooled systems. The systems can be as small as 1 MWe, with the optimal size for certain bulk electricity production applications in the 100- to 300-MWe range. Central receivers systems can provide the wide range of temperatures required for process heat applications.

Parabolic troughs are line-focusing systems that concentrate the sun's energy along a pipe (receiver) carrying a heat-transfer fluid. Either the collector or the receiver moves to keep the sun's image on the receiver. The transfer fluid carries the heat (400° to 600° F) from the receiver for applications in process heat, mechanical power, electricity generation, or total energy systems. System sizes can range from a few kilowatts to tens of megawatts.

For electricity applications, the parabolic dish module consists of four subsystems: the concentrator, the receiver, the power conversion unit, and an automatic control system to enable each module to track the sun. The concentrator collects solar energy from a large area and focuses it onto a very small area. The receiver, which is mounted at the focus of the dish, captures the concentrated radiation and converts the energy to heat in a working fluid such as hot gas. The working fluid transports the energy to the heat engine of the power conversion unit, which is mechanically linked to the electricity generator. In the simplest system configuration, shown in Figure 16-28, the power conversion unit is located on top of the receiver at the focus. The optical portion of the concentrator is a parabolic reflector, although lens concentrators also are being considered. To produce thermal energy for industrial, commercial, or agricultural applications, the power conversion unit is replaced with a receiver having flexible lines to conduct the working fluid to a heat-transfer network on the ground.

First generation parabolic dish system experiments currently operate at 750° F. Designs now being developed to achieve temperatures as high as 1500° F are more promising for small electrical (about 1 MWe), total energy, process heat, and possibly for utility applications.

A novel concept under development in the Solar Thermal Program is a concentrating configuration, based on a fixed reflecting hemispherical bowl. In this unit, a moving heat receiver is aligned within the bowl parallel to the direction of incident radiation.

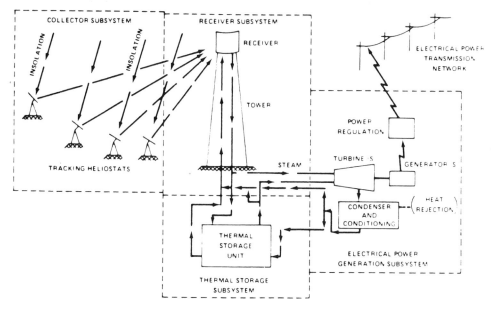

Figure 16-27. Central receiver solar thermal power system.

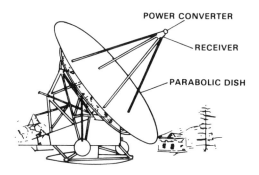

Figure 16-28. Dish Concentrator Power Converter.

Bowls can be used for mechanical power, process heat, total energy, and community electricity applications. Bowls can be operated at 600° to 1000° F and systems can range in size from tens of kilowatts to tens of megawatts.

## Program Strategy

The objective of the Solar Thermal Program is to establish the technical feasibility and cost readiness for its various market applications. The program strategy is to:

- Perform market studies and systems analysis to determine user requirements and cost goals and to identify important market applications

- Develop prototype systems to characterize performance and evaluate technology

- Perform R&D on critical components and subsystems and advanced technologies

- Conduct appropriate second-generation applications experiments and build engineering prototypes to establish cost readiness

The elements of this strategy are interconnected so that development in one can build on a base established in others. For instance, advanced R&D will depend primarily on results obtained in market studies and on technology development needs identified in engineering development. Similarly, the engineering development activity will utilize components and results from advanced R&D and market study activities, as well as feedback from the applications experiments.

Keyed to these activities are three levels of market size, as illustrated in Table 16-10: markets of early opportunity (1985), near-term markets (1990), and ultimate markets (1995 and beyond). Pilot plants using first-genera-

## Table 16-10. Market-Driven Solar Thermal Strategy.

| COLLECTOR TYPE | EARLY OPPORTUNITY | NEAR-TERM | ULTIMATE |
|---|---|---|---|
| Trough (400°F) | Large-Scale Solar Cooling | Solar Heating and Cooling | Low-Temperature IPH |
| Trough (600°F) | Mid-Temperature IPH Retrofit Enhanced Oil Recovery | Community-Scale Electric | Mid-Temperature IPH |
| Bowl | Small IPH | Small Irrigation | Large Irrigation |
| Dish* (Steam) | Small High-Temperature IPH | Industrial Total Energy | Industrial, Residential Total Energy (Dish only) |
| Dish (Gas) | Remote Electric or Pumping | Community-Scale Electric | Bulk Electric-Northern Latitudes, Fuels and Chemicals, Off-Site IPH |
| Tower (Steam) | Retrofit IPH Repowering | Intermediate C.F.**Bulk Electric-Southern Latitudes | High C.F. Bulk Electric-Southern Latitudes |
| Tower (Gas) | Community-Scale Electric | Small Cogeneration | Fuels and Chemicals, Off-Site IPH |

*(or small towers)                    **C.F.-Capacity Factor

Note: This is an example of current thinking subject to revision based on ongoing market and technology assessment studies and is not intended to be an exhaustive list of all possible markets

tion technology—and later, critical module experiments—will be conducted to establish experience and technical cost data bases in the context of near-term applications. Applications experiments using second-generation technology will be targeted toward markets of early opportunity. Advanced materials, component, and subsystem R&D will focus on requirements of systems tailored to ultimate large markets.

To ensure large-scale penetration of major heat, electricity and fuel markets and to direct solar thermal systems toward cost readiness, cost goals for all classes of collector systems have been established based on expected competition. As shown in Table 16-11, energy cost targets for the year 1990 are in the range of 72 to 180 mills-kWh for electricity generation and $6.00 to $6.60 per MBtu for industrial process heat. Cost goals on the order of 25% to 50% lower than 1990 levels are set for 2000.

These goals are based on 30-year life-cycle costing. To reach the energy cost goals, system and critical subsystem (e.g., concentrator) goals have been established (see Table 16-12).

These goals require that costs of heliostats, troughs, dishes, and other concentrators be reduced to $9 to $12/ft$^2$ (1980 dollars) by 1990, and be reduced by 25% to 50% by 2000. Cost goals for the Solar Thermal Program are under examination and will be adjusted based on recommendations of the Interlaboratory Solar Thermal Cost Goal Committee, chaired by the Solar Energy Research Institute. Performance goals have also been established for critical components and subsystems.

## Program Structure and Budget

The Solar Thermal Program consists of three elements: Central Receiver Systems, Distributed Receiver Systems, and Advanced Technology. They are described briefly below.

The overall objective of the Central Receiver subprogram is to establish the technical readiness of cost-competitive systems, beginning with initial applications in the mid-1980s. A phased technology development program will focus on near-term applications that will lead to increasing levels of fossil-fuel

## Table 16-11. Solar Thermal Energy Cost Goals (1980 Dollars)

| APPLICATION | 1990 GOAL | 2000 GOAL |
|---|---|---|
| Electric Generation | 72-180 mills/kWh | 50-60mills/kWh |
| Industrial Process Heat | $6.00-$6.60/MBtu | $4.00-$5.00/MBtu |

Table 16-12. Solar Thermal Program: Major Cost Goals.

| SYSTEM CLASS | APPLICATIONS | COST UNITS | COST GOALS | | |
|---|---|---|---|---|---|
| | | | 1982 | 1985 | 1990 |
| Small Power Systems | Small remote systems | ($/kWe) | 6,000 | 1,800 | 1,200 |
| | Total energy systems | ($/kWe) | 12,000 | 3,600 | 2,400 |
| | Small IPH | ($/kWth) | 1,700 | 500 | 300 |
| Large Power Systems | Repowering | ($/kWe) | 4,000 | 1,500 | 1,200 |
| | Stand-alone | ($/kWe) | — | 1,900 | 1,400 |
| | Cogeneration | ($/kWe) | 8,000 | 3,000 | 2,400 |
| | Large IPH | ($/kWth) | 1,100 | 400 | 300 |
| Subsystems (Installed) | Trough | ($/ft²) | 19 | 12 | 10 |
| | Dishes | ($/ft²) | 25 | 15 | 12 |
| | Heliostats | ($/ft²) | 24 | 16 | 9 |

displacement as systems costs decrease and the technology is demonstrated. The coordinated development/commercialization effort will continue as more advanced systems applications are proven technically satisfactory and approach cost competitiveness.

Activities within the Distributed Receiver subprogram focus on linear concentrators (parabolic troughs), dish technology, and hemispherical bowls to decrease system costs, increase reliability, and increase the performance of collectors and receivers. This subprogram is comparable in scope and strategy to the Central Receiver subprogram, but deals with distinctly different solar collection systems; hence, it generally focuses on smaller applications.

The Advanced Technology program is developing lower cost, more durable, and more efficient materials, components, and subsystems that will result in significant improvements in performance and cost of solar thermal systems as they are adopted by manufacturers. An additional objective is to de-

velop and validate advanced concepts and processes that potentially could expand market opportunities for solar thermal systems. Activities that support the entire Solar Thermal Program, such as technology transfer, insolation resource assessment, etc., are also funded by this subprogram. The FY 1981 Solar Thermal Power Systems funding estimates are shown in Table 16-13.

## Program Detail

Central Receiver Systems. Central receivers are the Solar Thermal Program's most visible technical effort and the one that has received the most attention. The central receiver concept applies to moderate- to large-scale, high-temperature applications. However, recent studies by Sandia Laboratories and the Solar Energy Research Institute suggest good competitive prospects for the concept in smaller and lower temperature applications. This finding has significant im-

Table 16-13. Solar Thermal Power Systems Budget.

| SOLAR THERMAL PROGRAM | BUDGET AUTHORITY (DOLLARS IN THOUSANDS) | | |
|---|---|---|---|
| | APPROPRIATION FY 1979 | APPROPRIATION FY 1980 | ESTIMATE FY 1981 |
| OFFICE OF SOLAR TECHNOLOGY | | | |
| Central Receiver Systems | 53,034 | 65,000 | 37,250 |
| Distributed Receiver Systems | 31,822 | 34,000 | 46,250 |
| Advanced Technology | 13,444 | 22,000 | 34,000 |
| TOTAL | 98,300 | 121,000 | 117,500 |

plications for both program priorities and overall commercialization strategy.

Tasks related to central receiver development are:

- The 10-MWe pilot plant to be located near Barstow, California
- The development of systems for large-scale utility and industrial process heat applications and repowering of fossil-fuel-fired electrical power plants
- The verification of subsystems and components at the Central Receiver Test Facility near Albuquerque, New Mexico
- The development of lower cost and higher performance components, especially heliostats (guided mirror assemblies) and advanced heat receivers

*Central receiver applications projects.* The 10-MWe central-receiver pilot plant, now being constructed near Barstow, California, will provide much-needed data on costs, performance, operating conditions, and environmental impact, as well as experience in plant design with the most powerful analytical tools. The project reached a critical milestone during the fall of 1979 when the preliminary design effort was completed, cost estimates became available, and bids for heliostats were evaluated. Selection of the heliostat supplier was based on firm vendor bids and tests of preproduction prototypes at the Central Receiver Test Facility.

Response to an FY 1979 solicitation entitled "Solar Repowering/Industrial Retrofit Systems" included a number of potential industrial users. Project definition efforts have begun for repowering both industrial and utility boilers. Twelve 9-month contracts, awarded in September 1979, involved user/supplier teams, development of conceptual designs, and site-specific project plans. These efforts allow preliminary designs to proceed in both categories in FY 1981, subject to funding availability. The results of the central-receiver strategy analysis, completed in late FY 1979 by SERI, indicate that a near-term market exists to convert ("repower") existing oil- and gas-fired utility plants in the Southwest.

An experimental central-receiver system under construction in Spain was completed in FY 1981, when a 2-year test operation began. This 500-kWe, sodium-cooled, central receiver project is funded by the International Energy Agency. The data are to be shared by all contributing countries.

The FY 1981 funding estimate supports:

- Construction of the 10-MW pilot plant at Barstow
- Preliminary designs of two or more industrial retrofit or utility projects
- Operation of the International Energy Agency 500-kW project in Spain

Milestones and funding for this subprogram are given in Figure 16-29.

*Technology and development.* In parallel with the pilot plant and proposed retrofit projects, central receiver system and component R&D is funded at a level of approximately $25 million per year.

Management has been delegated to the DOE San Francisco Operations Office. Sandia Laboratories at Livermore provides technical management support and management of component (heliostat and receiver) development efforts. The DOE San Francisco activities include systems definition and evaluation related to major potential applications, as well as support of the 10-MWe pilot plant.

During FY 1979 several contracts were let to develop and validate mass-producible heliostat designs and were completed in FY 1981. Five advanced second-generation heliostats will be built in FY 1980 and tested in FY 1981. After the tests, selected second-generation heliostats may be installed at the Central Receiver Test Facility and at the Barstow pilot plant for additional long-term testing. Third-generation heliostat development, based on lightweight plastic structures, including a transparent plastic bubble to enclose the heliostat assembly, was begun in FY 1981.

The development and testing of advanced receiver concepts will be continued in FY

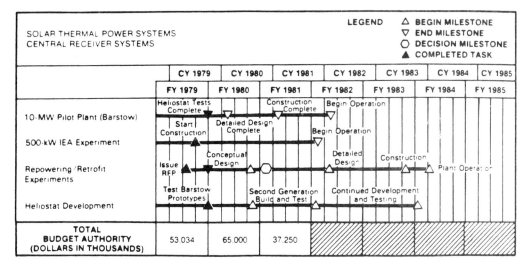

Figure 16-29. Solar Thermal Power Systems Central Receiver Systems subprogram milestones and funding.

1981. This will include water-steam, high temperature air, molten salt, and liquid metal designs that were selected for efficiency and overall improvement in system cost. Prototypes will be tested under near-operational conditions at the Central Receiver Test Facility. Materials development and materials compatibility testing will support receiver development.

Project definition efforts related to small-scale central receivers for total energy and cogeneration applications are also part of the DOE San Francisco Operations Office activities. They include site-specific conceptual designs to explore industrial cogeneration applications.

The latter effort can build upon gas receiver development funded by the utility industry through the Electrical Power Research Institute. Closely coordinated with DOE efforts, the Institute's program has produced two receiver designs with capabilities in the 1500° to 2000° F range. The first design was successfully tested at the Central Receiver Test Facility during FY 1979.

The FY 1981 funding estimate supports:

- Continued development of advanced receiver concepts
- Continued development of low-cost, second-generation heliostats

- Initiation of lightweight, third-generation heliostat development

**Distributed Receiver Systems.** The Distributed Receiver Systems subprogram focuses on the development of solar thermal technology for applications in which the energy supply system can be integrated at the point of use and for small, community-size electrical systems. Comparable conventional systems presently being used rely heavily on fuels such as natural gas, propane, and oil. Distributed receiver systems differ from central receiver systems in the type of concentrating collector used. However, they may not differ in the form of energy supplied. The thermal energy generated by the solar system can be used for process heat, and space-heating and cooling; it can be converted to direct mechanical energy or electricity; or it can be used in any combination of these.

Technologies under development in this activity include linear concentrators (such as parabolic troughs), fixed hemispherical bowls, and dishes. Linear collectors are developed for applications in the 400° to 900° F range and will be used for midtemperature process heat applications, total energy systems, and small remote electrical applications.

Experimental activities include the trough system being built at Almeria, Spain, through the International Energy Agency, and the midtemperature solar system test facility at Sandia Laboratories in Albuquerque. Currently, four different distributed receivers are being tested at Sandia, including second- and third-generation trough systems. In addition, the Coolidge, Arizona, irrigation experiment, which was dedicated in November 1979, will provide data on solar system performance and operations and maintenance costs for "real-world" trough systems. Several application experiments are being studied to establish the technical readiness of solar thermal systems for industrial process heat, total energy, and other applications. The DOE Office of Solar Applications is funding two preliminary designs of trough systems to provide steam for enhanced oil recovery.

The current cost of energy from these experimental systems is roughly $15 per 1 million Btu. The goal is $6 per 1 million Btu by 1990, through improved system performance and reduced manufacturing costs.

Dish technology uses a point-focusing collector with two-axis tracking, capable of achieving temperatures above 2500° F. Dishes are suitable for both total energy (electricity and heat) and modular electrical applications. Two systems experiments are being undertaken. The first is a 400-kWe total energy experiment at a knitwear factory in Shenandoah, Georgia, scheduled to operate in September 1981. The parabolic dish collectors used in this project will produce steam temperatures of 750° F. Design of the second experiment, a Small Community Engineering Experiment, is proceeding. The system will use Rankine-cycle engines mounted on 11-meter-diameter dishes. Site selection was completed in early FY 1980, and operation is scheduled for FY 1983.

To support the development of dish-mounted heat engine systems, the parabolic-dish test site is under construction at Edwards Test Station near Victorville, California, in the Mojave Desert. Planned capability is 10 to 15 kWe per dish module at temperatures of 1000° to 2000° F. Test bed concentrators have been installed and are being tested. In FY 1980, an air-cooled receiver and a Stirling heat engine were tested.

In contrast to dish technology, the bowl concept uses a fixed, hemispherical bowl-shaped reflector with a linear receiver tracking the sun. Construction on a 65-foot diameter bowl in Crosbyton, Texas, was begun in FY 1980. Information on construction, performance, and operating and maintenance costs will be used to evaluate the potential of this concept relative to that of other solar thermal systems.

The FY 1981 funding estimate supports activities in:

- Testing advanced trough components at the Mid-Temperature Solar System Test Facility at Sandia Laboratories near Albuquerque, New Mexico
- Preliminary designs of industrial/utility applications experiments
- Completion of construction of the Shenandoah total energy experiment
- Testing of dish systems at parabolic dish test site
- Operation of 20-meter bowl module at Crosbyton, Texas, and conceptual design of systems built from modules of this design

Milestones and funding for this subprogram are presented in Figure 16-30.

Advanced Technology. The Advanced Technology subprogram will develop lower cost, more durable, and more efficient materials, components, and subsystems that will result in cost-effective solar thermal systems in the mid 1980s. An additional objective is identification of new applications for solar thermal systems. Activities to support the entire Solar Thermal Program, such as technology transfer, insolation resource assessment, and university programs, are supported in this budget element. Activities included are:

- Technology assessment and support
- Materials technology
- Advanced components development
- Advanced systems and subsystems development

Review of the status of solar thermal tech-

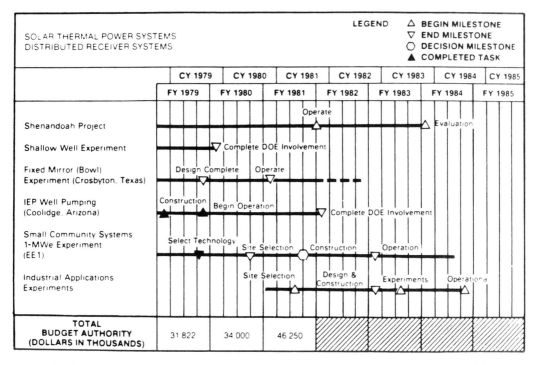

Figure 16-30. Solar Thermal Power Systems, Distributed Receiver Systems Subprogram, milestones and funding.

nology to identify Advanced Technology areas requiring new or additional emphasis was initiated early in FY 1980. Preliminary results indicate that improved technology is needed to provide more durable reflective surfaces, low-cost transmitting materials, thermal transport subsystems for industrial process heat applications, and thermal storage systems. This review was completed in late FY 1980 and provided direction for the Advanced Technology subprogram.

In materials technology, a pilot run of high reflectivity thin glass was achieved, and manufacturers' assessments of the potential for this glass in their equipment was completed in FY 1980. The materials activity investigates advanced reflective materials, long-lifetime polymers and glass, improved insolation absorption materials, and receiver materials. A major effort in FY 1980 and FY 1981 is to gain an understanding of the causes of mirror degradation. A coordinated, multilaboratory effort is under way to examine moisture, glass-silver adhesion, and other effects.

The advanced components development activity is aimed at providing higher efficiency

and more durable receiver, storage, transport, and heat-engine subsystems. Receivers capable of 2500° F operation are now being designed and may be tested in FY 1980. A detailed technology development plan for receiver technology will be completed in FY 1980. A similar plan for solar thermal storage systems was prepared jointly with the Energy Storage Program in FY 1979. During FY 1980, funds were transferred to the DOE Division of Energy Storage Systems to initiate coordinated development of second-generation storage subsystems, following the joint plan. This activity also supports the Advanced Components Test Facility at Georgia Institute of Technology, a full test facility for testing high-temperature materials for receivers and solar boilers as well as prototype subsystems.

The Advanced Technology subprogram also includes advanced systems and subsystems activity, which identifies attractive future systems and includes experimental development of components. Key subsystems for a parabolic-dish/Stirling-engine system are under development. During FY 1980, a receiver/engine subsystem manufactured by Fairchild and United Technologies will be tested. A number of feasibility experiments are under way to demonstrate the use of solar thermal energy to drive chemical reactions, such as coal gasification, shale retorting, and concentrated pyrolysis of biomass. The high heating rates available from solar thermal energy suggest a unique capability to produce liquid and gaseous fuels and other chemicals.

The FY 1981 funding estimate supports activities in:

- Assessment of Continued Advanced Technology program
- Assessment of high-performance thin glass
- Continued development of polymer protective enclosures for heliostats and dishes
- Continued multilaboratory mirror-durability assessment
- Testing of components for an advanced-dish/Stirling system

- Initiation of conceptual designs for solar thermal fuel production processes

Milestones and funding for this subprogram are shown in Figure 16-31.

## OCEAN SYSTEMS

Ocean energy systems have the potential for providing a renewable energy source for generating substantial amounts of baseload electricity and energy-intensive products. The potential resource extractable from United States waters using ocean energy systems is tens of quads per year. Systems currently under development include ocean thermal energy conversion (OTEC), wave energy, ocean currents, and salinity gradients. The latter three concepts are in the early development stage.

The OTEC concept is based on the use of ocean temperature differences between warm surface waters and cold water from the depths to operate a heat cycle that generates elec-

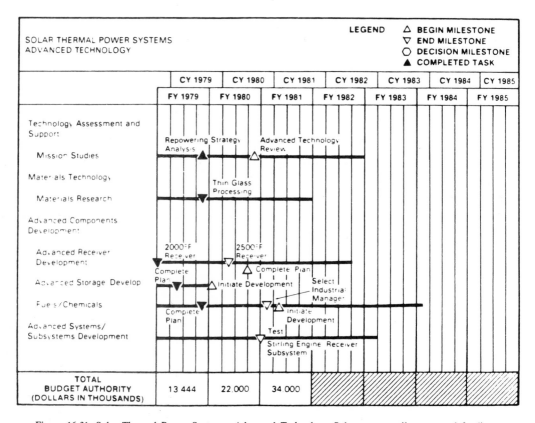

Figure 16-31. Solar Thermal Power Systems, Advanced Technology Subprogram, milestones and funding.

tricity. Two power cycles applying this concept are the closed cycle, which utilizes a fluid with a low boiling point such as ammonia, and the open cycle, which uses seawater as the working fluid. Figure 16-32 is a schematic diagram of the closed cycle. Hardware for the closed-cycle concept is closer to commercial availability than the hardware for the open cycle. Two ocean energy applications involve (1) moored platforms that supply electricity by submarine cable to shore and (2) floating platforms that "graze" for optimum temperature differences while generating electricity for onboard manufacturing of energy-intensive products.

Wave energy systems convert mechanical energy from ocean waves to electricity. Recent concepts that use wave energy by focusing the energy source appear to offer potential cost-effective advantages over other wave conversion systems that have been proposed. Ocean-current energy can be extracted by large-diameter hydroturbines suspended below the surface and driven by the current. Power may be produced from salinity gradients by using the energy potential across a selective membrane between two solutions of differing salinity. One approach uses seawater for the saline solution and freshwater for the lower concentration solution. Alternatively, brine from salt domes may be used for the saline solution and seawater as the low salinity source.

## Program Strategy

The goal of the Ocean Systems program is to develop a technology base and the institutional links that are required to catalyze a timely commercial exploitation of renewable ocean energy resources in an environmentally sound, economic manner.

Ocean thermal resources for OTEC plants tend to be available within 1 to 6 miles from United States tropical islands where water is 3000 to 4000 feet deep. American islands can be served by short underwater electrical cables and allow the use of alternating current for energy transmission. This cable requirement is technologically simpler than that used

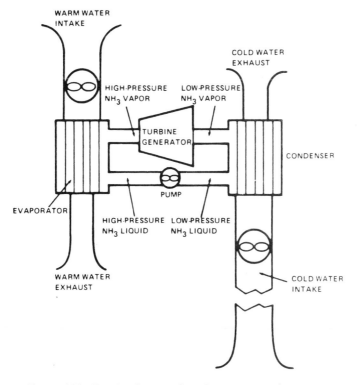

Figure 16-32. Closed-cycle ocean thermal energy conversion concept.

to transfer power from offshore OTEC plants to Gulf of Mexico markets, since suitable ocean thermal resources in the Gulf of Mexico are 80 to 150 miles from the coastline. These longer distances require direct current transmission and subsequent reconversion to alternating current onshore. Clearly, from the standpoint of underwater cables, United States island markets are more accessible at a reduced cost, compared to gulf coast markets, and tend to have significantly better thermal differences.

The identification of early competitive markets is an important adjunct to the commercial introduction of new technology. In the case of OTEC technology, it has become clear that electrical production for island users constitutes a market where even the first OTEC commercial plants can be cost-competitive with existing alternatives. The islands most likely to use ocean energy are those located in the tropics and subtropics, because of their proximity to a good ocean thermal resource. This is the case for United States islands such as Hawaii, Puerto Rico, the Virgin Islands, Guam, Micronesia, and American Samoa. Key factors making OTEC attractive for base-load electrical applications on United States islands are their present reliance on oil-derived electricity and their excellent thermal resources, conveniently accessible to floating OTEC plants via a few miles of electrical cable.

The above factors lead to the Ocean Systems Program strategy:

- Focus initially on markets that are subject to high-cost foreign fuels, supply interruption, and embargo, such as United States islands and military bases
- Perform further component and material R&D directed toward OTEC cost reduction to penetrate the United States mainland market in the 1990s
- Increase the capacity to distribute ocean energy cost effectively to a larger United States continental area than is accessible by undersea cable, through use of energy-intensive products such as ammonia
- Maintain close coordination with ocean energy R&D programs of other nations

Projections of future United States baseload electricity requirements have been modified downward in the past few years because of anticipated energy conservation activities, motivated in large part by higher fuel costs. Nevertheless, such projections still indicate increasing demand, especially in the southern states. By 1990, the United States islands will need an additional 3500 MW of electrical power and an additional 5000 MW by the year 2000. Similarly, the United States gulf coast will need about 50 gigawatts (GW) of base-load electric power by 1990 and about 100 GW by the year 2000. These requirements could be partially met by OTEC, if projected OTEC energy costs are shown to be competitive with those of coal, nuclear, and oil alternatives during pilot plant operation. In the near term, OTEC energy costs of $0.05 to $0.06 per kilowatt-hour are projected for 100-MW plants that could operate off Puerto Rico in the year 1990—an application that is less costly than new oil-fired plants—and perhaps displace existing oil-fired plants.

The preceding considerations suggest an "island strategy" for OTEC commercialization, whereby relatively small OTEC commercial plants can be introduced into United States island markets by 1990, with the United States gulf coast OTEC market commencing in about 1995. OTEC market penetration on an international basis would start in about the year 2000. Such a "market-pull" introduction of OTEC technology will allow time for the development of the submarine electrical cable technology required for OTEC penetration of the gulf coast market. The modularity and size flexibility of OTEC power plants are other factors conducive to initiating OTEC commercialization through such an island scenario.

Overall program responsibility and direction reside with the Ocean Systems Branch in the Office of Solar Technology. Detailed management is being decentralized, as shown in Figure 16-33.

## Program Structure and Budget

The Ocean Systems Program budget structure contains the following elements:

- Project Management

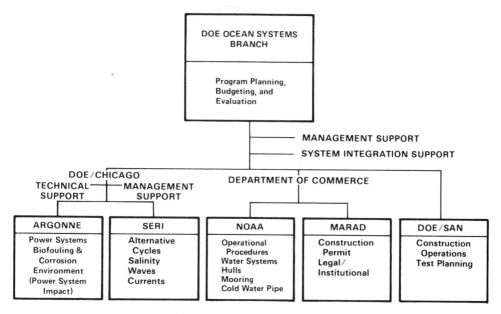

Figure 16-33. Ocean Systems program management.

- Definition Planning
- Technology Development
- Engineering Test and Evaluation
- Advanced R&D

This structure allows management to implement changes in program detail, based on definition planning and information from other program elements such as recent R&D results, including non-United States government efforts, analyses of changes in market conditions, and changing institutional considerations. The FY 1981 funding estimates are listed in Table 16-14.

## Program Detail

**Project Management.** Project Management includes funding requirements for DOE headquarters, field offices, other government agencies, and laboratories to manage the various projects and contracts for which they have been given responsibility under the decentralized management organization. Funding of this activity includes the cost of program support such as architectural and engineering services.

Activities under this element consist of necessary analyses, preparation of government cost estimates, data assessment, critique of contractor reports, review of contractor de-

signs, and management of major hardware programs. Technology Development plans, monthly and quarterly project reviews, and multiyear program plans are examples of the material that is generated. An initial multiyear program plan for ocean energy systems has recently been prepared and approved.

The FY 1981 budget request for this element supports management of program activities, with emphasis on the OTEC-1 engineering test facility and the pilot plant procurement.

**Definition Planning.** The Definition Planning activity includes systems and market studies, development of test program requirements, mission analyses, environmental impact studies, resource assessments and siting studies, and legal and institutional studies.

Cost projections and comparisons of projected OTEC energy costs to those of existing baseload electricity generating systems are continued. These studies show that OTEC power plants using existing state-of-the-art (i.e., enhanced shell-and-tube titanium) heat exchangers can be cost-competitive with existing oil-fired baseload capacity in island markets by 1985, and that improved OTEC systems can be competitive in United States mainland markets by the year 2000. Market integration studies of OTEC-generated elec-

## Table 16-14. Ocean Systems Budget.

| OCEAN SYSTEMS PROGRAM | BUDGET AUTHORITY (DOLLARS IN THOUSANDS) | | |
| --- | --- | --- | --- |
| | ACTUAL FY 1979 | APPROPRIATION FY 1980 | ESTIMATE FY 1981 |
| OFFICE OF SOLAR TECHNOLOGY | | | |
| Project Management | 3,728 | 4,655 | 4,520 |
| Definition Planning | 3,000 | 2,953 | 2,900 |
| Technology Development | 11,550 | 9,359 | 8,200 |
| Engineering Test and Evaluation | 19,247 | 19,546 | 20,000 |
| Advanced Research and Development | 3,620 | 3,487 | 3,380 |
| TOTAL | 41,145 | 40,000 | 39,000 |

tricity for islands, the mainland United States, and energy-intensive options such as the production of aluminum, hydrogen, and ammonia were completed in 1980. These studies bear on financing, legal, institutional, and technical integration factors.

A physical measurements program is continuing at several islands and the Gulf of Mexico to measure biological counts, chemical constituents, and physical parameters such as currents. Assessments of biological effects, environmental impacts, and resource availability are being pursued. The majority of potential OTEC utility users have organized an OTEC Utility Users Council for interfacing with the OTEC development program. Results from design studies and studies of electrical utility integration, energy-intensive options, environmental questions, and market assessment are being made available to the council and to industry.

The FY 1981 budget request supported continued economic evaluations and comparisons of OTEC with conventional power systems as well as legal and institutional studies and physical measurement programs. A study of financial incentives will lead to key policy recommendations in this area.

Technology Development. Technology Development activities for OTEC include the cold-water pipe, electrical cable, power systems, cleaning methods, development and utilization of test facilities for biofouling and corrosion studies, and development of open-cycle components. Activities for alternate ocean energy systems include developing membranes for salinity gradients, focusing devices for waves, mooring and low-speed rotating machinery for waves and currents, and water turbines for current systems.

In FY 1979, construction of the large 1-MWe capacity titanium state-of-the-art shell-tube heat exchanger was completed. This unit has design options that can simulate a 10-MWe heat exchanger through shrouding. It was deployed and operated in June 1980 as the first unit on the OTEC-1 Engineering Test Facility off the Kona Coast of Hawaii.

In separate competitive efforts for developing candidate heat exchangers, Westinghouse was selected to provide a smaller aluminum shell-tube unit, and Lockheed a titanium plate unit. The former unit was fabricated by September 1980 and the latter in June 1981. Funds for the completion of the fabrication of the Lockheed test hardware are provided in this budget.

Land tests on electrical cables will simulate electrical and mechanical stresses predicted for OTEC applications. Preliminary screening tests on six cables are to be conducted in FY 1980.

Results of ocean tests of a steel 5-ft diameter, 800-ft-long cold-water pipe off the coast of California will be correlated with predictions of analytical models. Cleaning tests of single-tube OTEC heat exchangers have been conducted during FY 1979 and FY 1980 at Panama City, Florida. By the time these tests are concluded in March 1980, they will have supplied valuable data including information from Hawaiian tests on the ability of various techniques to remove biofouling slime layers. So far, brush, ball, and intermittent chlorine injections have been successful in keeping the slime layer at acceptable levels for periods up to three months. By March, it is anticipated that longer term data will become available. A commercial countermeasure laboratory has been competitively selected for a 2-year program during FY 1980 and FY 1981 to define optimum biofouling cleaning methods.

Design of the Test Facility was completed in December 1979. Measurements at this facility will be conducted in two phases. Phase 1 put a warm-water facility in place, to be operational in July 1980. Continued operation of this facility is planned for FY 1981. Cold-water experiments are planned on the existing private OTEC platform or on OTEC-1 off the coast of Hawaii during FY 1980 and FY 1981 to augment the warm-water data from the Test Facility. Funds are provided in the FY 1981 budget for that activity. These tests will include single-tube, plate, and multitube experiments. Single-tube biofouling tests from a vessel off the coast of Puerto Rico were performed in FY 1980 and FY 1981.

Westinghouse completed a conceptual design study of an open-cycle OTEC power system that flashes the incoming warm water into steam and drives a turbine directly. Such a system mitigates problems associated with biofouling of heat exchangers.

Biofouling and corrosion studies are being continued, with primary emphasis shifting to

the first year of testing on OTEC-1 and at the Test Facility. Experimental research on corrosion will be consolidated by relocation at the La Que Center for Corrosion Technology.

The milestones and funding for the Technology Department subprogram are shown in Figure 16-34. FY 1981 activities supported in the funding estimate include:

- On-land tests of electrical cable prototypes for underwater cable
- Continued biofouling, corrosion, and cleaning studies at OTEC-1 and at the Seacoast Test Facility
- Continued studies and bench-scale tests on open-cycle OTEC components such as flash evaporators, degasification, condensation, and large steam turbine expanders

Engineering Test and Evaluation. The engineering test and evaluation subprogram includes tasks to develop and deploy a sea-based engineering test facility and to develop an OTEC pilot plant designed to lead to commercial OTEC systems capable of satisfying market requirements.

The sea-based engineering test facility, OTEC-1 (see Figure 16-35), was deployed in

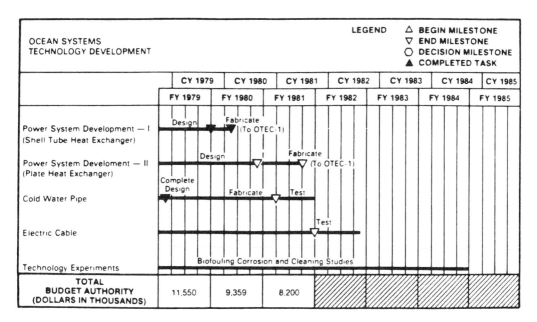

Figure 16-34. Ocean Systems, Technology Development Subprogram, milestones and funding.

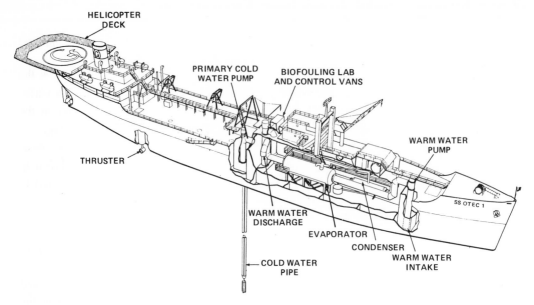

Figure 16-35. OTEC-1 engineering test facility.

June 1980, equipped with a TRW titanium heat exchanger, containing both smooth and enhanced tubes. Global Marine Development, Inc. converted an existing T-2 tanker for this purpose.

In FY 1981, the candidate heat exchanger designs on OTEC-1 were tested and the operation of a 2-year testing program was initiated. Based on results of this testing program, it is anticipated that certain heat exchanger concepts can be certified for commercial OTEC use.

During FY 1981, initial test operations on the 1-MW system will be completed. The vessel will be brought back to port for installation of more advanced heat exchangers, then redeployed. The funds provided in this category will provide for completion of test operations for the first deployment, modification of the vessel for new heat exchangers, and initial operations for the second deployment. Duration of the second deployment will be 2 to 2.5 years. Consideration will be given to installation of a turbine on OTEC-1 which is not provided for in current funding plans for the project.

A key activity to be continued in this period is the competitive design of OTEC pilot plants. These efforts were initiated in the last quarter of FY 1980 and completed in FY 1981. It is anticipated that detail design and construction can be started in FY 1982 and completed in FY 1985. The pilot plant will be of a scale sufficient to demonstrate the performance and reliability of all subsystems and of the integrated system. Its operation is necessary to permit an industrial decision on a subsequent demonstration plant. Several user organizations have indicated their desires to operate such a pilot plant by submitting unsolicited cost-sharing proposals to DOE. Government costs for managing the pilot plant design and construction project are also included in this subprogram. Milestones and funding are shown in Figure 16-36.

Advanced Research and Development. Advanced Research and Development activities concentrate on studies relating to advanced heat exchangers and advanced open-cycle concepts, including Claude and steamlift cycles. Results of these studies will be used in the design and development of future OTEC power systems. Alternative ocean energy concepts for utilizing salinity gradients, waves, and ocean currents as renewable resources are also being investigated. These concepts are being explored to determine their technical feasibility and economic potential for production of electricity or manufacture of energy-

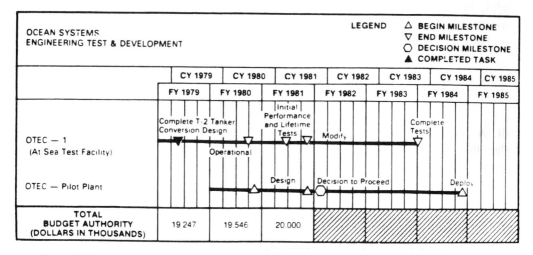

Figure 16-36. Ocean Systems, Engineering Test and Development subprogram, milestones and funding.

intensive products. Major activities are reviewed in the following subsections.

*Power systems.* Several heat exchangers will be core-tested at Argonne National Laboratories in FY 1980. During FY 1981, funds will be provided to fabricate cleaning systems for test units of these designs to be installed for the second deployment of OTEC-1. In addition, new candidate heat-exchanger concepts, such as direct contact designs, will be evaluated. There will be continued investigations of waterside and working-fluid-side heat transfer-enhancement systems. Experimental investigation of vapor-liquid entrainment interactions will be conducted in an effort to refine present analytical models. Effects of platform attitude and motion on the thermal performance of evaporators and condensers will be studied experimentally and analytically.

*Advanced open-cycle OTEC concepts.* Two advanced steamlift open-cycle systems are being evaluated. These approaches use the potential energy of the warm water to convert it into foam or mist, which then rises and is subsequently condensed. The resulting displacement rotates hydraulic turbines connected to electrical generators. Funds requested for FY 1981 will continue the laboratory efforts being performed on mist and foam open-cycle systems, including seawater tests to determine the impact of foam generation on the environment, and to study possible aperture plugging of mist generators.

*Salinity Gradients.* Based on current appraisals of the economic viability of this approach, no additional projects are planned during FY 1981.

*Waves.* Extraction of energy from ocean waves appears to be the simplest ocean energy system approach to exploit, and hence has attracted popular attention. However, it has not yet been shown to be potentially cost-effective. Recently, several promising approaches have been suggested to focus the wave front through diffraction, interference, and Fresnel phenomena.

The results of initial experiments on a focusing wave energy conversion device was completed in FY 1980. An analysis of this system and comparative assessments between alternative focusing approaches will be performed in FY 1981. These analyses and experimental results are expected to be instrumental in defining a prototype experiment. Ocean tests on a uniquely configured, United States wave energy conversion turbine, as part of an international program with Japan, the United Kingdom, and Canada were initiated in FY 1980 and continued into FY 1981.

*Ocean currents.* In FY 1979, special studies on ocean current systems were conducted, relating to the hydroelastic behavior of turbine blades, mooring and anchoring systems, and environmental influences. A study of a specific large ocean turbine system by Aero Vironment Corp. (California) is expected to provide results on which to base the FY 1981 engineering design of a 400-kW turbine, with the possibility of subsequently building such a unit. A system analysis will be performed to judge the overall cost effectiveness of the concept. Milestones and funding for this subprogram are presented in Figure 16-37.

## AGRICULTURAL AND INDUSTRIAL PROCESS HEAT

Approximately 40 percent of the total energy consumed in the United States each year is used in industry and agriculture. This currently represents about 30.0 quads/yr. of which up to 20.0 quads are used for industrial process heat applications such as process water heating, hot air for drying, and the production of process steam. The major share of this energy is derived from fossil fuels. Agricultural production consumes approximately 2.0 quads/yr. Approximately one-

quarter of agricultural consumption is provided by liquefied petroleum gas and used for low-temperature heat applications, such as for drying grains and crops and heating greenhouses and livestock shelters.

A large portion of President Carter's 20 percent solar goal is projected to be supplied by solar Agricultural and Industrial Process Heat (AIPH) systems. The Domestic Policy Review (DPR) has estimated that solar AIPH systems could displace 2.6 quads in the year 2000. This estimate translates into cumulative collector production requirements of 5.9 billion ft$^2$ by the year 2000. Solar system production rates to achieve the Domestic Policy Review target of 2.6 quads in the year 2000 are shown in Figure 16-38.

The major categories of solar agricultural and industrial process heat systems are:

- *Low- and intermediate-temperature industrial process heat (IPH) systems.* Low temperature systems 212° F or below will generally employ flat-plate collectors. Tracking line-focusing concentrating collectors or tubular collectors would be used for intermediate temperatures (212° to 550° F). Agricul-

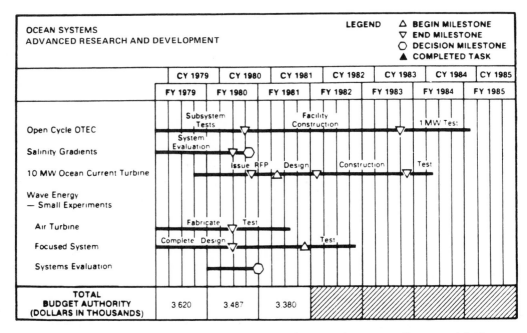

Figure 16-37. Ocean Systems, Advanced Research and Development subprograms, milestones and funding.

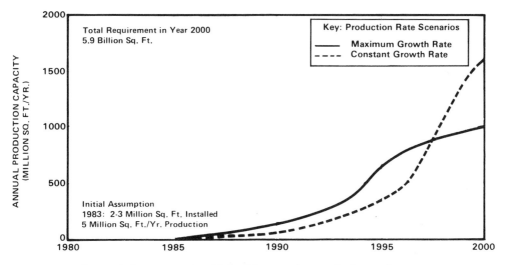

Figure 16-38. Agricultural and industrial process heat production requirements.

tural uses and industrial hot water, hot air, and steam generation are principal applications.

- *Large-scale low- and intermediate-temperature systems.* Having greatly reduced installation costs, solar systems to replace even small industrial boilers—which may be factory-produced modules of approximately 50,000 ft² each—are being developed. Field tests of these modules are planned for FY 1981. Industrial applications with steam temperatures up to 550° F are the principal near-term market.
- *Solar-enhanced oil recovery.* Solar-enhanced oil-recovery systems will initially involve the use of parabolic trough collectors to supply steam for enhanced oil recovery. Parabolic trough systems employ parabolic reflectors to

concentrate solar energy onto the working fluid on the absorber tubes. Figure 16-39 shows several line-focusing concentrator design concepts. It is aimed at demonstrating trough system capabilities in potential near-term applications. Later, enhanced oil recovery systems could use point-concentrating dish collectors or power tower systems, if these prove to be more economical.
- *Advanced applications including high-temperature parabolic dishes and central receivers in fuels, chemical, and total energy systems.* Figure 16-40 illustrates the configuration of a parabolic dish collector. The central receiver systems concept is applicable to large-scale, high-temperature applications. However, recent studies indicate that they may also be cost effective in smaller, lower temperature applications.

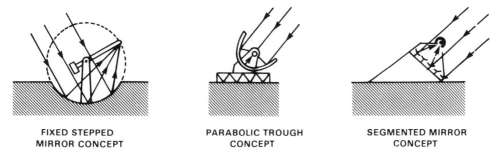

FIXED STEPPED MIRROR CONCEPT    PARABOLIC TROUGH CONCEPT    SEGMENTED MIRROR CONCEPT

Figure 16-39. Typical line-focusing concentrators.

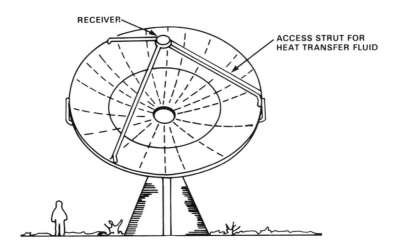

Figure 16-40. Parabolic dish collector.

Solar thermal systems can be used in numerous configurations for AIPH applications. Two configurations of operating industrial process heat systems are shown in Figures 16-41 and 16-42.

### Program Strategy

Due to differences in the stage of technology development and variations in requirements between the industrial and agricultural sectors, two program strategies are followed.

Industrial Sector Strategy. Solar systems for industrial processes are not presently economically competitive with conventional heat sources according to criteria used by industry for decisionmaking; consequently the strategy for achieving utilization in the industrial sector follows two parallel paths.

- Identify types of systems and markets that have the best prospects of early conversion to solar energy and focus most of the development efforts on those systems
- Determine the nature and extent of government incentives required to ensure substantial conversion

Recent studies suggest that mass-produced,

Figure 16-41. Solar industrial hot water system for concrete block curing, York Buildings Product Plant, Harrisburg, Pennsylvania.

Figure 16-42. Solar industrial process steam system for gauze bleaching, Johnson & Johnson Pharmaceuticals Plant, Sherman, Texas.

line-focusing, tracking collectors may be the most economical of systems under development for process heat applications at temperatures up to 550° F. The continued development of this technology will be a central industrial-process-heat program thrust. The program also will continue to support alternative technologies and to build and maintain a technology base for:

- Improved evacuated tube collectors and flat-plate collectors for low-temperature applications
- Reliable point-focus or central receiver collectors for high-temperature application
- Systems analysis and design to match technology to industrial needs
- Materials and engineering R&D to advance the technology

**Agricultural Sector Strategy.** The agricultural program will focus on solar applications in a market that is already fairly well defined. The main strategy, therefore, is to improve dedicated-use systems and develop integrated alternative technology and multiple-use systems for this market. The Office of Solar Applications will continue to work closely with the Department of Agriculture, its Science and Education Administration, and its Extension Service in order to draw upon its

farm experience and farm communication network. These will be utilized to aid in product development, market penetration and identification, and implementation of incentives.

**Program Management.** The Agricultural and Industrial Systems Branch in the Office of Solar Applications currently has the lead role for managing and coordinating the activities that implement the program strategy. Its activities are supported by the other branches in that office. Management responsibility for specific programs within the Office of Solar Applications will be decentralized, utilizing other government agencies and DOE field offices, laboratories, and regional offices. The Department of Agriculture will manage the agricultural programs, including R&D, field and market testing, and information dissemination. The industrial program will be managed by the DOE San Francisco Operations Office and the Solar Energy Research Institute. The Regional Solar Energy Centers, the DOE national laboratories, and other government agencies will provide support in technical and information-dissemination areas.

## Program Structure and Budget

The budget for Agricultural and Industrial Process Heat (AIPH) is part of the overall Office of Solar Applications budget. The approximate funding of AIPH activities is shown in Table 16-15. The budget is organized under the following functional headings:

- *Market Analysis* identifies the agricultural and industrial markets for solar thermal systems, both nationally and internationally. It investigates the infrastructure requirements, opportunities, and barriers for successful market penetration.
- *Systems Development* seeks to improve existing system performance, reliability, and durability. This is done through the design, development, and testing of components and subsystems; field testing to collect operating and cost data; systems analysis to establish cost and performance goals and to match systems with applications; and the development of simulation models, design tools, and handbooks for users.
- *Market Tests and Applications* consists of field tests in promising applications to assess technical performance, the users' interactions with and acceptance of solar systems, and institutional and marketing requirements.
- *Market Development and Training* stimulates interest in the use of solar energy for agricultural and industrial applications, develops the installation and maintenance skills necessary to support those applications, and promotes awareness of the potential of solar technology in these markets.

## Program Detail

This section describes the AIPH program in greater detail by each functional area: Systems Development, Market Tests and Applications, Market Analysis, and Market Development and Training. It is anticipated that DOE's involvement in the agricultural program will be reduced in FY 1984 in preparation for the Department of Agriculture's assumption of program responsibilities in FY 1985. In the industrial sector, emphasis will be placed on continuing Systems Development and on the testing and operational phase of programs begun or completed in the FY 1979 through FY 1980 time frame. The projects include continuation of current activities (Table 16-16), as well as several new projects aimed at specific market applications.

Overall milestones for the Agricultural and Industrial Process Heat Activities are given in Figures 16-43 and 16-44, respectively.

**Systems Development.** Over the next 5 years, the Solar Industrial Process Heat Program will seek to develop and test mass-producible, reliable, efficient, and cost-effective solar systems that can provide process heat in the low- to intermediate-temperature range (as high as 550° F) by the end of FY 1983 and in the higher temperature range (as high as 500° to 1500° F) beginning in FY 1984, as these technologies are transferred from the Solar Thermal Power Systems Program. Systems analysis will establish cost/performance goals and identify optimal systems for industrial ap-

## Table 16-15. Estimated Expenditures for Agricultural and Industrial Process Heat.

| AGRICULTURAL AND INDUSTRIAL PROCESS HEAT PROGRAM | BUDGET AUTHORITY (DOLLARS IN THOUSANDS) | | |
|---|---|---|---|
| | ACTUAL FY 1979 | APPROPRIATION FY 1980 | ESTIMATE FY 1981 |
| OFFICE OF SOLAR APPLICATIONS | | | |
| Market Analysis | 400 | 1,000 | 1,600 |
| Systems Development | 7,400 | 16,700 | 21,000 |
| Market Test and Applications | 13,468 | 16,300 | 14,800 |
| Market Development and Training | 500 | 1,000 | 5,600 |
| TOTAL | 21,768 | 35,000 | 43,000 |

## Table 16-16. Current Activities of the AIPH Program.

- Complete development of one or more mass-producible line-focus concentrating collectors to determine production requirements and costs

- Complete system costs/performance requirements for tracking the line-focus and point-focus collector systems and market incentives for economic competitiveness

- Complete prototype systems tests for flat plate and evacuated tube collectors and evaluate performance data

- Evaluate the technical and economic viability of salt gradient solar ponds

- Determine market potential for solar-assisted industrial heat pumps and solar-driven refrigeration units

- Bring on-line seven new high temperature (350–500°F) applications experiments averaging 10,000 ft² each

- Design (cost shared with industry) four high-temperature and four low-temperature large-scale (50,000 ft²) systems

- Complete design and begin construction of a solar enhanced oil recovery project of 250,000 ft²

- Provide current information on technology to industry to encourage solar energy use

- Promote changes in Federal, State, and local programs to provide financial incentives to encourage industry to make greater use of solar process heat and steam

- Continue 50 DOE/USDA research projects in agriculture for on-farm application in heating greenhouses and livestock shelters, drying crops, and processing grain and food

- Continue construction and operation of on-farm field tests for livestock shelters (90 sites), crop and grain drying (75 to 100 sites), and multiple use application (75 to 100 sites)

plications. The development and validation of system simulation models, design tools, and handbooks for various user sectors will be undertaken.

The agricultural program will focus on development of dedicated-use systems and integrated technology systems combining projects in solar thermal, biomass, and wind energy. Demonstrations of this technology will be conducted in limited field tests.

Existing solar industrial systems undergoing field tests for producing industrial-process hot water, hot air, and steam will be upgraded to improve instrumentation, data acquisition, and analysis. Uniformity of instrumentation, data logging, and thermal analysis techniques will permit comparison of various systems' performance and will provide a basis for component and system improvements from FY 1980 through FY 1983.

Based on improved data systems and performance experience, field tests of new improvements on existing systems will be conducted from FY 1981 through FY 1984. Tests will be directed toward determining reliability, reducing operating and maintenance costs, and providing a technical basis for continued improvement in component and system design.

In the agricultural program, a number of existing and workable systems for crop drying and the heating of livestock housing and greenhouses have been developed. Some are being demonstrated through the Department of Agriculture's Extension Service. An expanded program of field testing, demonstration, and preparation of educational materials for consumers and manufacturers will be undertaken to expedite widespread adoption.

A study of mass-producible collectors will be conducted to accelerate the development of improved designs for line-focusing, tracking, concentrating solar collectors, and for collector components and subsystems. Such designs should offer improvements in thermal performance, durability, and reliability, as well

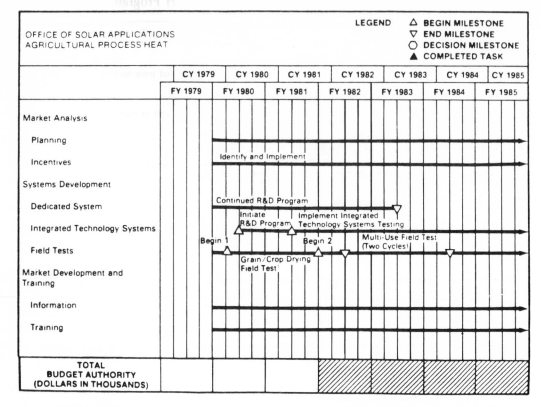

Figure 16-43. Office of Solar Applications, Agricultural Process Heat, milestones and funding.

as ease and economy of manufacturing and installation. The mass-producible system study will include the design, fabrication, and testing of a prototype system; development of a manufacturing plan; and identification of critical components.

Component studies will include design, test, and evaluation of system elements—including glass, structures, trackers, drive system, collector supports, materials, absorbers, distribution system, cleaning requirements, and procedures—and system security, with a view toward improved efficiency, lower costs, and extended lifetimes. Design and construction of site-built systems may be appropriate for agricultural applications. Development of potentially marketable systems that use readily available materials or that can be supplied in kit form for on-farm use will be investigated.

Subsystem development efforts will also include basic work on materials and engineering to keep advancing solar technology. This will apply to all types of systems, components, and collectors.

**Market Tests and Applications.** Following prototype development and laboratory testing, field test projects are undertaken in a user environment to assess technical performance (engineering field tests) and institutional/marketability requirements.

Market test projects will incorporate a balanced mixture of large- and small-scale tests. These include construction of three low- and three intermediate-temperature 50,000-ft$^2$ industrial process heat systems from the FY 1980 program, construction of one solar IPH system of 250,000 ft$^2$; amd designing two or three large-scale (100,000 ft$^2$) systems in diverse industries and locations. Because of the need to rank program activities and the high costs of field tests ($2 to $5 million each), it is necessary and appropriate to require cost sharing with industry.

**Market Analysis.** The Market Analysis program element will provide the supporting studies that are essential for successful commercialization of AIPH solar systems. These

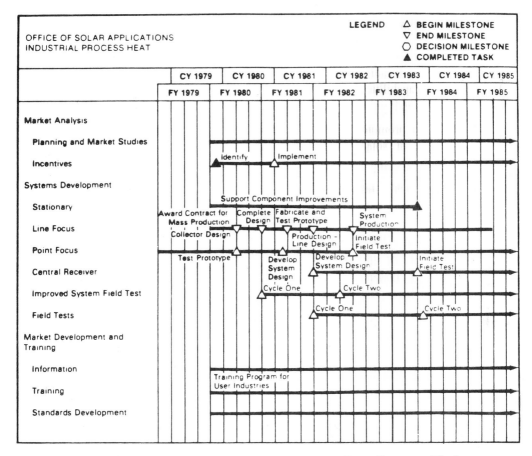

Figure 16-44. Office of Solar Applications, Industrial Process Heat, milestones and funding.

studies are aimed at identifying the opportunities, barriers, and extent of the various agricultural and industrial markets, both nationally and internationally. Estimating infrastructure requirements for successful market penetration and evaluating consumer demand patterns are also major objectives of this program element.

Specific Market Analysis activities in support of Agricultural and Industrial Process Heat include:

- Identification of financial incentives, laws, and regulations that will encourage commitment by industry to use solar systems. Emphasis will be placed on identifying improvements to or augmentation of existing laws and regulations, such as the Fuel Use Act, solar tax credits, fuel allocation preferences, and loan guarantees that are necessary

to promote large-scale use of solar AIPH.

- Development of site-specific market-definition data that rank specific sites by solar insolation, land availability, process load profile, competitive system costs and availability, enviromental constraints, state and local incentives (taxes, loans), and other market factors.

- Development of methods of meeting the capital required for large-scale use of solar systems by industry while minimizing the effect on productive capacity. Methods considered will include third-party ownerships that offer solar systems for lease or sale.

**Market Development and Training.** The primary objectives of the Training, Information, and Education program are to stimulate interest in the acceptance and utilization of solar energy for agricultural and industrial applications, develop the installation and maintenance skills and capabilities necessary to support the industrial and agricultural applications technology, and promote awareness of the potential benefits of solar utilization in these market sectors. Specific FY 1981 AIPH Market Development and Training activities include:

- Through regional solar energy centers, development of an information program, working to disseminate data concerning new laws, regulations, and incentives that could make the use of solar process heat systems more attractive.
- Workshops and economic analyses in areas where solar AIPH systems are most competitive
- Identification of training barriers to large-scale solar production installation and operation
- Through regional solar energy centers, identification and removal of state and local barriers to wide-scale solar use by industry, and evaluation of low-temperature, market-ready applications
- Development of codes and standards for tracking line-focus and point-focus collectors

## ACTIVE SOLAR HEATING AND COOLING

Active solar heating systems are capable of attaining temperatures of 150° to 220° F. They are therefore a good thermodynamic match to the requirements of heating buildings and providing hot water. Widespread application of such systems could save fossil fuels for other uses, such as chemical feedstocks or transportation. The estimated contribution of active heating and cooling systems to President Carter's 20 percent solar goal is 2.0 quads/yr from solar systems installed in more than 20 million homes. In the short term an estimated 0.2 quad of energy could be provided by active solar systems by 1985. Active solar heating systems are being sold commercially today; it is estimated that 80,000 hot-water heating systems combined hot-water/space-heating systems are now installed throughout the nation. Active space-cooling systems are generally considered to require additional technology and systems development to become cost-competitive. The DOE's Active Solar Program objectives to meet these challenging goals are summarized in Table 16-17.

The active systems shown in Figure 16-45 provides heating, hot water, and cooling to a single-family home built by students at the

## Table 16-17. Active Solar System Contributions Required to Meet the 20 Percent Solar Goal.

| SYSTEMS & YEAR | NUMBER OF SYSTEMS INSTALLED (MILLIONS) | PRIMARY ENERGY DISPLACED (QUADS/YEAR) |
|---|---|---|
| Solar Water Heating Systems | | |
| 1985 | 1-1.5 | 0.05 |
| 2000 | 20-26 | 0.65 |
| Solar Space & Water Heating Systems (Includes Solar Heat Pumps) | | |
| 1985 | 1.5-2.0 | 0.13 |
| 2000 | 8-9 | 0.75 |
| Solar-Driven Heating and Cooling Systems (Includes Water Heating | | |
| 1985 | 0.3-0.5 | 0.02 |
| 2000 | 2-3 | 0.60 |

Figure 16-45. Active solar heating system, Lawrenceberg, Texas.

273

Lawrence County Vocational Center, Lawrenceberg, Texas. A combination of active, passive, and hybrid solar systems supplies approximately 80 percent of the space-heating and hot-water requirements of the commercial building Walpole, New Hampshire (see Figure 16-46).

The purpose of the Active Solar Heating and Cooling Program is to accelerate the development and commercialization of five major types of systems. These are:

- *Direct solar space- and water-heating systems.* Solar energy is used to heat air or a liquid, which is then used to heat a building and/or hot water. These systems usually include provisions for storing the heated fluid for use at night or during cloudy days.
- *Combined solar/heat-pump systems.* A heat pump is a device that absorbs heat from outside a building and transfers it inside, thereby heating the building. It is actually an air-conditioner operating in reverse, and can also cool by collecting heat from inside and expelling it to the outside. A solar-assisted heat pump uses solar collectors to augment the pump with a heat source other than outdoor ambient air.

- *Solar-driven absorption heating and cooling systems.* An absorption device is essentially a vapor-compression machine that transfers heat by expansion and compression of a liquid such as water or ammonia in combination with an absorbent. It is similar in principle to the operation of an air conditioner. The absorbent is a liquid that combines chemically with the water or ammonia and releases heat while combining. Solar-driven absoption systems use sunshine instead of electricity as an energy input.
- *Solar-driven Rankine heating and cooling systems* use solar collectors to supply heat to vaporize an organic fluid, which in turn is expanded through a turbine to produce mechanical energy. The heat transfer process is similar in principle to a steam turbine. The turbine is coupled to a compressor in a vapor compression air-conditioning unit, which is similar to an ordinary air-conditioner or refrigerator. Rather than using an electrical motor to compress the organic fluid, solar Rankine systems use solar energy to drive a turbine that supplies the work to compress the organic fluid.

Figure 16-46. Combined active and passive system, Walpole, New Hampshire.

- *Solar-drive desiccant heating and cooling systems* provide heat transfer by dehumidification of air, using either a liquid or solid desiccant. Solar energy is used to dry the desiccant material, which in turn absorbs moisture from the conditioned space. This dry air is then humidified using an evaporative cooler to produce the cooling effect.

## Program Strategy

The Active Solar Heating and Cooling Program includes a few relatively well-developed technologies with some established commercialization histories (e.g., domestic water heaters) as well as several technologies still in an early development stage. The Office of Solar Applications periodically evaluates these technologies to determine when program emphasis should shift from systems development to market development. Thus, the program strategy has two primary thrusts:

- *Systems Development* emphasizes systems integration and design improvements that lead to required cost and performance levels for specific applications
- *Market Development* emphasizes gathering and distribution of appropriate information to groups of users to reduce solar systems barriers and to encourage their acceptance

For direct solar space-heating and water-heating systems (the active systems with the greatest near-term potential for consumer acceptance), the strategy emphasizes market development efforts. Component and systems reliability, durability, and cost-performance characteristics still have significant potential for improvement. An expanded materials R&D of cheaper or more efficient materials and components, as well as improved systems integration, could have a major impact on reducing costs. Also needed are control subsystems that help to improve system performance, permit load management, and permit the monitoring of system integrity.

For combined solar-heat-pump systems and the less mature, combined solar-heating-and-cooling systems, the strategy emphasizes systems development activities. However the emphasis is expected to shift to market development efforts for combined solar-heat pumps in 1981–82 and for combined solar heating and cooling systems in 1983–85. Novel collectors, chillers, and ground coupling for heat pumps are some of the components needing further development. Systems integration issues, including controllers for dual-source heat pumps, load leveling, load management, and storage strategies will require resolution through systems development activities. The most urgent market-related development needs are realistic cost-performance, reliability, and maintainability goals upon which sound systems-development objectives for all active solar applications can be based.

To implement this strategy, the activities of the Active Solar Heating and Cooling Program are structured along the Product Development Process used by industry to bring products to the marketplace. Table 16-18 shows areas of federal interaction in the Product Development Process. The overall objective of the Active Solar Heating and Cooling Program is to plan and implement appropriate federal activities to accelerate technology and market development and acceptance and utilization of active systems, and to thereby reduce national dependence on nonrenewable forms of energy.

The general activities of the DOE active solar system program include:

- Determining the cost, performance, reliability, and maintainability of state-of-the-art active solar systems for residential and commercial applications and setting future cost, performance, reliability, and maintainability goals needed to meet the 2.0-quad goal by the year 2000
- Comparing active solar systems with the most cost-effective and energy-conserving nonsolar systems and assessing required incentives for active solar systems in light of these comparisons
- Supporting the development of im-

Table 16-18. Federal Functions for Product Development Process.

| EIGHT-STAGE PRODUCT DEVELOPMENT PROCESS | REPRESENTATIVE FEDERAL FUNCTIONS |
|---|---|
| 1. Basic and Applied R&D | • Select candidate research projects<br>• Directly fund selected projects<br>• Select projects for subsequent development |
| 2. Market Research and Product Justification | • Develop cost/performance goals<br>• Perform market research<br>• Conduct analyses of barriers<br>• Analyze policy and incentives |
| 3. Component Development & Test | • Assess component cost/performance characteristics<br>• Determine needed improvements<br>• Directly fund selected projects |
| 4. Prototype System Development and Test | • Analyze systems marketability<br>• Identify system cost/performance improvements<br>• Assess systems compatibility<br>• Directly fund selected projects |
| 5. Field Testing | • Select apropriate test systems and applications<br>• Directly fund selected projects on a cost-shared basis<br>• Provide feedback for program improvement<br>• Disseminate technical information |
| 6. Test Marketing | • Analyze product suitability for marketing<br>• Assist in focusing market effort<br>• Underwrite selected marketing costs |
| 7. Production | • Analyze production needs<br>• Assess production capabilities, methods, and costs<br>• Support selected projects |
| 8. Sales and Marketing | • Promote product acceptance and utilization<br>• Remove institutional market barriers<br>• Support training and education programs<br>• Assist development of market infrastructure<br>• Provide user information<br>• Define needed market incentives |

proved materials, components, systems, design, installation techniques, and service capabilities for active solar systems

• Increasing consumer acceptance of active solar systems by supporting information, education, and training programs

Specific goals for the program are to:

• Develop active heating and cooling systems which can be commercially distributed and utilized to displace 2.0 quads of fossil energy by the year 2000

• Bring into mass production a minimum of six generic types of domestic water-heating systems that can be cost-effectively utilized in 50 percent of the geographic areas of the United States by 1985, and displace 0.05 quad of fossil energy with these systems by 1985 and 0.65 quad by the year 2000

• Provide development support to industry to ensure commercial availability of at least 10 cost-effective, residential-size space-heating systems and at least 6 cost-effective commercial-size space-heating systems by 1985, to displace 0.13 quad of fossil energy with these systems by 1985 and 0.75 quad by the year 2000

- Continue R&D activities in materials, components, and system configurations leading to at least two generic types of cost-effective cooling systems by the year 1990, and displace 0.6 quad of fossil energy with these systems by the year 2000

A large number of federal, state, and local agencies are involved in the Active Solar Heating and Cooling Program.

## Program Structure and Budget

The budget structure of the solar applications program contains the five subprograms listed below:

- Market analysis
  1. To determine the type of products needed for given market sectors and their cost goal
  2. To identify market barriers
  3. To prepare recommendations for action, such as incentives and policy to remove or mitigate them
- Systems development
  1. To conduct R&D resulting in technically proven components, design, and systems
  2. To project equipment and installation costs along with operating costs
  3. To gather reliability and performance data and analyze it via prototype and field testing
  4. To provide essential feedback to the product improvement process
  5. To provide resource assessments and urban and community solar systems planning
- Market test and applications
  1. To conduct large-scale market testing and product support activities for technically proven designs and products
  2. To induce manufacturers to invest in mass production capacity to reduce per-unit price to the customer, using federal funds (on a cost-shared ba-

sis) to assist the private sector and federal purchases to create demand

- Market development and training
  1. To gather and disseminate information in a format useful to customers and to those who influence the pace of new product development and sales
  2. To develop a trained contractor and installer work force in cooperation with industry
  3. To involve the program with other federal, state, and local agencies and private institutions to accelerate solar commercialization
  4. To remove uncertainty and enhance consumer confidence in new solar products
- Solar international programs: to export United States solar energy products and services to the industrial and developing world for the mutual benefit of each

The estimated expenditures for the Active Heating and Cooling Program are summarized in Table 16-19.

## Program Detail

This section describes the Active Heating and Cooling Program according to major budget categories. Figure 16-47 summarizes activities for the five principal heating and cooling technologies.

Market Analysis. The Market Analysis subprogram collects and analyzes market data to establish reliability, maintainability, cost, and performance goals for the Systems Development subprogram. Activities include analyses of economic and financial incentives; utility interfaces; legal, regulatory, and institutional issues; employment and manpower issues; consumer response; and domestic and international market potential.

Ongoing activities for FY 1980, which will continue during FY 1981, include:

- Direct solar space and water heating
  1. Continuing analysis of the effects of

Table 16-19. Estimated Expenditures for Active Solar Heating and Cooling.

| ACTIVE SOLAR HEATING AND COOLING PROGRAM | BUDGET AUTHORITY (DOLLARS IN THOUSANDS) | | |
|---|---|---|---|
| | ACTUAL FY 1979 | APPROPRIATION FY 1980 | ESTIMATE FY 1981 |
| OFFICE OF SOLAR APPLICATIONS | | | |
| Market Analysis | 1,013 | 1,200 | 2,300 |
| Systems Development | 27,500 | 20,100 | 20,000 |
| Market Test and Applications† | 60,200 | 35,000 | 11,400 |
| Market Development and Training | 1,700 | 11,000 | 15,400 |
| International | — | — | 3,000 |
| TOTAL | 90,400 | 67,300 | 51,700 |

†Includes Federal Buildings activity.

utility practices on water- and space-heating systems commercialization

2. Monitoring actual market penetration on a regional and national basis
3. Development of reliability, maintainability, cost, and performance goals for solar water- and space-heating systems and components on a regional basis
4. Continuing computer modeling, market surveys, and other methods of projecting market penetration and the impacts of federal and state financial initiatives on the solar market
5. Collection and analysis of performance, cost, reliability, and maintenance data of test-marketed systems
6. Collection and analysis of information concerning public response, industry involvement, and overall market acceptability of test-marketed systems

• Solar-assisted heat pump systems
1. Continuing accelerated market planning activities
2. Reliability, maintainability, cost, and performance criteria development for solar-assisted heat pump systems on a regional basis
3. Market projections for solar-assisted heat pumps with emphasis on the effect of present and projected regional utility rates
4. Analysis of the impact of various federal programs on the solar-assisted heat pump market

5. Monitoring market penetration progress regionally
6. Analysis of long-term economic factors that affect the future market for solar-assisted heat pumps

• Solar absorption heating and cooling systems
1. Continuing market planning activities with emphasis on systems for commercial buildings
2. Reliability, maintainability, cost, and performance goal development, in order to achieve competitive status with existing commercial heating and cooling equipment
3. Market-penetration predictions on a regional basis for a range of sizes of cooling equipment
4. Evaluation of federal and state initiatives
5. Monitoring the market penetration of absorption systems in commercial buildings
6. Analysis of the effects of long-term economic factors

• Solar-driven Rankine-cycle heating and cooling systems. The Market Analysis activities outlined for absorption systems are directly applicable to Rankine-cycle systems as well.

• Solar-driven desiccant heating and cooling systems. The Market Analysis activities outlined for absorption systems are also applicable to desiccant systems. The desiccant systems development program emphasizes hybrid (active-passive) systems for the residen-

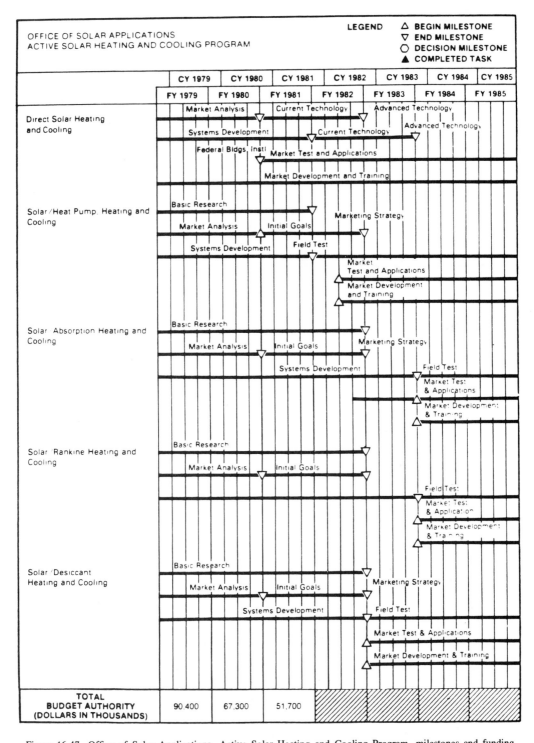

Figure 16-47. Office of Solar Applications, Active Solar Heating and Cooling Program, milestones and funding.

tial market. This is, of course, reflected in the Market Analysis activities.

**Systems Development.** The Systems Development subprogram focuses on engineering and developing solar products. The subprogram delivers engineering-tested components that are technically proven and ready for systems integration and prototype development; develops and tests prototype systems; conducts field tests to verify engineering performance; collects reliability and performance data; and assists in the development of manufacturing methods, product specifications, and applications.

Major activities by principal solar technology are listed below:

- Direct solar space and water heating
  1. Continued development and testing of advanced and improved system control methods and devices
  2. Continued development and publication of reliable engineering designs, information, and data; materials properties, selection guidance, and economic data; for designers and manufacturers of collectors, thermal energy storage devices, and system controllers
  3. Continued testing and evaluation of state-of-the-art water-heating systems
  4. Development and testing of advanced pumped and natural-convection water-heating systems
  5. Verification of improved design and installation practices for air systems
  6. Continued support of engineering field tests of advanced space-heating systems
- Solar-assisted heat pumps
  1. Continued systems analysis activities to define optimum heat-pump system configurations and control schemes to interface with the electrical and utility grid; development of simplified analysis and design tools
  2, Evaluation of existing collectors to reduce cost at operating temperatures appropriate for heat-pump operation
  3. Conducting low-cost collector development activities
  4. Evaluation of the earth's potential as a storage medium for solar-assisted heat-pump systems in a variety of soil types and climates
  5. Evaluation of the potential of other large-volume, long-term storage techniques (such as ponds) for residential and commercial buildings

6. Continued field testing of state-of-the-art, solar-assisted heat-pump systems
- Solar absorption and Rankine-cycle heating and cooling systems
  1. Continued systems analysis in order to identify optimum configurations, control schemes, attractive areas for component and system performance improvements and to develop appropriate analysis design and application tools
  2. Development of high-efficiency non-tracking and tracking mass-producible collectors
  3. Improved application method development for sensible heat-storage devices, improved phase change-material storage media for hot-side and cold-side storage applications at appropriate storage sizes, charge and discharge rates
  4. Advanced components and systems development solicitations (three cycles are anticipated—the first cycle concentrates on commercial-scale systems)
  5. Component testing of residential- and commercial-scale chillers currently under development
- Solar-driven desiccant heating and cooling systems
  1. Solid desiccant systems
     a. Analyses and experiments to identify required desiccant reactivation temperatures, preferred desiccant materials storage and system configurations, allowable system parasitic power requirements, and optimum hybrid (active-passive) systems
     b. Performance testing of existing integrated prototype collector-dryer panels for a range of field conditions
     c. Development of improved desiccant dryers with reduced parasitic power requirements and reactivation temperatures
  2. Liquid desiccant systems
     a. Determine required heat and mass-transfer rates, preferred

component configurations, liquid desiccant containment options, and allowable system parasitic power requirements

b. Prototype and engineering field testing of advanced solid and liquid dessicant systems

Major activities in Systems Development planned to be supported in FY 1981 are listed below:

- Solar direct heating—residential and commercial
  1. Completion of prototype testing of advanced water-heating systems and selection of the most promising systems for field tests
  2. Publication of information and engineering data generated by the collector materials program in a comprehensive handbook for collector designers and manufacturers
  3. Development of improved test methods for evaluating and rating solar heating and cooling systems
- Residential solar-assisted heat pump. A 1-year engineering field test and comparative evaluation of residential-size, solar-assisted heat pump systems.
- Commercial solar-assisted heat pump systems
  1. A 1-year engineering field test and comparative evaluation of four commercial-size, solar-assisted heat pump systems
  2. Evaluation of commercial-size solar ponds with heat pumps
- Solar absorption systems emphasis
  1. Field tests of commercial-scale chillers and systems currently being developed
  2. Tests of an advanced ammonia/water absorption chiller with significantly improved coefficient of performance
- Solar-driven Rankine-cycle heating and cooling systems
  1. Field tests of several chillers and systems currently being developed
  2. Prototype fabrication and test of a new fossil-fuel-boosted, solar-steam

Rankine chiller with high potential for increased efficiency
- Residential desiccant systems
  1. Construction and laboratory tests of a passive dehumidifier
  2. Construction and laboratory tests of a prototype liquid desiccant cooler

**Market Tests and Applications.** The major objective of this subprogram is to conduct large-scale market testing and product-support activities for designs and products that have been proven technically sound during systems development. This activity focuses on the five principal technologies under development, although solar desiccant systems are not expected to be tested until FY 1982. Activities include:

- Market tests of solar systems that have successfully completed engineering field tests
- Ongoing design review and construction assistance to federal agencies in purchasing, installation, and operation of active solar heating and cooling systems in federally owned buidlings
- Determination of overall market response resulting from legislated programs
- Assessment of marketability and appropriate federal assistance for industry-developed solar products

Activities focus on the following system types:

- Direct solar space heating and water heating
- Solar-assisted heat pump
- Solar absorption and Rankine-cycle heating and cooling systems

The Market Tests and Applications subprogram has had responsibility for management of the commercial phase of the joint DOE/Department of Housing and Urban Development heating and cooling demonstration program, and includes the following activities for FY 1980:

- Direct solar space and water heating
  1. Continued operational testing of commercial heating and cooling systems funded under the heating and cooling demonstration program
  2. Regional planning of market testing of improved and/or advanced systems
  3. Evaluation of market response, industry involvement, and overall market acceptability of solar products
- Solar-assisted heat pumps. The Market Test and Applications activities outlined for direct space- and water-heating systems are applicable also to solar-assisted heat-pump systems.

Major Market Test and Applications activities planned in FY 1981 include:

- Market testing of improved direct space- and water-heating systems
- Continued operational testing of commercial heating and cooling systems funded under the heating and cooling demonstration program

**Market Development and Training.** The major objective of the Market Development and Training subprogram is to gather information resulting from other solar program activities and disseminate it in a format useful to consumers and to others who influence the pace of new product development and sales. Information-dissemination activities were given high priority in President Carter's message to Congress. Activities include:

- Support of consumer protection and consensus-based product-performance standards and codes
- Joint programs with local building authorities, the Federal Housing Administration, and utilities
- Support to renewable-resource development and solar programs administered by other public and private agencies
- Information programs to serve the needs of builders, engineers, lenders, farmers, homeowners, industrial and utility decisionmakers, government officials, and the general public
- Training programs for contractors and installers, and development of educational programs for vocational-technical and secondary schools and colleges

Market Development and Training focuses on the five principal solar system types. Activities for FY 1980, which will continue in FY 1981, are listed below:

- Direct solar space and water heating:
  1. Promotion of general public awareness of the benefits of direct solar space and water heating through mass media, consumer education, and other programs aimed at creating a constituency and market demand for these systems
  2. Establishment of appropriate instructional programs to enable heating, ventilating, air-conditioning, electrical, mechanical, and construction contractors to install and maintain direct solar space- and water-heating systems
  3. Assessment of the need for and training of code, inspection, licensing, and planning officials associated with space- and water-heating installations
  4. Definition of the requirements for and implementing primary, secondary, and college educational programs, in direct solar space- and water-heating technology
  5. Establishment of a program of workshops, seminars, and educational materials on the financial, legal, environmental, and technological issues of solar space- and water-heating technology
  6. Dissemination of technical and marketing information on solar space and water heating to manufacturers, distributors, and potential users
  7. Development and dissemination of

information on direct solar space- and water-heating initiatives, incentives, and financial programs to the building industry, the financial community, and users

8. Establishment of technical and marketing assistance programs to consumers, designers, installers, distributors, and others in the supply infrastructure

9. Development of an International Market Information Program to stimulate foreign demand for direct heating systems

10. Support of the development of component standards, a comprehensive approach to systems standards and certification, and the development and adoption of model solar codes

11. Continued support of component testing

12. Development of consumer assurance programs

- Combined solar-assisted heat pump systems. Training, information, and education programs as outlined for direct solar space and water heating are applicable for solar-assisted heat pump systems.

- Combined solar absorption and Rankine-cycle heating and cooling systems. Training, information, and education programs as outlined for direct solar space-heating and water heating are applicable for solar absorption and Rankine-cycle systems.

- Solar-driven desiccant heating and cooling systems. Training, information, and education programs as outlined for direct solar space-heating and water-heating systems are applicable for solar/desiccant systems.

International Solar Programs. The major objectives of the subprogram are to enhance American solar technical capabilities and to increase exports of United States–manufactured solar equipment, in order to contribute to the United States policy goal of reducing its—and global—dependence on imported oil and encouraging the transition from depletable fuels to alternative energy sources.

An international solar commercialization working group has been created to coordinate solar export development activities of the federal government. The activities of this group support United States industry in marketing solar hardware and expertise abroad. The benefits of increased foreign sales will be the reduction of unit costs of solar systems, expanded United States manufacturing capability, and overseas project experience.

Technical cooperative R&D projects are under way both bilaterally and multilaterally. Bilateral agreements have been initiated with Saudi Arabia, Italy, Israel, Spain, and Mexico, and other arrangements are in the planning stages. Multilateral projects are generally carried out under the aegis of the International Energy Agency. Joint technical collaboration serves to fill gaps in the domestic R&D program, reduce costs to the domestic program, accelerate the achievement of program objectives, and broaden the pool of technical talent, approaches, and ideas.

## PASSIVE AND HYBRID SOLAR HEATING AND COOLING

Passive and hybrid solar buildings employ designs and products where energy transfer into, out of, and within a building relies primarily on natural processes—conduction, convection, and radiation. Minimal dependence is placed on mechanical equipment such as fans, pumps, and compressors unless they can be used effectively to augment natural energy flows for control purposes or when capital costs and operating energy are justified by improved system performance. When other solar technologies (active heating systems or photovoltaics) also are integrated into the design, the result is considered a hybrid solar application.

The Solar Energy Domestic Policy Review (DPR) estimated that passive techniques could displace as much as 1.0 quad of conventional energy in the year 2000. However, the DPR found that passive design techniques had not been adopted by the building industry because designs and systems are not well understood by the industry and general public and because builders are unwilling to take risks with unconventional designs.

Given that nearly 80 percent of the energy consumed in residential and commercial buildings is used for space heating and cooling, lighting, and water heating, passive and hybrid solar energy systems can make an important contribution toward reaching this objective.

Based on the concept of energy delivery to the space, passive heating systems may be divided into three broad categories:

- Direct heating: solar radiation enters the space directly to heat the interior surfaces and contents
- Indirect heating: an intermediary surface external to the space absorbs and converts the radiation into heat
- Isolated heating: the external surface is isolated from the space so the temperature of the space can be regulated

Figure 16-48 illustrates some of the basic

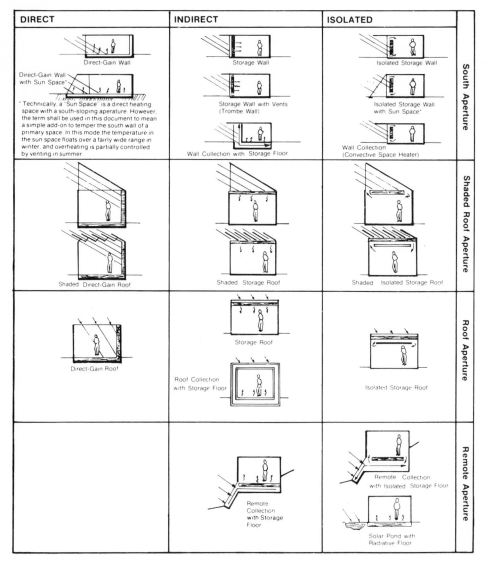

Figure 16-48. Examples of passive solar heating systems.

design possibilities for passive heating, plus collection techniques. The four collection techniques are:

- South aperture
- Shaded roof aperture
- Roof aperture
- Remote aperture

Combinations of the four collection techniques with the three delivery choices account for the varied design possibilities being stud-

ied by the DOE program. DOE is also performing research on and demonstrating how these techniques can be applied to larger, multistory buildings.

Passive cooling involves designing processes to discharge heat through natural means. The sky, atmosphere, ground, and water are potential heat sinks for these systems. Figure 16-49 illustrates varied design concepts for passive cooling systems. As in space heating, roles for passive cooling in more general energy management schemes for large-scale

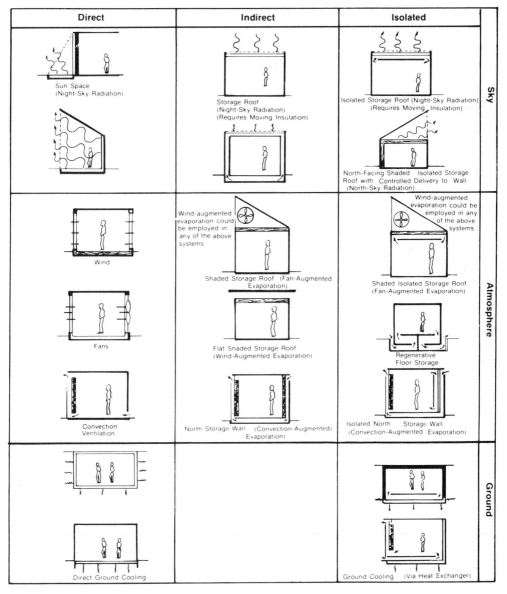

Figure 16-49. Examples of passive solar cooling systems.

commercial buildings are being identified and evaluated.

Passive and hybrid systems are projected to provide 1.0 quad of energy in the year 2000. An additional quad could be saved by integrating passive and hybrid solar technologies into large-scale, multiple-building applications in cities and towns. Architects and builders are now beginning to offer passive heating of homes to the public. However, most installations are in relatively expensive, custom-built homes. There has been little activity in applications in housing development and retrofits to existing buildings. The home shown in Figure 16-50 uses a large expanse of glass for a greenhouse that provides comfortable temperatures on even the coldest winter days. This configuration provides 90 percent of the annual space-heat load. The house in Figure 16-51 uses a large expanse of south-facing windows, an overhanging roof, and thermal storage collectors in the roof to reduce conventional fuel consumption by 40 percent.

## Program Strategy

To meet an established short-term goal of saving 0.1 quad by 1985 through the use of passive and hybrid systems, approximately 12 percent of new residential and commercial buildings (about 1 million) and 59 percent of existing buildings (about 360,000) must be passively solar-heated to a significant degree (30 to 80 percent of the energy requirements). The long-term goal is 1 to 2 quads in the year 2000, if 1985 goals are met.

The Passive and Hybrid Systems Program strategy has two principal thrusts:

- Introduction and dissemination of passive systems design and construction technologies into the building industry

Figure 16-50. Maximum use of direct solar energy.

Figure 16-51. Integrated passive solar home.

- Development of new materials products and systems for buildings and commercial systems

Because passive solar technology is essentially a design procedure employing passive components, the essence of the overall program strategy is (1) to create design procedures in a form amenable to application by architects, engineers, and builders; (2) to encourage the profession to apply the procedure; and (3) to inform both the consuming market and the financial, utility, and real estate sectors of the advantages of these procedures and the underlying technical approaches. Successful commercialization will accomplish energy-conserving, cost-effective buildings and community systems. Development of passive solar products and systems will be a key to widespread use and increased energy savings from advanced systems. The implementation of this strategy is shown in Figure 16-52.

The emphasis of the Passive and Hybrid

Systems Program has been divided into two functional areas: Systems Development and Market Development. The program is divided further into three application categories: building systems, communities, and agriculture. Since the principal impediment to passive system adoption is information, the thrust of the program is to make applications visible throughout the nation. Specific goals for federally supported installations in each application are listed below.

- Buildings
  1. Service hot water: 1000 installations by 1985 using thermosyphon, breadbox, and other systems
  2. Space heating: 2000 single-family and 1500 multifamily units by 1983, plus construction of 500 commercial buildings by 1983 using direct-gain wall, sun space, storage wall, convective loop, and other systems
  3. Space cooling: 800 residential and 200

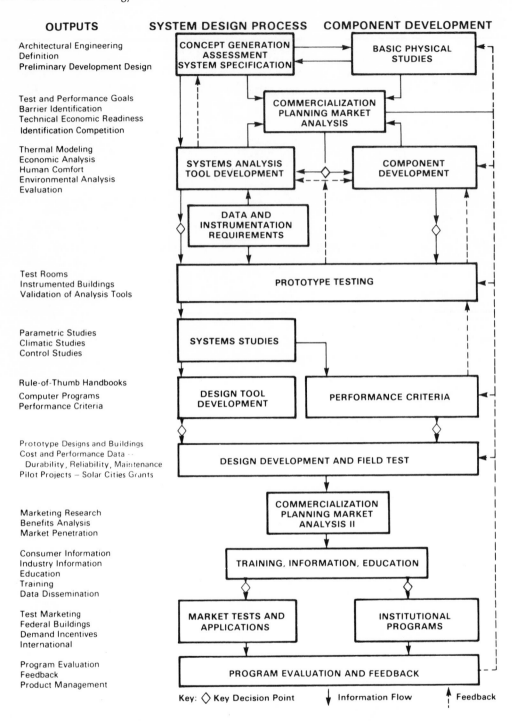

Figure 16-52. Passive and hybrid program implementation.

commercial buildings by 1983–85, using storage roof and storage wall, ventilation, evaporation, ground, desiccant, and other systems

4. Hybrid systems: 500 residential and

commercial thermal and electric application for 1981–87 using combined active, passive, photovoltaic, and other systems

• Communities and cities: 150 pilot projects in cities and towns will extend over 10 years, beginning in 1980–81. They will concentrate first on passive and active solar systems, and later on the mixed use of other conservation and solar technologies

- Agriculture: 500 projects are planned between 1981–85 for grain drying, greenhouses, and shelter heating and cooling.

At DOE headquarters, the Passive and Hybrid Systems Branch in the Office of Solar Applications provide general management for the program. In addition, the Department of Housing and Urban Development and the Department of Agriculture have important roles. The implementation of program elements and project management involves DOE field offices, the Solar Energy Research Institute, the national laboratories, regional solar energy centers, and key support contractors. Approximately 80 percent of the work will be performed by businesses, universities, nonprofit organizations, and state and local governments. A regional management focus is required for this program because the use and characteristics of passive solar systems vary according to local conditions.

## Program Structure and Budget

Table 16-20 summarizes the estimated funding for the Passive and Hybrid Systems Program. Expenditures for Market Development and Training represent a $10.9 million increase over the FY 1980 authorization.

## Program Detail

The DOE Passive and Hybrid Systems Program has activities in Systems Development, Market Tests and Applications, and Market Development and Training. The two principal products of the Systems Development Activity in passive systems are either a specific product or design and information regarding design on actual buildings that demonstrates proven performance of a product. These products are developed through research, testing, and assessment.

Eight technology applications have been identified using passive and hybrid systems for both multiple use (buildings and communities) and agricultural markets:

- Service hot water: involve thermocirculation hot-water systems using flat-plate liquid- or air-heating collectors, breadbox water heaters, and integrated site-built systems. The systems for this application are not fully developed.
- Space heating; includes 15 single-zone systems that can be used singly or in combination for multizone residential and commercial building applications, including multiple large-scale buildings.
- Space cooling: includes 17 single-zone systems using convective, evaporative, conductive, and radiative heat exchange processes for cooling. These systems can be used in combination for multizone residential, commercial, and community applications.
- Mixed technologies: integrates solar

## Table 16-20. Estimated Expenditures for Passive and Hybrid Systems.

| PASSIVE AND HYBRID SOLAR HEATING AND COOLING PROGRAM | BUDGET AUTHORITY (DOLLARS IN THOUSANDS) | | |
| --- | --- | --- | --- |
| | ACTUAL FY 1979 | APPROPRIATION FY 1980 | ESTIMATE FY 1981 |
| OFFICE OF SOLAR APPLICATIONS | | | |
| Market Analysis | 400 | 1,900 | 2,150 |
| Systems Development | 5,100 | 16,200 | 15,000 |
| Market Test and Applications† | 18,000 | 10,000 | 9,000 |
| Market Development and Training | 400 | 5,800 | 16,700 |
| International | — | — | 1,000 |
| TOTAL | 23,900 | 33,900 | 43,850 |

†Includes Federal Buildings Activity.

technologies into urban multiple building applications to be carried out under the Solar Cities and Towns projects.

- Hybrid thermal and electric: integrates solar thermal and electrical systems into residential and commercial buildings. Systems development and analysis are being undertaken to identify workable combinations.
- Space heating for shelters and greenhouses: includes farm and animal shelters, as well as small and commercial, attached and detached greenhouses. The systems for these applications are being studied by the Department of Agriculture and the Solar Energy Research Institute.
- Drying of agricultural and industrial products.
- Cooling for agricultural and industrial processes, including applications for animal and produce storage.

Systems Development. DOE has funded Systems Development activities since 1976. Most support has gone to determine the thermal requirements of various kinds of buildings and the performance characteristics of passive heating systems. A smaller amount has supported passive cooling research and market development activities.

The Solar Cities Project provides grants to cities and towns to become more energy self-sufficient by integrating solar and energy-conscious designs into their policy, planning, and development processes.

The status of major activities in heating buildings is:

- A computer simulation model for thermal storage walls is computed and in use. Mass storage-wall (with and without vents) and water storage-wall computer modeling approaches have been validated by comparison with test room results. Models for direct gain, sun space, and roof storage systems are being developed. Three thermal storage-wall heating systems are being analyzed using hourly solar and weather data from 29 cities.

- Twenty-five passive-solar-heated buildings are now instrumented; 2 to 4 months of winter data have been accumulated for each building. Fifteen 5-by-8-by-10-ft test rooms have been constructed.
- Twelve building experiments have been funded to demonstrate passive and hybrid solar heating and cooling systems involving a range of systems: direct gain (south and shaded roof apertures), storage walls, storage roof, sun space (greenhouse and solariums), and ponds.
- Seven assembly component studies have been funded to develop, monitor, and assess various passive system elements, including heat pipes, control techniques, collector assemblies, reflectors, lightweight storage, and thermic diodes.
- A manufactured buildings program has been initiated to investigate how passive techniques apply to factory-built residential, commercial, and agricultural buildings. Twenty-seven manufacturers are involved in this program.
- Forty-two commercial building prototype projects are being supported to integrate passive heating, cooling, and daylighting into commercial structures.
- Investigations of the macro- and microeconomics of residential applications of passive techniques are ongoing, including the development of a passive system costing technique and an economic system optimization procedure.
- A program has been initiated to identify and evaluate possible roles for passive solar energy within more general energy management schemes for large-scale commercial and industrial buildings.
- Systems studies are under way to investigate various passive solar domestic-hot-water options.
- A passive solar simulation capability is being integrated into public building energy-analysis computer programs.
- Two user-oriented handbooks for residential design were completed in FY 1979.

The status of major activities in passive and hybrid cooling is:

- One test cell has been funded to test and analyze a combined passive and hybrid night-sky radiation and convection cooling system.
- Data collection and evaluation are under way for a roof pond passively heated and cooled house in Las Cruces, New Mexico.
- Computer simulations have been used to calculate the net rate of heat dissipation by the combination of radiation, convection, and evaporation from horizontal surfaces at fixed temperatures in 58 American cities. An assessment of the net cooling rates for tilted surfaces has been conducted for four cities representing a wide range of climates.
- A detailed spectral radiometer has been developed and four instruments have been deployed in various parts of the United States.
- A contract has been awarded to refine movable-insulation schemes for thermal storage roof systems. An alternative storage roof system using fixed insulation and moving water has been built and monitored.
- A 5-year joint United States–Saudi Arabia program in passive cooling was begun in FY 1979.

The status of activities in the agricultural applications area is:

- Basic physical studies are under way in solar heating and greenhouses, direct drying, and cooling.
- Materials, components, and systems development and assessment activities are also under way for solar heating and greenhouses and direct grain drying and cooling.
- Prototype testing and design development and field testing are under way for solar heating and greenhouses.

Activities in passive heating that are planned for FY 1981 include the following:

- Complete the first cycle of 42 engineering field tests of commercial buiding projects
- Develop computer design tool capability for heating and initiate studies for cooling
- Develop performance criteria for passive heating systems
- Complete the first cycle of manufactured buildings field tests
- Collect performance data for residential heating systems

Under the Systems Development Program, the following physical studies are planned for FY 1981:

- Basic R&D will be conducted on physics (radiation, ground properties, etc.) and materials for glazing solar control films and selective absorption surfaces.
- Residential space-cooling studies (dehumidification, evaporation, etc.) will be initiated in FY 1981 and completed in FY 1983.
- Shelter heating and greenhouses for agricultural applications will be completed in FY 1983.

Data implementation, side-by-side prototype testing, test rooms, and instrumented buildings should be completed in FY 1981 in:

- Space heating for single-family residential buildings; multifamily to be initiated in FY 1981 and completed in FY 1982
- Service hot water for residential and commercial buildings (to be completed in FY 1982).

Systems analysis and design tools should be completed in FY 1981 for:

- Two handbooks on space heating in single-family residential buildings
- Three commercial buildings computer programs
- Service hot-water and cooling design tools

Design development and field tests should be completed in FY 1981 in:

- Service hot water for residential and commercial buildings
- Multifamily project completion delayed until FY 1982
- Product development, expected to be completed to permit development and field testing of 30 additional marketable passive and hybrid products

The Solar Cities Project should be completed in FY 1981 for:

- Three to five demonstrations

Market Analysis. The market analysis activities support the Passive and Hybrid System Program through analysis of the types of products needed for given applications, the resulting cost goals, identification of barriers, and recommended actions for barrier removal, such as incentives and policies. It also provides commercialization support.

The passive program plan for commercialization planning and market analysis for space heating in single-family residential buildings and for service hot water for both residential and commercial buildings, was completed in 1980. Market development activities have begun in most areas.

Market development activities planned for FY 1981 include commercialization planning and market analysis in:

- space heating in single-family residential buildings
- space heating in commercial buildings
- space heating in city and town multiple buildings
- shelter heating and greenhouses in agricultural applications
- direct grain drying

Market Development and Training. Commercialization of passive solar systems is distinctly different from commercialization of hardware-oriented technologies. In the latter case, the components and systems are typically visible entities, and their contribution to the welfare of the consumer is easily deter-

mined or readily apparent. Passive solar systems, on the other hand, generally are integral parts of the building design. The passive solar components are not easily distinguishable from the remainder of the structure; they often consist of materials that already are common in the building industry; e.g., glass, masonry, and insulation.

Only limited passive commercialization activities exist throughout the country, and most of them are nonfederally funded. A few of these activities are listed here.

- The Department of Housing and Urban Development has sponsored, in conjunction with DOE and the Solar Energy Research Institute, a passive residential design competition resulting in 262 awards.
- The states of Minnesota and Illinois have sponsored residential design competitions.
- The Tennessee Valley Authority, Wisconsin Power & Light, Pacific Gas & Electric, and many other utilities are investigating the peak load sharing potential of passive systems and consumer loan programs to finance solar system costs.
- New Mexico, California, Oregon, and Colorado have income tax credits for passive systems.
- Virginia has a property-tax-exemption provision for passive systems.
- California, New Mexico, and several other states have initiated passive-solar education courses for builders and designers.
- Four regional solar energy centers have been established by DOE to assist in the accelerated commercialization of solar energy technology.
- Certain localities (e.g., Davis, California) have modified building codes and zoning regulations to accommodate passive systems.
- The Solar Energy Research Institute has completed a technology readiness assessment of passive solar heating and cooling, focusing primarily on economic performance projections.
- Several publications and pamphlets on

passive systems have been completed for designers and builders, including a passive solar bibliography and a survey of passive solar buildings.

- Training, information, and education for service hot water in residential buildings. Space heating and cooling was delayed 1 yr.
- Institutional programs for service hot water in residential and commercial buildings.

Market Tests and Applications. The objective of Market Tests and Applications is to test passive solar technologies under actual market conditions and to monitor tested installations to obtain performance information to be used for systems development and policymaking. FY 1980 activity provided funds for the design of passively heated buildings from 27 different manufacturers, on a regional basis, and continued manufacturing of previously authorized installations.

- The FY 1981 budget request supports continued tests of 100 manufactured buildings utilizing passive design.

- Initiation of 175 new single-family dwellings using passive techniques. Additional residential activity will occur in FY 1982.
- There will be dissemination of cost and performance information from earlier projects.
- Information and education programs for builders and consumers will focus on residential heating applications.

International Solar Programs. A number of international bilateral cooperative activities are under way with Israel, Italy and Mexico, which involve passive solar research, development, and demonstration projects. A multilateral program sponsored by a NATO pilot study program, entitled "Committee for the Challenges of Modern Society" (CCMS), supports a passive solar applications group. Approximately 15 countries participate in RD&D activities (i.e., United States, Great Britain, France, Israel, Spain, Mexico, Denmark, Switzerland, Greece, and others) aimed at advancing the state-of-the-art of passive solar applications.

# 17. Solar Program Description by Function

This chapter describes the DOE Solar Energy Program by crosscutting function and provides backup for the FY 1981 Office of Solar Applications budget request. Details on solar activities by technology (for example, solar heating and photovoltaics) are provided in Chapter 3.

A number of DOE organizations are involved in the solar program, in addition to the Office of Solar Applications and Solar Technology under the DOE Assistant Secretary for Conservation and Solar Energy. The DOE Assistant Secretary for Environment supports solar energy in environmental research and technology assessments, and the DOE Office of Energy Research supports a number of basic research and solar-specific activities in advanced materials, photochemistry, and advanced photovoltaic materials, as well as the concept evaluation of the solar power satellite. DOE's Energy Storage Division, under the Deputy Assistant Secretary for Conservation, is developing a number of storage options that specifically support solar tech-

nologies, and the Division of Electric Energy Systems (under the Assistant Secretary of Resource Applications) is cooperating in integrations of solar technologies into the electrical utility grid.

This chapter describes the Solar Energy Program by functional breakdown. In many respects, this duplicates the technology-specific material presented in Chapter 3; however it presents it in functional rather than topical terms.

These functional activities can be grouped into a logical sequence, as follows in Table 17-1.

The overall DOE funding for solar-related programs was $707 million for FY 1981. A functional breakdown of the solar-related budget is shown in Table 17-2.

## TECHNOLOGY BASE

Technology Base activities include basic and applied research and exploratory development.

Table 17-1. Functional Crosscut of Overall DOE Solar-Related Activities.

| FUNCTION | DESCRIPTION |
| --- | --- |
| Technology Base | Basic and applied research and exploratory development |
| Environment, Health and Safety, and Social Impact | Research and procedural compliance in environment, health and safety, and social aspects |
| Market Analysis | Characteristics of energy demand and use |
| Technology Development | Advancements in development of technology integral to providing an energy service |
| Systems Development | Advancements in design of cost-effective systems to provide energy |
| Market Tests and Applications | Gathering data on user experience and acceptance of particular solar systems |
| Market Development and Training | Dissemination of information on solar systems to potential users and provisions of worker training |

## Table 17-2. Functional Breakdown of FY 1981 Solar Budget.

| FUNCTION | BUDGET AUTHORITY* (DOLLARS IN THOUSANDS) | | |
| --- | --- | --- | --- |
| | APPROPRIATION FY 1979 | APPROPRIATION FY 1980 | ESTIMATE FY 1981 |
| Technology Base | 71,326 | 93,173 | 114,713 |
| Environmental Analysis | 8,500 | 9,200 | 6,900 |
| Market Analysis | 4,700 | 8,400 | 14,500 |
| Technology Development | 170,263 | 202,384 | 178,042 |
| Systems Development | 170,111 | 214,742 | 235,845 |
| Market Tests and Applications | 106,700 | 72,400 | 76,000 |
| Market Development and Training | 2,800 | 20,500 | 40,900 |
| International | — | — | 15,000 |
| Program Direction/SERI Facility | 9,141 | 16,765† | 17,300 |
| TOTAL | 543,541 | 637,564 | 699,200 |

*Estimate. Includes activities in the Office of the Assistant Secretary for Conservation and Solar Energy, Office of the Assistant Secretary for Resource Applications (Electric Energy Systems Division) and the Office of Energy Research (Basic Energy Sciences Program and the Solar Power Satellite). In FY 1981, this includes $186.1 million from the Office of Solar Techonology, $465.9 million from the Office of Solar Technology, $32.6 million from the Office of Energy Research, $2.8 million from the Assistant Secretary for Environment, $7.0 million from the Electric Energy Systems Division (Resource Applications), and $4.8 million from the Energy Storage Division (Conservation).
†Includes purchase of a scientific computer for SERI.

They encompass efforts to increase knowledge and quantitative understanding of both natural phenomena and the environment; problem-oriented research in energy-related sciences, with no clear-cut applicability to specific projects; and, finally, research efforts focused on ultimate application to a particular system, product, or process. These activities often span several scientific disciplines and lay the groundwork for innovation in particular energy technology areas.

## Goals

A major function of the Technology Base effort is to conceive of and explore technology options from a fundamental point of view. This results in the evolution of a broad R&D base that can provide the necessary underpinnings for the development of specific projects to carry promising technology options forward. Specifically, the primary goal of exploratory development is to bring candidate energy systems to a point where they can be rigorously screened for technical, environmental, and commercial viability.

An additional function of Technology Base activities is to provide a resource for attacking technological problems inherent in ongoing development programs, as well as unforeseen technological problems; e.g., from externally changed constraints. A final function of the Technology Base effort is to provide the technological backup necessary for the performance of the DOE's non-R&D functions, such as environmental protection activities under the National Environmental Policy Act, or technical aspects of resource assessment, policy development, and regulation.

## Budget

The budget for Technology Base activities (shown in Table 17-3) includes specific programs related to the five technologies in the DOE Office of Solar Technology, as well as solar-related research efforts in the Basic Energy Sciences Program of the Office of Energy Research.

## Program Thrust

Biomass Energy Systems. The major portion of current biomass usage is generated by the forest products and pulp-and-paper industries for their own use from mill residues and process streams, with 0.2 to 0.3 quad used for residential heating. In order for biomass to contribute a total of 6 to 10 quads/yr of energy after the year 2000, new techniques must

Table 17-3. Estimated Expenditures for Technology-Base Activities.

| ACTIVITIES | BUDGET AUTHORITY (DOLLARS IN THOUSANDS) | | |
|---|---|---|---|
| | ACTUAL FY 1979 | APPROPRIATION FY 1980 | ESTIMATE FY 1981 |
| OFFICE OF SOLAR TECHNOLOGY | | | |
| Biomass | 573 | 2,462 | 15,000 |
| Photovoltaics | 34,630 | 47,000 | 50,000 |
| Wind | 3,909 | 5,600 | 5,453 |
| Solar Thermal | 3,613 | 4,825 | 8,350 |
| Ocean Systems | 5,311 | 4,386 | 11,610 |
| OFFICE OF ENERGY RESEARCH | | | |
| Basic Energy Sciences† | 16,700 | 20,400 | 27,100 |
| TOTAL | 64,746 | 84,673 | 117,513 |

†Operating Expenses only.

be developed, and increased understanding of biomass energy conversion processes and types of fuel products derivable from them must be obtained.

The objective of the biomass Technology Base activities is to identify and investigate areas of research that promise to increase the resource base of biomass, the rate of conversion into energy and energy-related products, and the types of fuels and petrochemical substitutes that can be derived from biomass.

Major Technology Base efforts are under way in three areas:

- Land-based aquatic biomass production
  1. The University of California at Berkeley is evaluating the production and harvest of microalgae in a 1000-m² pond. The 48-hr settling batch process involves bioflocculation of the algae prior to sedimentation. The objective is to develop a prototype integrated wastewater-treatment/biofuel-production system that includes the anaerobic digestion of the microalgae to methane gas.
  2. The Woods Hole Oceanographic Institute is determining the maximum rates of biomass production of selected species of seaweeds and freshwater plants in 0.25 acre ponds at Fort Pierce, Florida. Sustainable yields of 25 ton/acre/yr for freshwa-

ter hyacinth have been demonstrated in smaller test facilities.

- Production of hydrocarbon-bearing plants. The Lawrence Berkeley Laboratory has experiments under way to analyze hydrocarbons as well as to optimize extraction techniques from the *Euphorbia tirucalli*. The plant produces a milky juice called *latex*, one-third of which is composed of hydrocarbon-like materials that may be suitable as refinery feedstocks. Included in the projects are genetic selection designed to improve the yield of the latex.
- Biophotolysis. The University of California at Richmond has demonstrated the continuous production of hydrogen by blue-green algae through biophotolysis. Current studies are focused on improving the rate of hydrogen production in outdoor cultures, and on comparing the hydrogenase systems in vivo and in vitro.

In addition to these ongoing activities, a program of basic and applied research and exploratory development is being carried out in such areas as advanced biochemical conversion techniques (e.g., biophotolysis), plant breeding, biomass farming, pyrolysis, fermentation, and distillation. A major focus is on identification of biomass feedstocks with significant production potential.

Photovoltaic Energy Systems. The Technology Base for phtotovoltaics is already quite broad. Crystalline silicon technology is well advanced. Other cell materials and design approaches are being vigorously pursued through a program of advanced R&D. This effort is designed to ensure a broad and advancing base of knowledge and to demonstrate technical feasibility of various new materials and components. The specific objective is increased cell efficiency and reduced cell costs. It has already resulted in the following accomplishments:

- Obtained about 9 percent efficiency in two thin-film cells (9 percent on cadmium sulfide and 9.5 percent on polycrystalline silicon)
- Obtained 12 percent efficiency on 5-element (III-V) compound cell
- Obtained 28 percent efficiency on a photovoltaic concentrator split-beam cell
- Accelerated the silicon thin-film effort

Photovoltaic Technology Base activities were pursued in FY 1981 in the following areas:

- Research support and fundamental studies in photophysics, photochemistry, optics, and materials
- High-risk items: advanced concentrators, eletrochemical cells, innovative concepts, emerging materials.
- Advanced materials and cell research: amorphous silicon (Si), gallium arsenide (GaAs), cadmium sulfide (CdS), and polycrystalline silicon (Si).

Wind Energy Conversion Systems. The Technology Base of the wind program includes subprogram work in research and analysis to investigate innovative concepts; analysis of wind system economics and modeling of wind systems in utilities; and a subprogram to develop techniques for wind prospecting and wind resource assessment.

The status of this work and accomplishments to date are described below.

- Innovative concepts
  1. The innovative concepts currently being investigated are the diffuser-augmented wind turbine, the tornado wind system, electrofluid dynamic driven generators, the Madras rotor power plant, and the wind/electric power transducer.
  2. The generic studies consist of evaluations of augmented horizontal and vertical axis systems, highlift devices, vortex extraction devices, and sailwing-type systems.
- Wind energy prospecting
  1. Large-area survey techniques were developed and successfully tested in the Pacific Northwest.
  2. Contracting for regional resource assessment was initiated in FY 1979 with all regions completed in FY 1980.
  3. The National Wind Resource Atlas was published in FY 1981. This provides estimates of the resource and its uncertainty across the nation on a 15-mile grid. For areas that appear to have good wind resources but high uncertainty, specialized assessments were made with a revised/updated atlas to be published in FY 1982.

The major wind energy system technology base activities to be pursued in FY 1981 are as follows:

- Innovative concepts
  1. Development of a cost estimation methodology applicable to innovative wind concepts by the Solar Research Institute (SERI). (The methodolgy will be applied to systems under investigation within SERI-managed subcontracts and will be used in the evaluation of future innovative concepts. The methodology will also be used in identifying those

concepts that should receive greater support under the evolving technologies program element.)

2. Investigation of several innovative concepts and generic studies under the auspices of SERI.

- Information dissemination. Conducting a newly developed type of decision-oriented workshop for utilities and other user groups; briefings; site visits; face-to-face contacts to articulate the various reports, workbooks, case studies, computer models, and other available analytic tools.
- Wind energy prospecting
  1. Preparation of a National Wind Resource Atlas using the data obtained in the regional assessments.
  2. Preparation of simple, efficient siting methodologies.
- Electrical utility analysis
  1. Existing tool designs have been acquired by SERI and their adequacy for modeling wind systems is being addressed.
  2. In parallel, technical requirements for improved models are being developed.

**Solar Thermal Power Systems.** Technology Base activities are an integral part of the solar thermal energy system development strategy. By providing improved, more durable materials and higher performance, lower cost subsystems, systems, and processes, the competitiveness of solar thermal energy is advanced. In addition, potential new applications for solar thermal energy are identified and supported through exploratory development.

Ongoing Technology Base programs are exploring advanced system designs and have already identified major application categories where solar thermal technology can penetrate additional national energy market sectors (i.e., advanced electricity, fuels and chemicals, and industrial process heat at temperatures greater than 1000°F). Advanced subsystems and components are also being studied, as are a variety of materials configurations and characteristics.

*Major activities.* The solar thermal Technology Base activities are divided into two areas: evaluation of advanced systems and applications, and advanced materials. The advanced systems and applications evaluation activity emphasizes development and understanding of systems/applications beyond the capabilities of the solar thermal technology currently being developed. These systems will provide higher performance and lower cost alternatives. New applications that can provide additional demand for solar thermal systems and subsystems include advanced electricity, fuels and chemicals, and industrial process heat (greater than 1000° F). The advanced electricity subelement is aimed at demonstrating the high performance potential of small (15 to 20 kWe) heat engine-alternator combinations operating at the focus of two-axis, tracking dish concentrators. Fuel and chemical activities will strive to adapt existing processes, such as coal gasification, to a solar heat input and to develop new processes that take advantage of the characteristics of solar heat. The industrial process heat subelement will look into requirements for high-temperature solar thermal industrial process heat systems and identify needed development work.

There are five material functions relevant to solar thermal technology: absorbing, reflecting, transmitting, structural support, and heat transfer fluids/containment. Near-term emphasis is on:

1. identifying a low-cost durable silver/glass mirror system

2. evaluating thin glass for reflector and transmitting material applications

3. identifying low-cost stable plastics for reflectors and transmitting materials

4. establishing the temperature stability of black chrome and black cobalt selective absorbers

5. evaluating cellular glass, plastics, wood/paper products and composites as support materials for advanced heliostats and concentrators

6. evaluating ceramics for high-temperature receivers

7. developing appropriate measurement techniques and formulation of a detailed solar materials property data base

8. understanding degradation mechanisms

Intermediate-term R&D emphasis is on intermediate-temperature selective absorbers (705° to 1300°F), compatible fluid/alloy containment combinations operative at temperatures greater than 1100°F, composite structural materials, and detailed understanding of degradation mechanisms leading to accurate lifetime prediction.

Ocean Systems. The major element of the ocean systems Technology Base effort is advanced R&D. The ocean systems program is supporting four main technology areas for estimating energy from the oceans: ocean thermal energy conversion (OTEC), salinity gradients, waves, and currents.

Of the four main technology areas, OTEC is the most advanced and receives approximately 95 percent of ocean systems funding. The OTEC program is at the stage where preliminary design for a pilot plant were initiated in FY 1981, with subsequent construction and deployment by the end of FY 1985. The other three technologies require further research, technology, and systems definition of efforts to determine the more promising candidate systems for commercial-scale conceptual designs.

In OTEC, investigations of water-side and working-fluid side-heat transfer-enhancement systems for closed-cycle heat exchangers have been performed. Also, two advanced open-cycle systems are being evaluated. These approaches use the potential energy of the warm water to make a foam or mist from this water, which is then lifted and condensed. Hydraulic turbines are used to convert the energy.

In the wave energy area, experiments have been conducted in wave tanks during FY 1980, and comparative economic assessments have been made between alternative wave-focusing approaches.

For salinity gradients, the cost analysis of a 50-kWe osmotic plant were completed in FY 1980 and a small-scale solar, stratified pond/ dialytic battery experiment also was conducted. Preliminary cost projections are not competitive with other ocean technologies, but if final results look attractive, then a reassessment of priorities within the advanced concepts will be made. This priority reassessment would be required, since current funding plans do not provide for further salinity gradient work.

In ocean currents, an engineering design of a 400-kW prototype turbine including grid, cable, and mooring requirements was completed in FY 1980.

The major ocean systems Technology Base activities that were pursued in FY 1981 are listed below.

- *Mist and foam*. Continuation of the laboratory effects being performed on mist and foam open-cycle systems. These will include seawater tests to determine the impact of foam generation on the environment and plugging of the apertures in mist generators.
- *Waves*. On the basis of results of the wave tank experiments and assessments of alternative wave-focusing approaches, pilot experiments were initiated in FY 1981. Tests on a uniquely configured wave turbine will be continued as part of an international program with Japan, England, and Canada.
- *Ocean Currents*. Based on the engineering design referred to above, cost-effectiveness projections and environmental assessments for ocean current energy conversion will proceed in FY 1981.

Basic Energy Sciences. The Basic Energy Sciences (BES) Program in the DOE Office of Energy Research is charged with the responsibility of conducting basic research in disciplines underlying the energy technologies. Thus, the Basic Energy Sciences research program is developed carefully to be relevant to the needs of the energy technologies, while maintaining scientific flexibility and engineering developments.

The Basic Energy Sciences strategy for solar development is two-pronged. First, technical areas of greatest need or greatest promise are identified through workshops, studies, and discussions. Second, unsolicited proposals related to solar energy are welcomed and encouraged from all disciplines and classes of R&D institutions, and the best of these proposals are selected for funding through careful peer review and coordination with the applied programs.

Several solar-energy-related workshops have been held under the auspices of the Basic Energy Sciences Program. Reports resulting from these meetings include, for example, "Polymer Materials Basic Research Needs for Energy Applications, August 1978," and "Thin-Film Problems and Research in Energy Systems, January 1979." The former dealt with such areas as polymer development and use in solar components, such as glazings, concentrators, receivers, converters, and structural components, as well as polymeric applications in heat transfer and energy storage. The latter included specific needs of solar thermal and photovoltaic energy systems for improvements in thin-film technology. Both workshops included presentations of research needs by DOE program representatives, descriptions of the state-of-the-art in relevant scientific and engineering areas, and analyses and discussions leading to recommendations concerning necessary and promising future research directions.

In addition, the Basic Energy Sciences Program sponsored a workshop in May 1979, "Assimilate Partitioning in Green Plants," which examined the current understanding of mechanisms of how plants distribute photosynthetically fixed carbon. These mechanisms are critical determinants of biomass productivity, an understanding of which could lead to improved strategies for biomass production.

The Basic Energy Sciences Program has five discipline-oriented subprograms, and an Advanced Energy Projects subprogram. The former sponsor energy-related research in nuclear science, materials science, chemical science, engineering, mathematics and geosciences, and biological sciences. The latter supports research on high-risk, high-payoff scientific concepts and technology systems that might not otherwise receive DOE support.

Examples of solar-related research supported in the disciplinary programs include:

- Materials research on semiconductors, polymers, and ceramics for applications in solar thermal and photovoltaic systems, as well as solar collectors and energy storage systems
- Research on heat transfer materials for photothermal energy conversion
- Studies of mechanisms of stress corrosion cracking and other materials failure modes relevant to safety of large solar systems, such as wind energy conversion systems, solar thermal systems, and ocean thermal energy conversion systems
- Research in solar-related photochemistry, chemical physics, and catalysis that can help make possible the eventual development of advanced solar technologies such as fuel-making solar cells
- Research in algae and higher plant photosynthesis as model systems for creating artificial photosynthesis

## ENVIRONMENT, HEALTH AND SAFETY, AND SOCIAL IMPACTS

An understanding of the full spectrum of Environmental, Health and Safety, and Social Impacts of solar energy technologies, as well as methods to effectively mitigate critical problems, is essential to the national solar program. Consequently, the DOE is engaged in extensive environmental assessment and control R&D activities for each solar technology. These activities are compiled and summarized for convenience in this chapter. First, overall objectives and program management strategy are discussed. Then major environmental, health, and safety concerns and associated program activities are described for each technology area. Finally, a crosscutting discussion of social impact issues, including community-level technology assessments, is presented.

When reviewing the concerns and program activities described below, it is best to remember that, for solar energy, there is little commercial experience. Thus, it is not possible to anticipate all adverse or beneficial impacts in the early stages of the assessment process. Rather, it is expected that reliable and detailed impact assessments will be possible only when specific system designs and test installations are available. Therefore, the assessments described in this section should be viewed as evolutionary rather than as final and definitive. It is conceivable that the cumulative effort of a large number of dispersed systems could result in adverse impacts that would otherwise be too small to be significant. For these reasons, community-level technology assessments are also being pursued, both to evaluate the relative acceptability of various "mixes" of solar technologies in a variety of model communities and to project the cumulative effects of each technology mix on communities and their energy needs.

The objective of DOE's environmental, health, and safety activities is to ensure that those energy technologies that it develops, demonstrates, and brings to the commercialization phase by DOE are environmentally acceptable. They must not only meet the present requirements of the law (e.g., impact assessment for federally funded projects, in compliance with the National Environmental Policy Act), but they must also identify potential environmental, health, and safety problems as well as measures that will avoid or mitigate adverse effects of widespread and long-term use of these technologies.

The DOE impact assessment strategy consists of the systematic development and analysis of information on the environmental, health, and safety aspects of each energy technology through the performance of a variety of research efforts, studies, and assessments, and the use of this information to develop environmentally acceptable technologies. At the heart of this process is the technology-specific Environmental Development Plan, which is timed to precede key technology program decisions as the technology moves from the exploratory development stage to an engineering development or technology development phase.

The Environmental Development Plan (1) identifies and evaluates environmental, health, and safety concerns; (2) defines research and assessment needs to resolve these concerns; (3) provides a schedule for environmental, health, and safety research that is coordinated with the technology development program; and (4) indicates the timing of related environmental management documents. These related documents are of three types: Environmental Readiness Document; Environmental Assessment/Environmental Impact Statement; and Safety Analysis Report.

The Environmental Readiness Document (ERD) is an independent assessment report prepared by the DOE Assistant Secretary of Environment (ASEV) with the full cognizance of technology program line managers. It provides a state-of-the-art assessment of the environmental status of a technology at the phase of development being considered for management decisions, usually at the key decision point of the management system. The document presents the results of a critical review and analysis of environmental research results carried on in the preceding phase; it provides further definition of concerns and research needs for the ensuing period. The ERD provides the basis for the DOE Assistant Secretary for Environment position on the environmental readiness of a DOE technology at each key decision point.

The National Environmental Policy Act of 1969 established the statutory requirements for Environmental Impact Statements. For all programs and projects that may require an impact statement, an Environmental Assessment may be prepared. An Environmental Assessment is a document that provides the information on which to base a determination of the necessity for an Environmental Impact Statement or a finding of no significant impact. Environmental Assessments may be prepared for any action and at any time to assist departmental planning and decisionmaking. Environmental Assessments and Environmental Impact Statements are prepared by program offices according to DOE procedures and are reviewed by the DOE Office of Environment and the Office of the General Counsel.

A Safety Analysis Report is prepared by program managers according to DOE procedures early in the design phase of a proposed facility that DOE intends to procure and operate. The use of a safety analysis (to identify hazards, eliminate and control identified hazards, assess residual risk, and document management authorization of a DOE operation) applies equally to Technology Base activities and energy system acquisition project. The purposes are to limit risks to the health and safety of the public and employees and to adequately protect property and the environment.

The Environmental Development Plan and the Environmental Readiness Document provide management overview of the environmental research program; National Environmental Policy Act and safety compliance are provided through the performance of Environmental Assessments, Environmental Impact Statements (where necessary), and Safety Analysis Reports. These efforts are coordinated through the Solar Environmental Coordinating Committee, constituted under authority of DOE Order 5420.1, and composed of representatives of various DOE offices. The group assists in the implementation of the DOE Environmental Development Plan System. The primary functions of the Solar Environmental Coordinating Committee are to monitor and oversee the status of environmental R&D programs, ensure that the intent of the Environmental Development Plans is achieved, and promote regular information exchanges and coordination between offices responsible for environmental research and development. Specifically, the Solar Environmental Coordinating Committee, through appropriate subcommittees:

- Participates in the preparation and review of Environmental Development Plans and identifies the need for and recommends revisions in the plans
- Maintains a collective awareness of the content, status, and results of environmental research efforts and apprises management periodically of status and issues
- Advises management of gaps, redundancies, and potential conflicts in the R&D efforts and recommends corrective options for management considerations

- Coordinates between performing offices those necessary physical and institutional arrangements required for the conduct of respective research efforts

DOE environmental, health, and safety research is jointly funded and managed by the Assistant Secretaries for Environment and for Conservation and Solar Energy, with the funding mix determined to optimally utilize the skills, expertise, and resources of each organization for each technology area. In addition, DOE coordinates its environmental activities with those of other federal agencies (e.g., Environmental Protection Agency) and with state and local governments where appropriate. The Environmental Development Plans describe all completed, current, and planned research projects of DOE and other organizations by purpose, performer, and sponsor.

## Program Thrust

Biomass Energy Systems. Major environmental concerns are associated with biomass production, collection, harvesting, and conversion processes. The primary considerations for production, collection, and harvesting are the effects on soil, erosion, soil quality, water quality, water and land use, and emissions and pollutants. The primary environmental concerns associated with biomass conversion are potential effects on air and water quality. These concerns are listed below.

- Soil erosion and quality
  1. Erosion on harvested farmland
  2. Impaired productivity
  3. Increased particulate pollution
  4. Reduced fertility through increased water runoff
- Water quality
  1. Increased sedimentation from erosion
  2. Eutrophication
  3. Biomass residual pollution
  4. Process water disposal
  5. Impaired subsurface water quality from settling ponds and biomass collection points

- Water and land use
    1. Disturbance of terrestrial and water habitats
    2. Land requirements for energy farms
    3. Loss of ecosystems from whole-forest lot harvesting
- Emissions and pollutants
    1. Fugitive dust, particulates, and combustion products
    2. Sulfur oxides and nitrous oxides
    3. Liquefaction byproducts, i.e., tars, chars, and carcinogens

A detailed environmental research plan to address the issues described above has been developed. The Argonne National Laboratory has the lead responsibility for organization, coordination, and analysis of environmental research in biomass conversion technology. The Oak Ridge National Laboratory leads the research on production and harvesting impacts. The environmental research plan sets forth a cohesive set of research projects coordinated with the biomass technology research schedule in the areas of:

- Regional environmental impacts
- Loss of ecological resources
- Air and water quality and soil fertility effects of biomass production
- Air and water pollution from biomass conversion processes
- Health effects of emissions

The following Environmental Assessments are planned or in progress:

- Regional Environmental Assessments of biomass energy systems for the Northeast, Northwest, Southeast, Midwest, and South Central regions (FY 1979–1983)
- Programmatic Environmental Assessments of biomass energy systems
- Site-specific Environmental Assessment for the 500-acre biomass energy farm scheduled at the Savannah River Plant site
- Environmental Assessments for all silvicultural energy farm proposals selected by DOE
- Site-specific Environmental Assessments for each preliminary design of a 300-ton/day, medium-Btu biomass gasification facility

The environmental goals and milestones for the biomass program, with reference to the associated technology development goals, as well as a more detailed description of the environmental program, can be found in the Biomass Environmental Development Plan (DOE/EDP 0032, September 1979).

**Photovoltaic Energy Conversion Systems.** Photovoltaic energy conversion typically involves relatively few hazardous residuals at the point of use. Some heat is released to the atmosphere by the collectors. However, major potential impacts are associated with chemical and thermal releases under accident or overheating conditions or during the mining, cell manufacture, or disposal/recycle stages of photovoltaic system production and use. The key environmental concerns are threefold.

- Release of toxic gases during system operation or malfunction. Decomposition or combustion of photovoltaic cells can result in outgassing or the production of a number of toxic materials. Silicon cells are inert, but polymer materials used in concentrators (e.g., methyl methacrylate) can burn, producing potentially hazardous products. Arsenic compounds from gallium arsenide cells and cadmium particulates from cadmium sulfide cells pose potential carcinogenic, respiratory and renal hazards.
- Inhalation of toxic gases and dusts by photovoltaic industry workers. Hazardous materials to which photovoltaic industry workers may be exposed include silicon dust, doping agents such as phophine and boron trichloride, cadmium dust or salts, and arsenic and its oxides. Considerable industrial experience in dealing with these materials exists, and most gases are covered by OSHA standards. However, large-scale production of photovoltaic cells may raise new questions about worker safety

and may make new demands on environmental control technology.

- Solid-waste disposal and effects of gases released during materials mining, manufacture, and disposal. Potential problems include surface and ground-water contamination from open pit mining, silicon distillation, and mercuric and acidic effluents and sludge from gallium production. Solar arrays disposed of in landfills, pits, or open dumps could also lead to leachate problems. The nature of the Resource Conservation and Recovery Act and the Toxic Substances Control Act implementation by EPA will significantly affect the photovoltaic industry as the production expands.

The management of the environmental research program in photovoltaics is decentralized, with two lead laboratories:

- Jet Propulsion Laboratory, operated by the California Institute of Technology, has responsibility for the National Environmental Policy Act and safety compliance activities, siting criteria, and control technology development. Researchers are in the process of developing a detailed environmental criteria document for evaluation of proposed photovoltaic deployment sites.
- Brookhaven National Laboratory has responsibility for research on health and environmental effects. A workshop on the health effects of photovoltaic technologies, held from July 31 to August 2, 1979, brought together experts from industry, the environmental community, and health effects research institutions. A research program built upon the insights obtained at this workshop is in the developmental stage. A second workshop devoted entirely to environmental issues was held in FY 1980.

Environmental Assessments have been or are being pursued in two areas by the Office of Conservation and Solar Energy. First, an Environmetal Assessment of the Federal Photovoltaics Utilization Program (implementation of title V, part 4, of the National Energy Conservation Policy Act) was published in June 1979. This assessment resulted in the determination that the probable environmental impacts of the program will not significantly affect the quality of the human environment and, thus, that an EIS would not be required. Second, an Environmental Assessment of the recently issued Program Research and Development Announcements for photovoltaic installations is presently being carried out.

The environmental goals and milestones for the photovoltaic program, with reference to the associated technology development goals, as well as a more detailed description of the environmental program, can be found in the Photovoltaic Environmental Development Plan EDP (DOE/EDP 0031, September 1979).

Wind Energy Conversion Systems. Small wind energy conversion systems are based on a proven technology with a long history of use in diverse environments. No major environmental problems have been encountered in the manufacture, installation, or operation of these systems. There is much less operational and environmental experience with large wind energy conversion systems. Concerns exist with respect to safety, electromagnetic interference, and noise associated with these machines.

- Safety: including (1) structural failure of tower, blade shaft, or hub and (2) risk of tall structures to public
- Electromagnetic interference: disruption of television and microwave reception
- Noise: residential siting limitations

Environmental concerns associated with wind energy conversion systems (WECS) are being addressed through design and siting studies. A safety plan has been developed and design analyses, testing, and safety review continues for each generation of WECS. The NASA Lewis Research Center is the lead laboratory for the safety research program. A comprehensive technical study of electromagnetic interference potential from wind machines is in progress at the University of

Michigan. Noise at WECS sites will be monitored. Bird impacts are being monitored for future analysis at the NASA Plum Brook test facility and other government installations. Siting of wind machines in rural areas and away from migratory bird routes should minimize the impact of noise, interference, and bird collisions. In general, with good design and careful siting, it should be possible to mitigate all of these problems.

A variety of Environmental Assessments have been performed by DOE for wind systems. An EIS was published in 1978 for the Wind Turbine Generator System on Block Island, Rhode Island. Environmental Assessments have been prepared for 17 candidate sites for experimental wind generators, and a programmatic Environmental Assessment of WECS has been submitted for management review and approval.

The environmental goals and milestones for the WECS program with reference to the associated technology development goals, as well as a more detailed description of the environmental program, can be found in the wind energy conversion systems Environmental Development Plan (DOE/EDP 0030, July 1979).

Solar Thermal Power Systems. Solar thermal power systems are of two principal types. Large-scale centralized systems are land- and capital-intensive. Sites for early development probably will be in remote areas of southwestern deserts. The primary environmental concerns will be onsite worker safety and offsite ecological impacts. Smaller scale, dispersed-receiver solar thermal systems are likely to be located near areas already developed for agriculture or industry. The primary environmental concerns will be the protection of health and property.

Handling and disposal of working system fluids are key concerns for both types of systems. Accidental or emergency release or flushing of working and storage fluids (such as liquid sodium, sodium hydroxide, hydrocarbon oils, and eutectic salts) could cause fires and explosions, contaminate drinking water supplies, increase soil salinity, affect terrestrial and aquatic communities, and reduce the effectiveness of sewage treatment. Major concerns for the central receiver system include the following:

- Ecological and microclimate effects
  1. Unknown effects on desert ecosystems
  2. Disruption of native plant and animal communities
  3. Disruption of surface and groundwater patterns
- Misdirected solar radiation
  1. Severe eye injury
  2. Fire hazard
  3. Disrupting nearby air and ground traffic

The lead laboratory for environmental research on solar thermal power systems is the University of California at Los Angeles (UCLA) Laboratory of Nuclear Medicine and Radiation Biology. Research is under way at both UCLA and other institutions addressing:

- Failure mode analysis for alternative solar thermal power system configurations, and ramifications for worker and public safety
- Ecological effects of accidental and mismanaged fluid releases (e.g., toxicity of specific working fluids, effects on desert ecosystems)
- Onsite impacts of solar thermal operations, including monitoring of local conditions at the 10-MWe Barstow site, experiments on managing vegetation grown under heliostats, and studies of microclimate changes under heliostats at a test site near Tempe, Arizona, conducted by Arizona State University

A programmatic Environmental Assessment for solar thermal systems has been submitted for DOE approval. Environmental Assessments have been and will continue to be prepared for each of the solar thermal test sites.

The environmental goals and milestones for the solar thermal program with reference to the associated technology development goals, as well as a more detailed description of the environmental program, can be found in the

Solar Thermal Power Systems Environmental Development Plan (DOE/EDP 0035, August 1979).

Ocean Systems. Studies to identify environmental concerns of technologies for converting energy from waves, currents, and salinity gradients are in the beginning stages and EDP's have not been developed yet. The major environmental concerns associated with OTEC systems fall into three categories: (1) effects on oceanic properties of the redistribution of cold deep water and warm shallow water, as well as of effluents; (2) chemical pollution; and (3) impact of the large OTEC structure and supporting systems. The major environmental concerns are summarized as follows:

- Ocean property change: disturbance of natural thermal structures, salinity gradients, nutrients, dissolved gases and turbidity; entrained biota in an OTEC system; and change of local climate by sea surface-temperature change
- Chemical pollution: biocides used to reduce fouling, leakage of working fluids, and erosion and corrosion of metal particles from OTEC
- Platform inputs: discharges from life-support systems

A comprehensive environmental research program aimed at resolving key uncertainties and providing the input needed for the development of effective impact mitigation strategies is being conducted. The Lawrence Berkeley Laboratory is the lead laboratory serving as the focus of program planning and technical direction. It is supported in the measurement and analysis areas by the NOAA Atmospheric and Oceanic Measurement Laboratory and in the hydrodynamic and numerical modeling areas by Argonne National Laboratory. The research program involves both laboratory and onsite testing components. Existing OTEC test sites are in the Gulf of Mexico, Hawaii, Puerto Rico, and the south Atlantic.

A number of Environmental Assessments of OTEC have been or are being carried out. Both a programmatic Environmental Assessment and an assessment of the 10-to-40-MWe facility, are planned for completion in FY 1980.

The environmental goals and milestones for the OTEC program with reference to the associated technology development goals, as well as a more detailed description of the environmental program, can be found in the OTEC Environmental Development Plan (DOE/EDP 0034, August 1979).

Agricultural and Industrial Process Heat. Environmental analysis done to date suggests that solar agricultural and industrial process heat (AIPH) systems have a relatively benign effect on the environment. However, there are several environmental concerns that are being studied to minimize uncertainties concerning their significance. These are primarily associated with (1) the toxicity of substances released through collector overheat or fire or the handling and disposal of system fluids and wastes, and (2) potential contamination of foodstuffs and other products through bacterial and fungal growth or contact with biocides. The major concerns are:

- Collector overheating and fire: toxic fume release and biocides, fungicides, and anticorrosion compounds released during overheat or fire
- Handling and disposal of systems fluids and waste: intentional flushing or leakage causing eutrophication, impairing sewage plant operation, or contaminating soil and water
- Microbial contamination of food: air and water storage temperatures that may be conducive to growth of fungi and bacteria
- Chemical contamination of products: spillage, leakage, or carry over of fluids and fluid additives into heating or drying areas or into air used for direct heating that may cause contamination of food and other products such as textiles and tobacco; the use of biocides in working fluids that may present another contamination problem.

Considerable research has already been carried out on the environmental and fire hazards of materials used in solar heating and cooling systems at the National Bureau of Standards and at Sandia and Los Alamos laboratories. The AIPH program will utilize these results, together with an ongoing research program that is developing toxicological data on materials not previously studied, to develop appropriate control strategies. Proper design and operation of the solar AIPH systems, based upon these studies, should minimize these impacts. Work on the characterization and measurement of microorganisms will support the food contamination control effort. Techniques such as periodic flushing of working fluid equipment; indirect heating in secondary working fluid systems; and careful monitoring of working fluids, air and water quality in the processing area, and the product itself will be employed to avoid chemical contaminations of products. Finally, specific research is under way to develop nontoxic heating systems and components.

The environmental goals and milestones for the AIPH program with reference to the associated technology development goals, as well as a more detailed description of the environmental program, can be found in the AIPH Environmental Development Plan (DOE/EDP 0033, September 1979).

Active Solar Heating and Cooling. Active solar heating and cooling systems have relatively small environmental impacts, compared to fossil fuel-fired alternatives. However, a number of environmental concerns are undergoing research and analysis to resolve uncertainties and allow the development of systems designed to minimize environmental, health, and safety hazards. The key concerns are:

- Potential hazard of toxic working fluids and storage media: contamination of potable water and direct human impacts from inhalation or contact
- Improper handling and disposal of working fluids and storage media: contamination of waterways, groundwater, and soil
- Collector overheat and fire (see p. 000)
- Degradation of living-space air quality:

stagnant air and accumulation of airborne contaminants; and storage system buildup of mold, fungus, and bacteria

DOE has an active program aimed at commercializing solar heating and cooling system devices that mitigate the environmental concerns just described. Considerable research on the environmental and fire hazards of solar heating and cooling system materials has been conducted at the National Bureau of Standards and at the Sandia and Los Alamos laboratories. Ongoing DOE-funded environmental research is focused on:

- Toxicity evaluations of solar heating and cooling system materials
- Control measures to mitigate working fluids leakage
- Ecosystem impacts of working fluids
- Transport and effects of fluids and additives
- Pathways into water resources
- Disposal and recycling of hazardous chemicals from solar heating and cooling systems
- Methods for acceptable disposal of liquid wastes
- Atmospheric transport and potential seepage into inhabited areas of toxic substances emitted by solar heating and cooling systems during overheating or fires

In addition, monitoring of air quality in solar buildings and design studies are being actively pursued to address those environmental concerns associated with building safety and the indoor environment. Inferior air quality in energy-efficient buildings has become a concern. The environmental goals and milestones for the SHAC program with reference to the associated technology development program, as well as a more detailed description of the environmental program, can be found in the Solar Heating and Cooling Environmental Development Plan (DOE/EDP 0029, september 1979).

Passive and Hybrid Heating and Cooling. Passive solar heating and cooling systems (see

Figure 17-1) pose few environmental problems. Concerns are limited to the potential degradation of interior air quality (as measured by temperature, humidity, and air circulation problems), potential compromise of fire integrity of the building structure by passive system air passages, and increased hazard from glass breakage associated with the use of large expanses of glass in passive system designs. The toxicity of materials used in passive solar heating and cooling systems also must be evaluated.

The major environmental research problem associated with passive solar heating and cooling is indoor air quality. DOE has an ongoing program on building thermal envelope design; DOE also plans to monitor air quality conditions in the living and working space of passive heating demonstrations as a complement to EPA research on indoor air pollution. Toxicity studies carried out in support of solar hot-water and active solar heating and cooling can be expanded where necessary to provide needed data in this area. The issues of fire safety and glass breakage

are similar to those experienced in traditional building activities and are being addressed through design studies; education programs for designers, builders, and consumers; and building codes and standards activities.

The environmental goals and milestones for passive solar heating and cooling of buildings with reference to the associated technology development and commercialization programs, as well as a more detailed description of the environmental program, can be found in "Solar: Hot Water and Passive, Commercialization Phase III Planning" (DOE/EDP-0010, September 1978).

## Social Impact

The use of solar energy technologies will lead to a number of social, economic, and community impacts, some of which have been identified in the Environmental Development Plans; others will become apparent as the technologies evolve in the marketplace. A comprehensive "Technology Assessment of Solar Energy," initiated by the DOE Assistant Sec-

Figure 17-1. Solar Heating and Cooling installed in the Georgia Town elementary school in Atlanta, Georgia.

retary for Environment, investigated the long-range socioeconomic (and environmental) impacts of distributed solar energy systems, including analysis at the national, regional, and community levels. The DOE technology development and commercialization programs will utilize the results of the technology assessment as well as the research programs from the Environmental Development Plans to design strategies that avoid or effectively mitigate (where possible) social and economic impacts.

Of particular interest is a set of research efforts in the solar program to examine the problems and opportunities of moving toward increased reliance on local, renewable energy sources. Technology assessments, scenarios for future energy development, and a number of supporting studies associated with self-sufficiency planning at the community level are being pursued. These technology assessments and energy self-sufficiency studies address important social and institutional impacts and how governments and other community groups at the local and state levels will be affected. The studies also assess modes of community response to mitigate negative impacts and to control implementation of renewable resources in specific geographic areas. They deal with the cumulative impacts of individual systems, as well as the impacts of specific "mixes" of technologies within a limited geographic region. Because the choice of solar technologies and the nature and magnitude of effects associated with the "mix" will differ by location, an effort has been completed to perform community-level studies (with a high level of community involvement) that are easily replicable by other communities without complicated research tools or extensive resources.

Work has been completed at the *local* level—where direct activity to implement renewable energy resources will be the greatest and where the resulting social impacts will be felt most intensely. A community-level Technology Assessment Program is in process to assess the social, political, institutional, and life style effects of using decentralized solar technologies. Using this information, researchers could develop a process by which communities can conduct their own formal technology assessments. One set of studies provides background information on the creation of prototypical scenarios for solar energy use at the community level. Scenario development studies are being pursued in several communities, and the results from these efforts will be disseminated as part of the Technology Assessment Program in the form of case-study reports and guidebooks.

The central activities of the Technology Assessment Program are the community-level technology assessments. A variety of demographic units and geographic regions have been chosen to plan prototype solar futures based on individual community energy needs, economic activities, housing patterns, and local availability of solar and renewable resources. In each local area, a working team has been established; it includes community members and technical consultants to evaluate plausible solar futures and their consequences, based on local expert judgment. The consequences for every possible aspect of community life must be considered, including infrastructure requirements, environment, quality of life, self-reliance, convenience, health and safety, and citizen desires.

One of the local technology assessments, a study of three counties in south-central New York has been completed. The final report from this project provided a renewable energy plan for the counties involved, described in detail the projected use of conservation and alternative energy sources, and outlined the potential impacts associated with the use of each. State legislation and local regulations affecting the energy development process were identified. The most significant finding of this research is that the region could reduce its use of energy supplied from outside the region (most of the energy used) by 50 percent or more within the next 10 years. The other communities involved in the study are following similar programs and anticipate comparable findings during the next year.

As the program has proceeded, it is apparent that the Technology Assessment Program is evolving in most of the communities toward an energy planning and implementation process, in an attempt to achieve the preferred future energy state. The program is following and evaluating these efforts as an extension of the technology assessment process and as an opportunity to verify the results of early work. While the program evaluation efforts have been minimal to date, it appears that not only are the limited goals of the program being met but that the program has also served to mobilize interest in other aspects of energy within the community. The program has generated considerable interest from communities beyond the defined study area. Future plans for the program include outreach activities to share the cumulative experience and findings of the local studies and the support of model implementation studies.

Additional support studies have been conducted to deal with special issues and examine greater policy implications. These have provided valuable background information on institutional and legal barriers and social costs related to the adoption of solar technologies. Special studies of specific employment effects are also under way.

The results of these programs dealing with the social context of decentralized solar technologies have provided valuable insight into the marketing problems that are likely to occur as technologies are broadly applied. Areas of consumer resistance to particular configurations of technologies for local application have been identified, and the social settings in which all technologies must conform and compete have been clarified.

In other research, the major social issues affecting solar energy technologies that are identified in the Environmental Development Plans include:

- Land and resource use: large land requirements, and resulting conflicts with other uses, and scarce resource utilization
- Utility impacts: intermittent nature of solar technologies, load management problems, and excess power purchase by utilities
- Aesthetics: large-scale wind and solar installations and residential design requirements

- Institutional impacts: sun rights, building code development, and international environmental laws for OTEC siting and operation

## MARKET ANALYSIS

Market Analysis includes those activities that evaluate potential markets for all solar technologies and develop market strategies for those technologies. This activity involves determining the type of products needed for given market sectors and setting cost goals for systems R&D of particular products. Additionally, Market Analysis identified nontechnical barriers to commercialization of new products and prepared analyses and recommendations for government action to remove or mitigate their adverse effects.

Rather than pushing market development of all these technologies in all markets, it is necessary to identify the markets that show the most promise and then to focus on those systems most applicable to the markets. Since market acceptance cannot be achieved through technological improvement alone, a parallel effort is necessary to determine the nature and extent of government incentives to ensure substantial conversion to solar energy. This is the approach used in the Market Analysis activities.

## Goals and Purpose

In general, Market Analysis efforts collect and analyze market data to establish cost and performance criteria for all solar technologies. Thus, the goals for the Market Analysis subprogram are:

- Identification of markets
- Selection of technologies for intensive development
- Identification and evaluation of barriers to the development and commercialization of solar energy (economic, legal, and social)
- Recommendation of government policies to remove barriers and accelerate solar commercialization

The Market Analysis effort determines national and international market needs and measures current product potential and readiness against those needs. In addition to this central purpose, several other factors motivate Market Analysis activities, including the needs to:

- Coordinate market research
- Evaluate progress of solar market penetration
- Identify needs for technical development, financial incentives, and regulatory actions
- Provide information and backup analyses for inquiries from the White House and Congress
- Identify potential government actions to develop markets; deliver technical, financial, and marketing services; and

promote the establishment of the industrial infrastructure

## Budget

The Market Analysis budget request for the Office of Solar Applications is shown in Table 17-4. These activities support the technologies shown in Table 17-5.

Appropriate Market Analysis expenditures in the Office of Solar Technology are $2.0 million in FY 1979, $2.4 million in FY 1980, and $7.2 million in the FY 1981 budget request.

## Program Detail

Status. The status of Market Analysis activities in the Office of Solar Technology is

### Table 17-4. Major Crosscutting Market Analysis Activities.

| | BUDGET AUTHORITY (DOLLARS IN THOUSANDS) | | |
|---|---|---|---|
| | APPROPRIATION | APPROPRIATION | ESTIMATE |
| ACTIVITIES | FY 1979 | FY 1980 | FY 1981 |
| Economic Modeling and Analysis | 1,000 | 1,200 | 1,300 |
| Economic and Financial Incentives | 300 | 500 | 1,400 |
| Public Utility/Solar Energy Interface | 200 | 500 | 1,300 |
| Legal Issues | 400 | 1,000 | 400 |
| Employment and Manpower Assessment | 400 | 1,000 | 500 |
| Marketing and Consumer Response | 300 | 1,600 | 2,100 |
| International Market Analysis | 100 | 200 | 300 |
| TOTAL | 2,700 | 6,000 | 7,300 |

### Table 17-5. FY 1981 Market Analysis Budget.

| | BUDGET AUTHORITY (DOLLARS IN THOUSANDS) | | |
|---|---|---|---|
| | APPROPRIATION | APPROPRIATION | ESTIMATE |
| TECHNOLOGY | FY 1979 | FY 1980 | FY 1981 |
| Passive Solar | 400 | 1,900 | 2,150 |
| Active Solar | 1,013 | 1,200 | 2,300 |
| Agricultural and Industrial Process Heat | 400 | 1,000 | 1,600 |
| Photovoltaics | 300 | 1,000 | 400 |
| Wood/Biomass | 300 | 500 | 850 |
| Wind Systems | 300 | 400 | 0 |
| TOTAL | 2,713 | 6,000 | 7,300 |

described in Chapter 16 of Part III and is not repeated here. The present status of the Market Analysis program in the Office of Solar Applications is summarized here in terms of the work completed in each activity, listed below with the major accomplishments.

- Solar economic modeling and analysis
  1. MITRE SPURR model
  2. ERDA-MITRE report, "An Economic Analysis of Solar Water and Space Heating"
  3. The white paper, "An Analysis of the Current Economic Feasibility of Solar Water and Space Heating"
  4. Studies of economic viability of solar technologies other than solar heating and cooling; e.g., wind energy conversion systems, biomass, and photovoltaics
- Economic and financial incentives
  1. Policy options documents dealing with financial incentives
  2. Numerous analyses of the costs and effects of different types and levels of financial incentives for solar technologies
  3. Battelle Pacific Northwest Laboratory study of federal incentives for energy production
  4. Estimates of costs to state and federal governments of solar tax credits
  5. Estimation of the impacts of National Energy Act solar incentives
- Solar energy/public utility interface
  1. Sections of policy options documents dealing with the solar/utility interface
  2. Study of interface strategies
  3. Study entitled "Impact of Passive and Active Solar Energy Systems on Utility Load Factors"
  4. Series of four workshops on the solar/utility interface problems
  5. Analysis of utility laws affecting the solar/utility interface
- Legal, regulatory, and legislative issues
  1. Appropriate sections of policy options documents

2. Legal barriers to the solar heating and cooling of buildings
3. Preliminary assessment of effects of changes in the federal tax structure on solar energy feasibility
4. Solar access and land use: state of the law, 1977
5. Analysis of building codes as a barrier to dispersed solar applications
- Employment and personnel issues
  1. Appropriate sections of the policy options documents
  2. Assessment of the personnel and employment effects of the National Energy Plan goal of 2.5 million solar installations by 1985
  3. Evaluation of federal efforts in the labor and personnel areas relating to solar energy

Major Activities. In addition to the accomplishments listed above, there are many other activities underway in the Market Analysis subprogram. Ongoing and planned activities are listed below by program element.

- Solar economic modeling and analysis
  1. Complete development of MITRE-SPURR model.
  2. Complete development of Booz, Allen & Hamilton solar energy market penetration model.
  3. Study SHAC private market data base.
  4. Develop prototype solar heating and cooling data collection, analysis, and evaluation system.
  5. Organize system simulation and economic analysis working group.
  6. Develop Oak Ridge National Laboratory residential conservation/solar model.
  7. Assess cost effectiveness of active versus passive systems.
  8. Determine the net capital requirements of solar energy systems, as compared with those of other energy technologies.

9. Continue analyses of the economics of near-term solar applications.

- Economic and financial incentives
  1. Update Battelle Northwest incentives study, and extend findings to include impacts.
  2. Assess incentives applicable to passive systems.
  3. Hold workshops on Battelle incentives study.
  4. Develop structure for monitoring and evaluating effectiveness of federal and state solar incentives.
  5. Expand analyses to include other near-term solar applications (e.g., photovoltaics, wind energy conversion systems, biomass).
  6. Identify incentives applicable to photovoltaic systems.

- Solar energy/public utility interface
  1. Analyze the impact of utility policies on solar system design and sizing.
  2. Perform utility interface modeling.
  3. Perform case study of Wisconsin to develop "model" solar/utility policies.
  4. Analyze the impacts of pricing policies of different utilities on dispersed solar energy systems.
  5. Study financial accounting aspects of utility involvement in onsite solar energy systems.
  6. Analyze impact of dispersed solar energy systems on an electrical utility grid.
  7. Develop special accounting model for various solar/utility options.
  8. Develop strategies for interface with different types of utilities (electric, gas, combined, co-op, federal, municipal, etc.).

- Legal, regulatory, and legislative issues
  1. Publish the *Solar Law Reporter* through the Solar Energy Research Institute.
  2. Develop comprehensive analysis of impact of federal tax policies on economic feasibility of solar applications.

- Employment and personnel issues
  1. Assessment by Battelle-Columbus of the personnel implications of solar energy devlopment.
  2. Assessment of personnel impacts on the near-term solar technologies.
  3. Analyze the skilled, semiskilled, and unskilled labor components of dispersed solar applications.
  4. Develop a general energy/employment impact model.

- International market analysis
  1. Study industry incentives used in Japan.
  2. Assess the export market for dispersed solar applications.
  3. Hold international solar energy commercialization workshop.
  4. Study the export market potential for photovoltaic systems.
  5. Participate in NATO's Critical Concerns for Mankind and Society work.

## TECHNOLOGY DEVELOPMENT

Technology Development is the systematic use of the knowledge and understanding gained from research in order to achieve technical feasibility of a concept and to gauge economic and environmental potential of energy concepts, processes, materials, devices, methods, and subsystems. It bridges the Technology Base activities and systems development and includes development of components and subsystems, energy system concept formulation, comparative and trade-off analysis of alternative concepts, and development and testing of laboratory-scale engineering feasibility models. Although these activities do not comprise a separate budget item in the Office of Solar Technology, they do represent a significant functional component of the effort in each technology area, focused on a common objective; then are compiled and summarized for convenience in this section.

### Goals and Purpose

Technology Development takes promising solar technologies and concepts identified through Technology Base activities and develops them to the stage at which they are ready for engineering field testing. The emphasis in this stage is on developing components and subsystems needed for product improvements or subsequent systems development activities. As the major link between research and full-scale systems tests and demonstrations of solar technologies, Technology Development is a key component in achieving the goals of the United States Solar Program.

## Budget

All Technology Development activities in Conservation and Solar Energy are funded through the Office of Solar Technology, and comprise approximately 38 percent of the total Solar Technology budget. The approximate expenditures for Technology Development activities in each of the five major solar technology areas are shown in Table 17-6.

## Program Detail

Biomass Energy Systems. Biomass Technology Development activities are designed to assist in moving a variety of promising technologies toward the marketplace. Since the focus is on those activities expected to have an impact within the next 5 years, efforts are directed primarily toward small-scale projects adapting existing technologies to the needs of such users as individual farms, farm coopera-

tives, and small utilities. The goal is to contribute 1.5 quads of energy by 1985 and 4.0 quads by 2000, in addition to the 1.5 quads already being produced by the forest products industry.

Major accomplishments include new development in direct combustion technology that could produce a near-term energy impact by using forest residues and other readily available biomass feedstocks—for example, by direct retrofitting without derating of oil- and gas-fired burners.

Technology Development activities related to energy from biomass include the following:

- On-farm anaerobic digestion
  1. The production of methane from animal manure may provide a substantial amount of fuel energy for individual farms and farm cooperatives and for concentrated areas of livestock operations.
  2. Researchers at Cornell University are trying to improve the economic and technical factors associated with anaerobic digestion systems for use in small-scale agricultural operations.
  3. The Ecotope Group has evaluated a 100,000-gallon anaerobic digestion system on a 350-cow dairy farm in Monroe, Washington.
  4. The Department of Agriculture at Cay Center, Nebraska, has developed design criteria for the optimal

## Table 17-6. Estimated Technology Development Expenditures.

| TECHNOLOGY | BUDGET AUTHORITY (DOLLARS IN THOUSANDS) | | |
| --- | --- | --- | --- |
| | ACTUAL FY 1979 | APPROPRIATION FY 1980 | ESTIMATE FY 1981 |
| OFFICE OF SOLAR TECHNOLOGY | | | |
| Biomass | 35,252 | 40,298 | 29,000 |
| Photovoltaics | 40,796 | 66,000 | 60,000 |
| Wind† | 30,094 | 30,780 | 25,767 |
| Solar Thermal | 48,188 | 55,545 | 52,685 |
| Ocean | 14,933 | 13,161 | 6,590 |
| TOTAL | 169,263 | 205,784 | 174,042 |

†Includes Wind Systems budget elements for Technology Development, Wind Characteristics, and other activities.

production of methane from manure and is developing methods to recover the digester residues, which have a high protein content.

- Alcohol production
  1. Sugar crops, mainly sweet sorghum and some sugar cane, are being investigated at a number of universities and USDA stations in Florida, Louisiana, Texas, and Puerto Rico. Sweet sorghum has the potential for a much wider geographic distribution than sugar cane. The development of harvesting, handling, and transportation systems began in FY 1980.
  2. Increased efforts on alcohol production systems are planned to improve the overall process efficiency. For example, new methods for sugar juice extraction are being evaluated, as well as methods for preserving the sugar feedstocks to lengthen the processing season. Planned activities also include the evaluation and development of new fermentation techniques, distillation units, and methods for processing the stillage produced during fermentation.

**Photovoltaic Energy Systems.** Technology Development for photovoltaics is aimed at developing arrays of photovoltaic cells with reduced cost, baseline technology improvement, and balance-of-system components and concepts needed to meet near-term program goals. These efforts have already achieved the following results:

- Overall array efficiencies greater than 10 percent are now ensured for both flat-plate devices and concentrators. The latter use silicon cells at concentration factors to 50 times normal sunlight.
- Progress with silicon is ahead of schedule, with promising results from several approaches.
- A study has been completed that addresses silicon supply during the next 5

years. Steps are being taken to avoid a possible shortfall in the 1982–85 period.

- The Advanced System Test Facility at Sandia Laboratory, Albuquerque, has been placed in operation, with two 10-MW arrays undergoing testing.
- Cadmium sulfide photovoltaic cells now appear sufficiently promising to warrant large-scale production in the early 1980s.

The major Technology Development activities in the photovoltaic area are:

- Development of advanced silicon technology and thin-film devices
- Baseline technology development pertaining to
  1. Flat-plate concepts, which use sunlight as it is received
  2. Concentrators, which employ optical devices to focus sunlight on a small area
  3. Total energy systems, which provide useful thermal energy as well as electricity
- Balance-of-system components and concepts, including everything other than array, such as control and power-conditioning system structural components, electrical storage subsystems and components, and any special equipment for interfacing with a utility.

**Wind Energy Conversion Systems.** Technology development for wind energy systems is focused on component development for large and intermediate systems and component development and testing for small systems. The Rocky Flats Test Center will have performed tests on more than 50 commercially available small wind systems by 1982.

Overall objectives of these efforts are to achieve wind system technical feasibility, to assess potential system improvements and supply the analytical tools and performance data needed to accomplish these improvements, and to identify and promote promising agricultural applications of wind energy.

The major wind energy technology development activities are managed in a decentralized fashion as follows:

- The Rocky Flats Test Center supports the existing wind energy industry through R&D testing of commercially available small wind systems.
- USDA is managing projects to identify design and performance requirements for farm and agricultural uses of wind systems.
- NASA is managing the development and testing of potential cost-reducing design options and components for intermediate- and large-scale wind turbines.

The wind systems program plans to develop and publish a set of handbooks dealing with design and siting issues for wind energy systems. The handbooks will draw upon the experience of demonstrations, test sites, Environmental Assessments, and other program data and expertise.

Solar Thermal Power Systems. Technology Development on solar thermal power emphasizes the provision of a sound technological and industrial base through reducing costs of key components. This includes developing prototype concentrator designs, characterizing the performance and verifying the design of subsystems, performing complete initial subscale system experiments, and developing mass-producible designs for concentrators and verifying their technological readiness through prototype testing. These efforts have already led to a number of concepts and subsystems now undergoing systems development for the Barstow test site and other ongoing projects.

The advanced subsystems and components element concentrates on advanced receivers, storage, transport, and heat engines, as well as on operation of the test facilities dedicated to technology-based activities. Advanced receiver development encompasses central receivers, thermal and chemical dish receivers, and generic receiver technology. The aim is to produce receivers with low capital and operating costs and high reliability that are well characterized and understood technically. Storage and transport development emphasizes second-generation subsystems offering cost and performance improvements, plus first-generation subsystems for solar thermal applications that have previously not incorporated storage. Early activities are concentrating on interfaces with water/steam, molten salt, liquid metal, and organic-fluid-cooled receivers. Heat engine activities include development of Stirling engines for use with dish concentrator systems; establishment of requirements and adaptation of Rankine cycle engines to solar thermal systems; and design, fabrications, and testing of high-efficiency Brayton cycle engines.

Ocean Systems. Technology development activities in the ocean systems program are aimed at improving the cost, performance, and reliability of materials, components, and subsystems. Included in the ocean thermal energy conversion (OTEC) program are the evaluation of numerous heat exchangers (condensers and evaporators) and their associated biofouling countermeasures. Other subsystems, now in the preliminary design phase, will advance to the final design and test phase in FY 1981, using the OTEC-1 test facility.

Under the salinity gradient concept, membrane and membrane module subsystems will be developed and tested. Technology Development in the wave concept will be directed at devices which focus wave energy, and the currents program will emphasize feasibility design of a rotary current device.

Technology Development for OTEC includes work on subsystems such as the cold-water pipe, electrical cable, and power systems; cleaning methods; development of the Seacoast Test Facility for biofouling and corrosion studies; and development of open-cycle components. From two procurements (one for shell-tube, the other for plate-type), two prototype, cost-effective 200-kW heat exchangers may be provided for test on the OTEC-1 oceangoing test platform. Several small 20-kW units may be used to evaluate configurations, materials, and heat-exchanger cleaning approaches. These tests will begin in FY 1981, and expected performance could reduce overall OTEC economics by $500/kW.

## SYSTEMS DEVELOPMENT

Systems Development focuses on the engineering phases of product development by supporting programs that result in technically proven components, designs, and systems. Equipment, installation, and operating costs are evaluated with the aim of minimizing costs. Reliability and performance data are gathered and analyzed via prototype and engineering field tests of integrated systems. The results provide important feedback to the product improvement process. Systems Development responsibilities in the Office of Solar Applications include technology assessments; systems analysis; and urban and community solar system planning, development, and applications programs.

### Goals and Purpose

The overall goals of the Systems Development subprogram are to support industry in bringing a new and diverse set of solar products, techniques, and services to market by:

- Supporting the development of improved components, systems, designs, and application procedures and practices for the residential, commercial, agricultural, and industrial process heat market sectors
- Conducting systems testing in the laboratory and in the field to determine operational performance, reliability, and maintainability characteristics
- Feeding these data back to the development process and to the outreach and legislative processes to define the development needs and state-of-the-art

The Office of Solar Applications, whose budget line item activities are described here, is responsible for near-term solar technologies and applications. The Office of Solar Technol-

ogy also supports activities that could be considered Systems Development for those solar technologies that have not yet reached market- or field-testing stage. The approximate budget and related activities for the latter office are briefly summarized in this section for completeness. Specific objectives for the Office of Solar Applications activities in Systems Development include the following:

- Agricultural and Industrial Process Heat. Bring into mass production reliable, durable, cost-effective solar applications capable of producing $6/MBtu within the next 5 years. First production on-line in 1983; capacity 10 million production on-line in 1983; capacity to double annually thereafter through 1988. Displace 2.6 quads of fossil energy by the year 2000.
- Active Heating and Cooling
  1. Develop active heating and cooling systems that can be commercially distributed and utilized to displace 2.0 quads of fossil energy by the year 2000.
  2. Bring into mass production a minimum of six generic types of domestic water-heating systems that can be cost-effectively utilized in 50 percent of the geographic areas of the United States by 1985. Displace 0.5 quad of fossil energy with these systems by 1985 and 0.65 quad by the year 2000.
  3. Provide development support to industry to ensure commercial availability of at least 10 cost-effective, residential-size, space-heating systems and at least 6 cost-effective, commercial-size space heating systems by 1985. Displace 0.13 quad of fossil energy with these systems by 1985 and 0.75 quad by the year 2000.
  4. Continue R&D activities in materials, components, and system configurations leading to at least two generic types of cost-effective cooling systems by the year 1990. Dis-

place 0.6 quad of fossil energy with these systems by the year 2000.

- Passive and Hybrid Solar Heating and Cooling. Introduce a range of marketable passive and hybrid solar designs, products, and systems for a variety of climates, building types, and community systems; stimulate use of passive and hybrid systems in 12 percent of new building construction by 1985; displace 1.0 quads of fossil energy by the year 2000.

## Budget

The estimated budget for Systems Development in the Office of Solar Applications and estimated expenditures in the Office of Solar Applications and estimated expenditures for the Office of Solar Technology are shown in Table 17-7.

## Program Detail

### Agricultural and Industrial Process Heat.

The objective of the AIPH Systems Development program is to develop and field-test increasingly more dependable and cost-effective

solar systems for application within the industrial and agricultural sectors. The budget request of $21 million represents an increase of $4.3 million over FY 1980. The FY 1981 program effort includes:

- Continued development and testing of dependable, cost-effective, mass-producible line-focusing collectors for low- and intermediate-temperature applications within the industrial and agricultural sectors
- Continued development and testing of point-focusing collectors for high-temperature applications within the industrial and agricultural sectors
- Verification of performance through field-testing, and systems refinement activities in support of low- and intermediate-temperature industrial process heat applications constructed in FY 1979 and FY 1980
- Verification, through field testing, of performance of 200 low- and intermediate-temperature applications within the agricultural sector, and R&D activities in support of these and future applications of solar energy to agriculture
- Materials and components R&D activities in support of stationary and tracking collectors, and systems for

### Table 17-7. FY 1981 Systems Development Budget.

| | BUDGET AUTHORITY (DOLLARS IN THOUSANDS) | | |
|---|---|---|---|
| | APPROPRIATION | APPROPRIATION | ESTIMATE |
| SYSTEMS DEVELOPMENT PROGRAM | FY 1979 | FY 1980 | FY 1981 |
| OFFICE OF SOLAR APPLICATIONS | | | |
| Active Heating and Cooling | 27,500 | 20,100 | 20,000 |
| Passive Heating and Cooling | 6,100 | 16,200 | 15,000 |
| Agricultural and Industrial Process Heat | 7,400 | 16,700 | 21,000 |
| Total Office of Solar Applications | 41,000 | 53,000 | 56,000 |
| OFFICE OF SOLAR TECHNOLOGY | | | |
| Biomass Energy Systems | 6,575 | 13,240 | 19,000 |
| Photovoltaic Energy Systems | 28,384 | 34,000 | 30,000 |
| Wind Energy Conversion Systems | 25,552 | 27,020 | 48,780 |
| Solar Thermal | 46,499 | 60,630 | 56,465 |
| Ocean Systems | 20,901 | 22,452 | 20,800 |
| Total Office of Solar Technology | 127,911 | 157,342 | 175,045 |
| TOTAL | 173,450 | 207,882 | 231,045 |

application within the industrial and agricultural sectors

## Active Heating and Cooling.

The objective of this program element is to support the development of active solar components, systems, and application techniques that are technically and economically feasible for domestic water- and space-heating and space-cooling applications. FY 1980 funding supported data collection and analysis of more than 75 active solar heating and cooling systems, plus the development of advanced collectors, thermal storage, and control strategies. Prototype testing was completed on two residential and two commercial heat-pump systems and two commercial solar cooling systems, developed in previous years. Cost, performance, and reliability goals were established as a basis for future development programs.

The FY 1981 budget request is $20 million, a decrease of $100,000 from FY 1980. However, beginning in FY 1981, this program element supports the data-collection activities of the National Solar Data Network. This will further reduce the funds available for hardware development activities. The requested funds will support:

- Development of cost-effective, total systems for active hot-water and space-heating systems
- Continued development of solar heat pump and solar cooling systems for residential and light commercial applications
- Development of low-temperature collectors and storage for solar heat pump systems
- Laboratory testing of two residential-sized desiccant cooling systems
- Prototype testing of one commercial-size absorption cooling system and one commercial-size Rankine cooling system
- Continued collection and analysis of data from 100 operational solar water-heating, space-heating, and space-cooling systems previously supported under

the Residential and Commercial Demonstration Program

The benefits from this program, in conjunction with other elements of the Solar Applications program, are expected to be the displacement of 0.2 quad by 1985, and 2.0 quads by the year 2000 in the buildings sector. FY 1981 funds were spent nationwide, primarily in the private sector.

## Passive and Hybrid Solar Heating and Cooling.

The objective of this program element is the rapid development and use of passive solar heating and cooling products and systems in residential and commercial buildings. The budget request for passive heating and cooling systems development is $15 million, a decrease of $1.2 million from the FY 1980 estimate.

The budget request is based on the need to develop marketable products and design tools for the building industry, so that approximately 2 million buildings incorporate passive solar measures by 1985. The funds requested provide for the following:

- Development and field-testing of 30 passive and hybrid marketable products, e.g., evaporative-radiative collector and storage assemblies, desiccant and dehumidification products, storage and control devices
- Development of 10 to 20 marketable designs and prototype buildings for passive-hybrid systems in residential heating, cooling, and hot-water applications
- Basic R&D of physical studies and materials for passive systems (e.g., glazing and solar control films, selective surfaces, and ground and sky properties and studies)
- Conduct of performance data collection for side-by-side testing, test rooms, and instruments buildings
- Development of design tools for passive and hybrid buildings (i.e., three handbooks and three computer programs)
- Completion of five pilot city and urban

planning projects as a part of the Solar Cities Program, using mixed solar technology
- Conduct of passive cooling studies and applied research in dehumidification, desiccant, ventilation and earth-coupled cooling process

The benefit of this program is the displacement of up to 0.1 quad of energy annually by 1985. Passive solar systems are usable in every part of the country. Funds will be spent nationwide. Field-testing will be conducted in prime market areas for each specific design. A major portion of the funds has gone to the construction industry and small businesses.

## Office of Solar Technolgogy Systems Development.

The Office of Solar Technology FY 1981 activities are briefly described below:

- In biomass, tests of improved conversion and production systems will be conducted.
- In photovoltaics, a moderate number of residential application experiments were conducted, and key international experiments were completed.
- In wind, the MOD-2 systems was operated, the development of improved small systems will be continued, and the design of advanced intermediate and large wind systems will be initiated.
- In solar thermal, the 10-MW pilot plant and 500-kW sodium central receiver will be operated; development and test of troughs, bowls, and dishes will be continued; the parabolic dish application experiment will be completed and operated; construction of repowering/retrofit application experiments will be initiated.
- In ocean systems, OTEC-1 will be completed and deployed, and the design of the 10 to 40 MW experiment will begin.

## Office of Solar Technology Systems Development.

The Office of Solar Technology FY 1981 activities are briefly described below:

- In biomass, tests of improved conversion and production systems will be conducted.
- In photovoltaics, a moderate number of residential application experiments were conducted, and key international experiments were completed.
- In wind, the MOD-2 system was operated, the development of improved small systems will be continued, and the design of advanced intermediate and large wind systems will be initiated.
- In solar thermal, the 10-MW pilot plant and 500-kW sodium central receiver will be operated; development and test of troughs, bowls, and dishes will be continued; the parabolic dish application experiment will be completed and operated; construction of repowering-retrofit application experiments will be initiated.
- In ocean systems, OTEC-1 will be completed and deployed, and the design of the 10 to 40 MW experiment will begin.

## MARKET TESTS AND APPLICATIONS

Following the Systems Development step in the product development sequence is the Market Tests and Applications activity in which products are tested in large-scale, real-world market applications in order to evaluate performance and user acceptability. Such activities are the responsibility of the Office of Solar Applications.

## Goals and Purpose

Market Tests and Applications (MTA) activities in the Office of Solar Applications program activities included both product support and large-scale market testing for technically proven designs and products. Federal funds are used on a cost-shared basis to assist the private sector in penetrating the buildings, industry, and utility energy technology markets. Through market test programs of significant size, federal technology purchases serve to stimulate demand. Both approaches induce manufacturers to invest in mass production capacity directly, with concomitant reductions in unit prices to consumers.

## Budget

As shown in Table 17-8, the FY 1981 budget for Market Tests and Applications activities in the Office of Solar Applications is $76 million, and includes efforts in buildings, agricultural and industrial process heat, photovoltaic systems, and federal buildings.

## Program Detail

Status. The activities support a variety of program efforts to meet the above objectives. These programs include the Solar Heating and Cooling Demonstration Program, the Solar Federal Buildings Program, the Federal Photovoltaic Utilization Program, and the International Solar Application Program. Overall management and direction stems from the MTA Planning and Implementation Program.

Specific accomplishments for the MTA activities include:

- The design, installation, and monitoring of solar heating and cooling systems in 6656 residential dwelling units
- The design, installation, and monitoring of 274 commercial solar heating and cooling projects
- The design and construction assistance for 400 to 500 federal building solar projects
- Completion of three cycles for Federal

Photovoltaic Utilization Program funding
- Establishment of the International Solar Commercialization Working Group
- Joint sponsorship of overseas trade shows and export market promotion with the Department of Commerce

Major activities completed in FY 1980 are listed below by subprogram.

*Solar Heating and Cooling Demonstration Program.* The Buildings Program assisted in the early establishment of solar industries for the design, manufacture, distribution, sales, installation, and maintenance of solar water-heating and space-heating and -cooling systems. Its scope includes:

- Residential, including all types of single-family and multifamily, nonfederal residential buidlings. This program was managed by the Department of Housing and Urban Development (HUD).
  1. Continue the design and construction of previously funded projects and HUD management (6656 of 11,375 units completed).
  2. Collect, analyze, and disseminate results on performance, operational reliability, maintenance, institutional problems, resale, and consumer acceptance.
  3. Identify problems and refurbish existing projects as appropriate.
- Commercial, including nonresidential buildings, some early demonstrations in federal buildings and offices, ware-

## Table 17-8. FY 1981 Market Tests and Commercial Applications Budget.

| MARKET TESTS AND COMMERCIAL APPLICATIONS PROGRAM | BUDGET AUTHORITY (DOLLARS IN THOUSANDS) | | |
|---|---|---|---|
| | APPROPRIATION FY 1979 | APPROPRIATION FY 1980 | ESTIMATE FY 1981 |
| OFFICE OF SOLAR APPLICATIONS | | | |
| Buildings (heating and cooling) | 55,000 | 36,750 | 18,000 |
| Agricultural and Industrial Process Heat | 11,000 | 14,000 | 20,800 |
| Photovoltaics | 15,000 | 10,000 | 35,200 |
| Federal Buildings | 25,668 | 11,750 | 2,000 |
| TOTAL | 106,668 | 72,500 | 76,000 |

houses, restaurants, hotel-motels, and other commercial buildings

1. Continue the design and construction of previously funded projects (134 of 274 projects are currently in operation).
2. Identify problems and refurbish existing projects as appropriate.
3. Continue funding support for passive commercial and manufactured buildings.

*Solar Federal Buildings Program.* The Solar Federal Buildings Program, authorized by title V, part 2 of the National Energy Conservation Policy Act (NECPA, P.L. 95-619), provided for the design, construction, and installation of solar heating and cooling equipment in existing and new federal buildings as a means of showing federal leadership the use of solar energy systems. For the three-year period of FY 1978–FY 1980, a total of $57,-418,000 was provided. The legislation also required that two rulemaking actions be completed prior to DOE's solicitation, evaluation, and award of funds for agency-proposed projects. Although the preliminary and final solar program rules were issued in April and October 1979, respectively, the final DOE rule for Life-Cycle Costing Procedures was issued early in FY 1980.

DOE has completed four solar design workshops for federal agency personnel, 11 regional preproposal conferences, a public hearing, and the establishment of technical design assistance to federal agencies preparing proposals for program submission.

*Federal Photovoltaic Utilization Program.* The objectives of the Federal Photovoltaic Utilization Program, in accordance with its legislative mandate under P.L. 96-619, title V, part 4, are to (1) accelerate the growth of a commercially competitive industry to make photovoltaic solar electrical systems available to the general public to reduce consumption of fossil fuels, (2) reduce fossil fuel costs to the federal government, (3) stimulate the general use within the federal government of methods for minimizing life-cycle costs, and (4) develop performance data on the program. The key activities are as follows:

- Conduct the Federal Photovoltaic Utilization Program (FPUP) Advisory Committee meetings.
- Prepare solicitation guidelines for Cycles 4 and 5.
- Recommend selection of proposed projects and participate in agency project reviews.
- Prepare annual report to Congress on the Federal Photovoltaic Utilization Program.
- Establish and implement a program for monitoring and assessing status of approved projects, including guidelines for project implementation.

*International Solar Applications Program.* The International Solar Applications Program provides support for international activities of the Market Tests and Commercial Applications Program. These projects will accelerate world utilization of solar energy and promote earliest application of United States solar systems and technologies in the international market. The key activities are as follows:

- Program plans and policy
  1. International Section of the National Plan for the Accelerated Commercialization of Solar Energy to be developed and updated
  2. International Solar Photovoltaics Plan to be entered and reviewed
  3. International Solar Commercialization Plan prepared
- Information systems
  1. Data base consolidated
  2. Dissemination systems established and coordinated
  3. Opportunity seminars exported
- Institutional support
  1. Financial institution defined and utilized
  2. Nontechnical issues identified and evaluated via workshops
- Trade promotion
  1. Solar trade program promoted
  2. International conferences supported and participated in
  3. Project implemented

These activities are funded directly by the

Market Tests and Commercial Applications Program and are not included in the FY 1981 solar international funding estimates.

Major Activities. Major activities supported in the FY 1981 budget include:

- Market-test 65 manufactured buildings and 120-single family residential projects using passive designs.
- Market-test 10 new passive commercial solar projects.
- Market-test eight active heating and cooling commercial and residential multifamily projects.
- Purchase 1.3 peak megawatt (MWp) arrays for 600 to 1000 photovoltaic systems for federal facilities.
- Complete design review assistance of FY 1980 projects; collect and analyze performance data derived from Federal Buildings Program.
- Market-test residential and light-industry wood systems to resolve site and area impacts ($6 million).
- Implement multiyear procurement strategy of 2 MWp of photovoltaic systems.
- Market-test 6 low- and moderate-temperature solar systems in industrial process applications.
- Market-test 1 intermediate-temperature solar system in a large AIPH project application.
- Develop 2 or 3 solar system designs for additional large-scale AIPH applications (100,000 ft$^2$ of collectors).
- Conduct approximately 100 feasibility studies for specific AIPH market-test applications.
- Conduct market test of solar-assisted heat pump AIPH applications.
- Market-test low-temperature solar collectors for AIPH applications.

## MARKET DEVELOPMENT AND TRAINING

The Market Development and Training Program in the Office of Solar Applications is responsible for gathering information resulting from Systems Development and Market Tests and Applications activities. The information is then disseminated in a format useful to customers and to those who influence the pace of new-product development and sales. This program also provides resources to develop a trained contractor and installer work force and links the DOE Solar Applications program with other federal, state, and local agencies and private institutions. The latter activity is aimed at ensuring that these agencies and institutions do not impede the introduction of solar products into the marketplace but, instead, that they complement and reinforce DOE activities to accelerate solar commercialization. Areas of program concern include financial incentives, regulatory programs, product standards, certification, and codes.

The Office of Solar Technology supports technical information dissemination on technologies under its purview and provides overall solar energy information support through the Solar Energy Information Data Bank at the Solar Energy Research Institute (SERI).

## Goals and Purpose

The ultimate purpose of the DOE Solar Market Development and Training effort is to remove any uncertainty impeding market acceptance of solar technologies and to build a high level of consumer confidence in new solar products. Major goals and objectives include:

- Providing current, unbiased technical and economic information to consumers and to those who play a major role in bringing new technology into the marketplace; e.g., architects, engineers, the financial community, builders, lawyers, state and local officials
- Providing training programs to stimulate industry and the private sector to develop a qualified work force to install solar systems and equipment
- Providing for consumer protection by accelerating the development of quality and performance standards for solar products by industry and standard-setting agencies
- Working with state and local governments to carry out a national effort to identify and remove all regulatory, in-

stitutional, and economic barriers that impede the marketing of solar technology

- Operating the National Solar Heating and Cooling Information Center and developing and disseminating information packages aimed at the general public through the regional solar energy centers

## Budget

The FY 1981 funding estimates for Market Development and Training are shown in Table 17-9. The Office of Solar Technology is responsible for the Solar Energy Information Data Bank located at SERI and described in Chapter 18.

## Program Detail

Status and Accomplishments. The Office of Solar Applications is engaged in a variety of programs to accomplish the Market Development and Training objective described above. These include:

- Coordination of a wide range of related activities of federal, state, and local organizations in such areas as technology and market information gathering and dissemination, standards development, equipment certification, and development and implementation of codes

- Operation of the National Solar Heating and Cooling Information Center and development and dissemination of information packages aimed at the general public through the regional solar energy centers

- Sponsorship of 75 to 100 workshops aimed at special groups (financial, industry, legislative, regulatory), both public and private, that can significantly influence critical decisions regarding solar marketing and commercialization. These workshops are an effective vehicle for the transfer of information and experience relative to the legal, financial, evironmental, and technical issues associated with solar technologies

- Development and implementation of contractor/installer training programs through existing apprenticeship organizations and mechanisms and through other institutions such as community colleges that train nonunion contractors and mechanics

- Development and dissemination of elementary/secondary school science, home economics, and industrial arts curricula dealing with solar energy technology and its application

- Support of various federal solar energy programs funded by other conservation divisions such as the federal Photovoltaic Utilization Program (FPUP), the Residential Conservation Program,

Table 17-9. Estimated FY 1981 Market Development and Training Budget.

| MARKET DEVELOPMENT AND TRAINING PROGRAM | BUDGET AUTHORITY (DOLLARS IN THOUSANDS) | | |
|---|---|---|---|
| | APPROPRIATION FY 1979 | APPROPRIATION FY 1980 | ESTIMATE FY 1981 |
| OFFICE OF SOLAR APPLICATIONS | | | |
| Active Heating and Cooling | 1,700 | 11,000 | 15,400 |
| Passive Heating and Cooling | 400 | 5,800 | 16,700 |
| Agricultural and Industrial Process Heat | 700 | 2,600 | 5,600 |
| Biomass (Wood) | 0 | 1,100 | 1,800 |
| TOTAL OFFICE OF SOLAR APPLICATIONS | 2,800 | 20,500 | 39,500 |
| OFFICE OF SOLAR TECHNOLOGY | | | |
| Solar Information Systems | 0 | 0 | 1,400 |

the Schools and Hospitals Program, and the Federal Energy Management Program

- Support of the Solar Energy Information Data Base development
- Support for a program to stimulate solar energy utilization, employment, and business opportunity in low-income communities, with emphasis in low technology applications

The Office of Solar Technology is funding technical information dissemination activities on biomass, photovoltaic, wind, solar thermal, and ocean systems, as well as the Solar Energy Information Data Bank at the Solar Energy Research Institute.

Major Activities. The major activities for FY 1981 in Market Development and Training included:

- Support for the development and consensus-based performance standards and codes to provide consumer assurance and enhance consumer acceptance of new products, with emphasis on solar heating and cooling materials performance, heating and cooling subsystems, solar thermal products, wood-combustion stoves, furnaces and boilers, and small wind-energy machines
- Joint programs with local building authorities, the Federal Housing Administration, and utility companies to eliminate institutional barriers and promote solar applications
- Support for renewable resource development in programs administered by other public and private agencies (e.g., Government National Mortgage Association financing, Small Business Administration loans, tax credits and incentives at the state level, Economic Development Administration development
- A limited effort to interest state regulatory commissions and energy offices in photovoltaic and wind machine development
- Information programs to serve the needs of builders, engineers, lenders,

homeowners, industrial and utility decisionmakers, government officials, and the general public

- Expanded and accelerated training and certification programs for contractors/installers and workshops for architects, engineers, and homebuilders to promote the use of passive-hybrid concepts in the construction industry
- Development of courses for vocational/technical schools, secondary schools, and colleges
- An expanded outreach program to promote consumer awareness of solar energy technologies, particularly passive solar heating and cooling
- Continued development of a comprehensive solar energy information data base
- Provision of technical, financial, and market services to the solar industry with emphasis on small solar businesses
- An expanded solar utilization economic development and employment program aimed at low-income communities and emphasizing low-technology applications

## INTERNATIONAL SOLAR PROGRAM

Nations around the world are increasingly looking to international energy cooperative activities in their search for energy supply alternatives. A principal reason for this trend is the increasing global problem with the cost and supply of petroleum fuels for economic development and national security. The widespread availability of substantial solar energy resources over a large part of the most populated areas of the Earth is leading other nations to develop solar plans; to undertake solar development programs; and to seek cooperation in solar R&D and implementation activities.

The United States has undertaken the largest and most aggressive national solar R&D and demonstration program. Because of this commitment and progress, many nations are interested in receiving information and assistance in their own efforts to evaluate and

326 Part III / Solar Energy

use their solar energy resources. In addition, the United States experience in implementing solar-energy-based systems commercially can have widespread ramifications for other national efforts, as well as for the international availability of reliable solar energy systems.

The United States solar program, organized in 1971, has grown rapidly since then. It is larger than the combined programs of all of the other nations in the world. The United States is fortunate to be one of the few highly industrialized countries to have substantial solar resources that are capable of providing a significant amount of its total energy needs.

President Carter's goal of 20 percent solar-derived energy usage in the United States by the year 2000 was set in 1979, on the basis of the Domestic Policy Review of Solar Energy.* In addition to a greater emphasis on development of solar energy technology for domestic energy needs, the report recommended a growing United States effort in international cooperation to enhance American technological progress and to assist other countries in the utilization of their solar resources. The International Panel's Final Report concluded that the United States should:

- Encourage the global transition from depletable petroleum supplies to alternative, renewable sources of energy
- Make energy an area of international cooperation
- Contribute to the economic and social advancement of developing countries by reducing energy-related obstacles
- Advance the state of United States technical energy programs
- Encourage the international use of alternative energy technologies developed by United States industry
- Avoid premature or excessive commitments to the use of nuclear energy
- Promote appropriate bilateral and multilateral scientific and technical cooperation

*International Panel. *Domestic Policy Review of Solar Energy, Final Report: Volume 1.* October 1978, p. 5.

As a consequence of the DOE mission to implement American energy policy, the management of United States international cooperative activities in solar energy RD&D is generally recognized to be a DOE responsibility. International agreements are developed and implemented in conjunction with the Department of State. Within DOE, the implementation of international solar activities is coordinated by offices reporting to the DOE Secretary for International Affairs, and managed by offices and laboratories reporting to the DOE Assistant Secretary for Conservation and Renewable Energy.

The DOE Secretary is the focus for coordinating DOE international agreements. All solar agreements are developed and managed by the Office of Technical Cooperation, and all official communications from DOE organizations to the State Department and other external organizations generally flow through this office. Policy issues in DOE international agreements are the responsibility of the Assistant Secretary for International Affairs, including required approvals by government departments and executive offices for specific agreements and projects. Staff members of the Office of Technical Cooperation work closely with staff from other DOE entities on all aspects of developing and implementing international agreements, including technical projects, information exchange, funding transfers, legal and patent issues, umbrella agreements, and project annexes. This office assists in negotiating solar energy agreements as well as agreements in many other areas of energy technology.

The DOE Assistant Secretary for Conservation and Solar Energy is responsible for the development and implementation of technical projects under the agreements in coordination with the Office of International Affairs. In general, funding of projects under the agreements is the responsibility of the Assistant Secretary and is justified in fulfillment of the objective of the technical program offices.

The Office of Science and International Programs (OSIP) under the Assistant Secretary for Conservation and Solar Energy provides a continuing interface between the technology program offices of the solar divisions and the Office of Technical Cooperation under the DOE Assistant Secretary for International Affairs. The OSIP is responsible for the development and integration of the solar international program. There are a substantial number of solar international activities under way.

In general, international cooperative activities of a technical nature originate in one of three ways: (1) in response to initiatives from foreign governments or other groups abroad; (2) as the result of high-level commitments of the administration; or (3) as the result of needs and opportunities identified by programmatic or policy elements within DOE. Some international solar agreements have been active since 1974, before the formation of either DOE or its predecessor organization, the United States Energy Research and Development Adminstration.

Some criteria for developing and evaluating potential international cooperative activities follow:

- Potential benefits to the domestic program
  1. Filling domestic program gaps
  2. Accelerating achievement of program objectives
  3. Reducing costs to the domestic program
  4. Utilizing overseas technologies and facilities
  5. Broadening the pool of technical talent, approaches and ideas
- Potential use of United States-developed energy technologies abroad and development of export market opportunities for United States industry
  1. Acquiring overseas project experience for United States industries
  2. Promoting United States technology for overseas needs
  3. Accelerating production to reduce equipment costs
  4. Identifying markets and export products
- Potential reduction in domestic and foreign demand for imported oil
  1. Accelerating United States deployment of new energy technologies

through increased production economies
  2. Stimulating the introduction and use of alternate energy systems abroad
- Potential support of United States energy policies
  1. Improving energy relationships with specific countries (including key oil exporters)
  2. Supporting international efforts to provide energy alternatives
  3. Promoting United States nuclear nonproliferation policies

## Goals and Purpose

DOE undertakes international solar activities to advance United States solar technology development, enhance the capabilities of American industry, and assist other countries in finding alternatives to the use of oil in their countries. More specific goals are listed below.

- To accelerate and expand solar energy use in the United States
  1. By increasing United States technical capabilities through international technical cooperative activities
  2. By strengthening United States industry through increased exports and other activities
- To accelerate and expand worldwide solar energy use
  1. Increasing the solar technical capabilities of other countries through international technical cooperation
  2. Promoting the use of solar systems overseas

Cooperative activities in information exchange, R&D projects, and applications analysis lead to new ideas, reduced R&D costs, and increased opportunities to study systems performance in a wider variety of user environments. The involvement of American industries in overseas projects and product adaptation enhances overseas business opportunities. Through sales in the international market, United States industry should be able to reduce the costs of domestically sold equipment more rapidly. Through information exchange, training citizens of developing countries, and demonstration projects, it is expected that renewable energy technologies can reduce the global demand for petroleum products more quickly. Worldwide cooperation and exchange of information on alternative sources of energy are important steps in achieving stability in the world energy market.

The DOE is the lead United States agency in undertaking major cooperative solar activities with international organizations such as the International Energy Agency, NATO's Committee on the Challenges of Modern Society, the World Bank, and the United Nations. These international organizations coordinate multinational efforts in information exchanges, conferences, surveys, R&D projects, application studies, and selected system demonstrations. Current areas of cooperation include information exchanges in biomass and wind technology areas; R&D projects in solar heating and cooling, thermal electric power,

and wave power; and conferences directed to the needs of developing countries.

One of the major cooperative projects under the International Energy Agency involves the construction and operation of two experimental 500-kWe solar thermal electric plants in Spain. One plant employs central receiver technology, and the other uses distributed receiver technology. The design phase has been completed, and the construction phase is under way, with initial operation of both plants scheduled for 1981. Ten countries are cooperating in the funding of this $45 million project. Figures 17-2 and 17-3 present sketches of these projects.

The Department of Energy has also been the lead United States agency in negotiating and carrying out bilateral solar energy agreements with many countries. The United States–Saudi Arabia Joint Solar Energy Project is the largest United States commitment to this type of international solar cooperation. Both countries will contribute $50 million over a 5-year period in a $100-million effort to advance the development and dissemination of solar energy technology. Major program

Figure 17-2. Artist's Concept of IEA Distributed Receiver Project at Almeria, Spain.

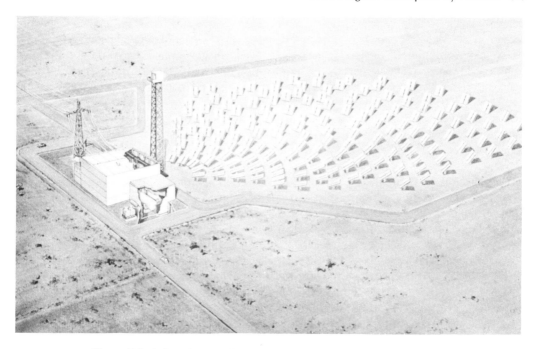

Figure 17-3. Artist's Concept of IEA Central Receiver Project at Almeria, Spain.

areas include urban, rural/agricultural, and industrial solar applications. A major project in the rural/agricultural program is a solar village power system to be built near two remotely located Saudi Arabian villages. The 350-kWe system uses photovoltaic technology and provides an opportunity to demonstrate new technology in the environment of a developing country. It will be in operation in mid-1982.

There are a number of other bilateral solar agreements in effect and several more in various stages of negotiations. Bilateral agreements for information exchange have been in operation for a number of years with France, Japan, the USSR, and Spain. Bilateral agreements have recently been initiated with Italy and Mexico and involve demonstration of solar systems in each of those countries. Other agreements are being negotiated for information exchange and cooperative projects.

While cooperative efforts with industrialized countries usually involve sharing technologies and expertise, the approach to working with developing countries is different. Since developing countries usually have not committed themselves to a broad, national, conventional energy base, they can pursue a development path that promises to meet their per capita energy consumption goals while, at the same time, providing for long-term reliability, increased self-sufficiency, and improved environmental quality.

In order to promote the accelerated commercialization of United States solar products abroad, the International Solar Commercialization Working Group was formed in early 1978. It provides interagency coordination of effort in solar international commercialization among the DOE, United States Agency for International Development, the Deparment of Commerce, and the State Department. Its goal is the accelerated commercialization of United States solar products abroad. An International Photovoltaic Plan was developed and submitted to Congress (in response to its request) early in 1980. The plan is aimed at accelerating the development of, and business opportunities for, United States industry in the 1980s through a combined international and domestic market development. An added international market can provide an extra, early impetus for production economies based upon the generally higher cost of petroleum-based power in developing country, rural economies.

The most immediate overseas market appears to be for relatively small, onsite solar technologies. Many development experts believe that decentralized solar technologies (e.g., small photovoltaic systems) can now economically provide many of the power needs for water pumping, communications, educational television and off-grid village power. In the near future, a market should develop for exports of specialty hardware and services; sale of system designs, engineering, and advanced technology under licensing; and royalty or joint-venture agreements.

In developing countries, there are large areas where there is no access to power grids. In other areas where electricity is generated, costs are often very high. As a result, on an economic basis, commercial application of solar electrical technologies could be possible earlier in developing countries than in the United States. There is a specific need in developing countries for low-cost, reliable, easily maintainable systems that do not require the construction of expensive transmission lines. A photovoltaic power system for a village in Upper Volta (Africa) is shown in Figure 17-4. Pilot testing of United States decentralized, renewable systems in these countries can be a preliminary step in improving the quality of life for large numbers of isolated villages around the world and in assisting American industrial efforts to serve those needs with United States–developed technology.

DOE manages much of its international solar activity through the International Program Division of the Solar Energy Research Institute in Golden, Colorado. The institute works with other DOE laboratories and industrial contractors to carry out the agreed-upon projects assigned to them by DOE headquarters staff. A very important Solar Energy Research Institute project for both the domestic and international solar program areas is the formation of an extensive international solar energy information data system.

## Budget

President Carter's FY 1981 budget request for DOE includes line items totaling $15 million for solar international activities under the Assistant Secretary for Conservation and Renewable Energy as shown in Table 17-10. The

Figure 17-4. Photovoltaic power system in Upper Volta, Africa.

Table 17-10. FY 1981 Budget for Solar International Activities.

| INTERNATIONAL ACTIVITIES | BUDGET AUTHORITY (DOLLARS IN THOUSANDS) | | |
|---|---|---|---|
| | ACTUAL FY 1979 | APPROPRIATION FY 1980 | ESTIMATE FY 1981 |
| Office of Solar Technology | — | — | 11,000 |
| Office of Solar Applications | — | — | 4,000 |
| TOTAL | — | — | 15,000 |

largest portion of the total line-item request ($11 million) is for activities developed by the Office of Solar Technology. A smaller portion ($4 million) is requested for international activities developed by the Office of Solar Applications.

These line-item funds would cover most of the firm commitments as of early 1979, of which the largest commitment is for the United States–Saudi Arabia Solar Agreement (approximately 80 percent of the total). The remaining funds are committed to agreements with the International Energy Agency and to the support of international activities at SERI. Some other program funds will be needed to continue a number of other existing bilateral agreements, e.g., those with Israel, Italy, and Mexico.

## Program Detail

The following sections present some of the considerations in evaluating the approaches and priorities in international solar applications.

Biomass Energy Systems. Developing countries have a particular need for biomass conversion equipment that is well adapted to their form of biomass collectibles and that can also produce good-quality fertilizer along with other derivatives. In contrast to industrial country needs, the development countries' market is oriented toward decentralized biomass conversion systems that use local materials and capabilities and minimize capital costs. The general objective is to operate experimental facilities, which, it is hoped, will make biomass fuel costs competitive.

Support for RD&D of decentralized biomass technologies, particularly those which produce high-quality fertilizer, is necessary to provide for the large market opportunities in developing countries. United States bilateral agreements involving biomass exist with Brazil and Israel.

Photovoltaic Energy Systems. There is considerable near-term market potential for a variety of decentralized photovoltaic applications, which will have an early market in developing countries. In December 1978, the DOE dedicated a 3.5-kW photovoltaic village system in an isolated Indian village at Schuchuli, Arizona. This photovoltaic system provides power to pump water, operate fluorescent lights and appliances, and provide refrigerated space for 15 houses and other buildings in the village of about 100 persons; the town is not connected to the electricity grid. A project of this type has great potential applicability in less developed countries with similar isolated villages that are either not serviced by traditional forms of electricity generation or are dependent on small-scale diesel engines to produce power.

A 350-kW peak output photovoltaic system is included among projects to be performed under the United States–Saudi Arabia Solar Energy Agreement. The system will be in operation in late 1981.

Current national photovoltaic goals seem modest, given the estimate that a photovoltaic market of 6 GW exists solely for water pumping in developing countries, with significant market penetration possible by 1981–82. However, many potential users are unaware of the potential benefits and unsure of the readiness of solar-cell power systems for their applications. While some photovoltaic systems are already competitive for many applications, overseas experimental programs by DOE may be necessary to satisfy potential users' concerns about their geographic compatibility and economic acceptability in specific settings of the users.

### Wind Energy Conversion Systems.

The increased emphasis of the DOE wind program on smaller dispersed systems for domestic power needs has a large potential spinoff for assisting developing countries in meeting selected power needs and for significant overseas commercialization activities for United States industry. More attention should be given to designing systems capable of efficiently operating in 6- to 10-mph winds, a lower average wind-speed regime than for the 10- to 15-mph winds common to temperate climates. Also, overseas demonstrations of wind-system applications for meeting the needs of developing countries may assist the United States in penetrating this market.

The two International Energy Agency wind energy conversion system agreements focus on the availability of data concerning larger wind systems appropriate for electrical utility and industrial needs. This is a reflection of the domestic program emphasis on reducing the costs of intermediate- and large-size wind energy systems. The United States has participated in a project with Denmark to obtain data on operating efficiencies of the Gedser windmill. This windmill was built by Denmark in 1961 to evaluate a wind turbine design. After a few years of initial testing, the facility was not operated for more than a decade until the recent renewed interest in wind turbine development emerged in both the United States and Denmark. The wind turbine was renovated in 1977 and has been operated to provide the United States and Denmark with performance data for an improved wind turbine design. Figure 17-5 shows the Gedser windmill. An additional bilateral wind project is under way with the United States in Spain.

### Solar Thermal Power Systems.

Competitive opportunities for small-scale solar thermal plants are emerging in remote areas where fuel and electricity costs are high and conventional fossil energy supplies are insufficient. In such areas in the United States and over-

Figure 17-5. Gedser windmill in Denmark.

seas, the best system approach may be the use of hybrid power systems in which dispersed solar systems are combined with fossil fuel plants as a backup system, or the use of hybrid systems as total energy systems producing both power and thermal energy for heating requirements.

Support of RD&D for small-scale thermal plants will raise the United States' visibility in this potential near-term market. International opportunities for developing solar thermal systems exist in multilateral (e.g., International Energy Agency) and bilateral (e.g., Italy) solar agreements. Such projects give American firms in the field some international market exposure, with possibilities of establishing joint commercial ventures with foreign firms. Another area of emphasis for overseas needs is in agriculture-based dispersed solar technologies. These small-scale systems can have early promise in developing countries because they can provide most of the thermal energy requirements for a farm that is not connected to a conventional electricity grid.

Commercialization of large-scale thermal plants overseas is expected to take considerably longer than that for small-scale plants. A large-scale thermal demonstration pant is not expected until at least 1986. A prime need is to reduce heliostat costs to about $14/ft² by the early 1980s and eventually to about $7/ft². At the latter price, thermal electrical plants are expected to be competitive with other electrical power plants in high insolation areas, such as in the southwestern United States and most tropical areas.

Ocean Systems. International commercialization considerations are particularly important for ocean systems projects as a result of the limited availability of operational sites near United States coasts. At present, the United States is participating in an International Energy Agency wave-power technical exchange agreement. This wave agreement should give the United States an opportunity to become more knowledgeable, at relatively low cost, about recent wave technology advances.

Planned commercialization of ocean thermal energy conversion systems may be accelerated by initiation of RD&D projects with countries that have promising test and operational sites and that are also potential customers.

Office of Solar Applications. The International Energy Agency cooperative agreement on solar heating and cooling was established to conduct joint tasks in addressing key technical problems (while avoiding duplication), sharing expertise, and making more effective use of resources. The solar heating and cooling collaboration has grown to seven projects in which 16 countries participate. The project on system performance has established common procedures for predicting system performance, measuring thermal performance and reporting system performance; in the coming year, work will center around validation of computer simulations and the study of solar-assisted low-energy dwellings. Having completed round-robin testing of flat-plate collector thermal performance, the collector testing project will concentrate its efforts on reliability and durability testing and the study of evacuated tubular collectors. Two of the projects involve meteorological support tasks designed to familiarize solar scientists with solar radiation and related meteorological data requirements and increase the meteorologists' understanding of the needs of the solar community. Another project has undertaken the study of the performance of a variety of systems utilizing evacuated collectors. The participants in a newly intitiated project will conduct a feasibility study and prepare site-specific designs for large-scale central solar heating plants with seasonal storage.

# 18. Solar Program Implementation

## PROGRAM MANAGEMENT

### Management Approach

Overall direction and planning for an integrated national solar program is accomplished at the DOE headquarters level with program and project management decentralized to lead field offices and lead technical centers. The objective of decentralization is to use research talent and facilities throughout the country and to establish centers of technological excellence. Field operations offices, national laboratories, SERI and regional solar centers, universities, private industry, and other federal agencies contribute to a consolidated national program.

The responsibility of DOE headquarters is to develop policy and long-range plans and to establish priorities to ensure coherent goal-oriented programs relevant to the National Energy Plan and legislative mandates. In addition, efforts are directed toward the development of the DOE budget, presentation of DOE programs to Congress, assessment of programs to meet goals and objectives, and coordination among key participants in the solar energy program.

Programs are being decentralized at three levels: subprograms, major elements of subprograms, and projects. Day-to-day decentralized management is the responsibility of operations offices, national laboratories including SERI, regional solar centers, other federal agencies, or a combination of key participants. The role of a "field" assigned a decentralized program or project is to manage resources to accomplish a given objective within prescribed funding and performance and schedule constraints, and to report progress to headquarters.

### Participants in Federal Solar Program

Table 18-1 presents the breadth of key participants in the solar program. Decentralization of program and project management is developing key centers of expertise in the field among a varied mix of performers, which are geographically dispersed throughout the country. Program managers in headquarters have an oversight and planning responsibility for the solar technologies. Field offices have a day-to-day management responsibility, providing support to technical groups at the field office, in onsite project offices, and at national laboratories. Contracts to industry and universities are awarded and managed by the field office as well as by the national laboratories. Laboratories with management responsibility for solar technologies have broadened their scope from performing in-house R&D to becoming lead technology centers with the charter to coordinate and integrate a spectrum of activities. Other federal agencies, many of whom have long been involved in solar areas, are bringing their experience to each of the solar technologies.

Each of the participants, as well as state and local governments, supports major efforts in industry, universities, and nonprofit groups. The nation's colleges and universities are significant resources in the solar program, involved in basic and applied research, exploratory and technology development, systems design and testing, and economic and social analysis. The commitment to colleges and universities in solar research and development is reinforced by the establishment of a specific university research program at SERI to stimulate basic and fundamental solar-related research.

General functions of key solar program participants are:

## Table 18-1. Major Participating Organizations in the Solar Program.

| PARTICIPANT TYPE | SOLAR THERMAL POWER | PHOTOVOLTAICS | BIOMASS/WOOD ENERGY | WIND ENERGY | OCEAN SYSTEMS | ACTIVE HEATING & COOLING | PASSIVE & HYBRID HEATING & COOLING | AGRICULTURAL/INDUSTRIAL PROCESS HEAT |
|---|---|---|---|---|---|---|---|---|
| DOE Headquarters | ST | ST SA | ST SA | ST SA | ST | SA | SA | SA |
| DOE Field Offices | SAN CHO ALO | SAN CHO ALO ORO | CHO RLO SAN | SAN CHO ALO RLO | CHO SAN | SAN CHO ALO | SAN CHO ALO | SAN CHO ALO ORO |
| Other Federal Agencies | NASA | NASA TVA BPA DOD DOI DOC AID DOS | DOD NBS TVA BPA USDA EPA | NASA TVA NBS DOD GSA USDA BPA DOI DOS | NOAA MARAD | NASA TVA CPSC NBS HUD AID USDA DOD GSA DOC DOS SBA HEW VA DOL DOI | HUD USDA NBS GSA NEA | USDA NASA NBS USAF |
| Solar Energy Research Institute (SERI) | √ | √ | √ | √ | √ | √ | √ | √ |
| Regional Solar Energy Centers | | | √ | √ | | √ | √ | √ |
| National Laboratories | SLA SLL | ANL BNL LBL LASL ORNL SLA PNL | ANL BNL LBL PNL ORNL | LLL PNL SLA RF | ANL LBL ORNL | ANL BNL LBL LASL | ANL BNL LBL LASL ORNL | SLA ORNL LLL LASL INEL ANL |

√  Denotes participation in the above program.

| | | | |
|---|---|---|---|
| AID | Agency for International Development | MARAD | Maritime Administration |
| ALO | Albuquerque Operations Office | NASA | National Aeronautics and Space Administration |
| ANL | Argonne National Laboratory | NBS | National Bureau of Standards |
| BIA | Bureau of Indian Affairs | NEA | National Endowment for the Arts |
| BLM | Bureau of Land Management | NOAA | National Oceanic and Atmospheric Administration |
| BNL | Brookhaven National Laboratory | | |
| BPA | Bonneville Power Administration | ORNL | Oak Ridge National Laboratory |
| | | ORO | Oak Ridge Operations Office |
| CHO | Chicago Operations Office | PNL | Pacific Northwest National Laboratory |
| CPSC | Consumer Product Safety Commission | | |
| | | RF | Rocky Flats SWECS Testing and Development Center |
| DOC | Department of Commerce | RLO | Richlands Operations Office |
| DOD | Department of Defense | RSECs | Regional Solar Energy Centers |
| DOE | Department of Energy | | (Mid American Solar Energy Complex) |
| DOI | Department of Interior | | (Western Solar Utilization Network) |
| DOL | Department of Labor | | (Southern Solar Energy Center) |
| DOS | Department of State | | (Northeast Solar Energy Center) |
| EDA | Economic Development Administration | SA | Office of Solar Applications |
| EPA | Environmental Protection Agency | SAN | San Francisco Operations Office |
| | | SBA | Small Business Administration |
| FTC | Federal Trade Commission | SERI | Solar Energy Research Institute |
| | | SLA | Sandia Laboratory Albuquerque |
| GSA | General Services Administration | SLL | Sandia Laboratory Livermore |
| | | ST | Office of Solar Technology |
| HEW | Health, Education and Welfare | | |
| HUD | Department of Housing and Urban Development | TVA | Tennessee Valley Authority |
| INEL | Idaho National Engineering Laboratory | USAF | United States Air Force |
| | | USDA | United States Department of Agriculture |
| LASL | Los Alamos Scientific Laboratory | | |
| LBL | Lawrence Berkeley Laboratory | VA | Veterans Administration |
| LLL | Lawrence Livermore Laboratory | | |

*DOE Headquarters*
- Planning
- Policy development
- Program development and coordination
- Fiscal control
- Evaluation of program results

*DOE Regional and Field Offices*
- Management of fiscal outlays
- Management support
- Project management

*Federal Agencies*
- Joint program management
- Technical support
- Support for initial market entry through purchases and incentives

*Solar Energy Research Institute (SERI)*
- In-house R&D
- National market development support
- Management support
- Technical support
- Project management

*Regional Solar Energy Centers*
- Regional market development support
- Outreach and information
- State and local government coordination
- Project management
- Technical support
- Evaluation of program results

*National Laboratories*
- Technical support
- Project management
- In-house R&D
- Evaluation of program results

*State and Local Agencies*
- Outreach
- Legislation
- Regulation
- Policy guidance
- Implementation of legislated programs

The Solar Energy Program involves a large number of universities, members of private industry, and research institutions.

# SOLAR ENERGY RESEARCH INSTITUTE

Since its inception in 1977, the Solar Energy Research Institute (SERI) has grown to employ 600 persons and is active in all phases of solar energy R&D, analysis, and commercialization. It has major R&D programs in:

- Photovoltaic systems
- Biomass energy systems
- Wind energy systems
- Solar thermal technology
- Industrial process heat
- Ocean energy systems
- Active solar heating and cooling
- Passive technology
- Energy storage systems
- Advanced solar energy research

In addition, SERI provides support to the DOE in planning and analysis, commercialization, and outreach programs. Specific programs include:

- Planning, analysis, and social science research
- Information systems
- International programs
- Academic and university research
- Commercialization activities

SERI was established to provide significant support to the national program of RD&D and deployment of solar energy technologies. The legislative authority for SERI was provided by Public Law 93-473, the Solar Energy Research, Development, and Demonstration Act of 1974, which calls for SERI to perform R&D and related functions as a part of the national Solar Energy Program.

## Program Overview

In accordance with this mandate, SERI is to provide the nation with a center of excellence dedicated to serving the needs of the public and industry in the development of solar energy as a major alternative energy source. SERI is an integral part of the Conservation and Solar Program. It has established continuing programs in research, analysis, and assessment; information and education; technology transfer; and commercialization in cooperation with other efforts on a national and international basis. SERI is necessary to accommodate these continuing programs by physically providing the environment for such research, data accumulation, and interaction of the management, scientific researchers, and staff.

## Photovoltaic Systems.

SERI has been designated as the lead center for advanced R&D in the DOE Photovoltaic Systems Program. The SERI program has four objectives. The first is to conduct research on promising new photovoltaic materials and conversion devices and to bring the most advanced concepts through the exploratory development phase of R&D to the point of technical feasibility. This includes improving the technical capability of the devices and developing, on a laboratory scale, promising cost-effective processes for their production. The second objective is to further develop the basic understanding of photovoltaic materials and phenomena so as to facilitate development of new and better photovoltaic devices. A third objective is to develop an understanding of photovoltaic related facets of the institutional, economic, and government policy areas to provide effective program planning and integration. The fourth major objective is to stimulate the formation and adoption of industry-established material, component, system, and subsystem performance criteria and standards for the design, application, and operation of reliable and safe photovoltaic power systems.

## Biomass Energy Systems.

SERI is developing the staff and facilities to conduct an effective R&D program composed of basic and applied research on selected topics; assessments of economic, social, and environmental factors; technology transfer to the private sector; and technical management and planning for assigned elements of the DOE national biomass program. SERI has major programs in biomass resource assessment to investigate the potential use of remote sensing for resource assessment and a bottoms-up assessment of forest resources in the United States. They also are analyzing production-supply relationships for agricultural products and scrap wood fuels and performing fundamental process research in thermochemical conversion and biomass production.

## Wind Energy Systems.

SERI has been assigned the lead mission in research and analysis activities of the DOE wind systems program. It conducts analyses to determine market, economic, environmental, and legal constraints that may affect wind systems development. SERI performs wind system studies to validate wind energy computer models, cost-estimating studies, television interference studies, and field measurements and noise studies. SERI also manages the advanced and innovative wind systems component of the DOE wind program.

## Solar Thermal Technology.

SERI has three major and several supporting objectives for this program. The major objectives are to provide effective management of the Advanced Technology Program that DOE has assigned to SERI to lead the component technology and materials development; extend the operating range of concentrator-receivers beyond 600° F; assist the development of nationally accepted standard performance tests for concentrating collectors; and support DOE in policy and programmatic decisions by analysis of solar thermal systems and their potential applications to establish cost and performance goals and to aid in the formulation of effective federal strategies.

The supporting objectives are to perform materials R&D to overcome component-limiting problems, such as low optical efficiency and thermal degradation; develop a comprehensive approach to quality assurance and standards leading to provision of interim performance criteria documents for the voluntary standards community; and disseminate technical information regarding solar thermal technology development and applications to both contractor and noncontractor audiences.

Ocean Energy Systems. The SERI program has three principal objectives. First, SERI will provide research data in the areas of biofouling and corrosion, heat transfer, and marine structures in direct support of the program's near-term objective to complete at-sea experiments with a closed-cycle power plant. Second, SERI will take the lead role in generating and assessing alternate power systems and conducting R&D on the most promising option. The goal of this effort is to generate enough data on the option so that a decision can be made on developing it in parallel or in place of the closed-cycle approach. Finally, SERI will develop social, environmental, legal, institutional, and market information that supports commercialization decisions relating to the alternative cycles.

Active Solar Heating and Cooling. SERI is developing solar heating and cooling systems that are increasingly cost-effective, efficient, reliable, safe, and durable. SERI provides, via its systems development activities, overall integration and direction to the DOE Systems Development Division's R&D effort, assists in focusing these efforts on the most effective aspect of product development and improvement, and provides the solar industry and others with methods to predict performance and to design and operate optimum solar systems. Specific objectives are to:

- Identify and consistently rank solar technologies and systems that are optimized to the intended end use
- Identify cost/performance goals for various solar technologies/systems that make them competitive with conventional systems
- Develop and validate system simulation models, and design tools and handbooks for various user sectors
- Develop improved control strategies and control systems to increase solar system performance

Passive Technology. The scope of the SERI passive technology program includes new and existing residential, commercial, and agricultural applications for heating, cooling, lighting, and hot water; energy conversion; and urban system analysis and design. The primary objectives of the SERI passive technology program are to develop predictive methods and design tools that promote energy-conserving, climate-adaptive structures and urban environments; verify the fuel-saving performance of passive/hybrid concepts in new and existing building and urban applications; and inform and motivate the public sector and the building community (including architects, engineers, land-use planners, builders, developers, manufacturers, and lenders) on the merits of passive/hybrid design concepts.

SERI's passive technology program is an integral part of the DOE National Passive and Hybrid Program, undertaking projects and activities that contribute to the accomplishment of national energy goals. It is designed to be responsive to high-priority tasks as identified in the national program for Passive/Hybrid Solar Heating and Cooling, the Commercialization Strategy for Passive Solar Heating, and the Multiyear Plan for Passive and Hybrid Systems.

Industrial Process Heat. The objective of the SERI industrial process heat (IPH) program is to accelerate the commercialization of solar energy for industrial applications by reducing system costs, increasing component reliability, understanding and predicting solar IPH system performance, simplifying industrial application/solar systems interfaces, and identifying potential markets. The main issue to be addressed by the near-term SERI IPH program is cost-effectiveness, with effort focused on low- to intermediate-temperature applications below 300° C. However, preliminary sytems and market analysis and efforts for industrial process applications above 300° C will also be initiated. A number of related issues will be covered. There is presently a great need for validated models to predict IPH solar system performance and for experimental testing of varying combinations of collector field types, storage, control strategies, and process loads. A field-test facility (SERAPH) aimed at these research needs is a priority of the SERI IPH program.

Solar Energy Storage. The overall objective of the Solar Energy Storage program is to gain a better understanding of advanced thermal storage technologies for solar applications and to obtain data and information that will allow thermal storage developers to select the most promising thermal storage technologies for specific solar applications. To this end, SERI will determine the value of thermal storage, develop concepts with the potential of low-cost storage, and identify the most promising storage concepts for specified solar applications. This is in support of the joint Thermal Storage for Solar Thermal Power Systems Plan between the DOE Division of Energy Storage Systems and the Division of Central Solar Technology. Specifically, the three main objectives are to:

- Perform R&D of advanced thermal energy storage by latent heat of fusion and reversible chemical reactions
- Evaluate thermal storage R&D activities and solar application storage requirements
- Conduct a comprehensive review of the thermal energy storage area

Advanced Solar Energy Research. The Advanced Solar Energy Research (ASER) Program is designed to identify, conduct, and support advanced research in areas not addressed by the existing DOE national solar technology programs. The ASER Program is comprised of four independent research tasks: photoconversion; materials research; energy resource assessment; and new concepts.

This program is oriented toward basic research, but directed to the ultimate goal of discovering and developing new solar energy conversion and materials options. It fits into the overall goals and mission of the DOE program because the ASER Program provides a mechanism by which new solar energy technologies can be identified, evaluated, and, if

warranted, developed into commercially feasible systems.

The energy resource assessment task will develop the methods and data required for assessing, characterizing, and forecasting solar energy resources and the associated impact on performance of solar conversion systems. The new concepts task will explore new solar energy conversion schemes that are not part of existing research programs. Its purpose is to evaluate new ideas and, when appropriate, to nurture them to a decision point where they may be incorporated into program areas or dropped. Such activity is essential to ensure the evolutionary development of solar technologies.

Planning Analysis and Social Science. The planning, analysis, and social science program area at SERI is intended to help DOE by providing analyses of past, present, and future government roles in solar technology development and application; information about solar technology performance, costs, and markets; information about progress toward Solar Energy Program goals; assessments of the social and environmental aspects of solar energy options; and assistance with R&D program planning and management. Planning, analysis, and social science analyses are characterized by their consideration of broad measures of social benefits and cost of alternative government actions, their assessment of alternative government actions that are not technology specific, and their use of comparisons among solar technologies or between solar and other technologies.

Information Systems. The information systems program focuses on the development, operation, and continued improvement of the Solar Energy Information Data Bank (SEIDB). The development of the data bank was authorized by Congress in the Solar Energy Research, Development, and Demonstration Act of 1974, Public Law 93-473, Section 8. The lead responsibility for the design, implementation, and operation of SEIDB was assigned to SERI by DOE.

SEIDB has been developed as a national

network with the direct participation of the regional solar energy centers and the National Solar Heating and Cooling Information Center. The network activities are coordinated closely with DOE information activities, particularly the Energy Information Administration and the Technical Information Center. Liaison with other appropriate federal information activities is maintained also to maximize utilization of existing information and data and to minimize duplication.

International Programs. SERI is chartered to serve as the focal point of America's international solar activities, aiding in the negotiation and development of bilateral and multilateral programs, project management, provision of information resources, and provision of aid to developing nations.

SERI supports DOE international programs and interacts with other DOE laboratories and other United States and international agencies in order to accomplish this objective. SERI has been asked by DOE to be the lead institute in collaborative agreements with Saudi Arabia, Italy, Spain, Australia, Israel, Mexico, and other countries. International solar activities are cooperative activities between the Office of the Assistant Secretary for International Affairs and the Office of the Assistant Secretary for Conservation and Solar Energy. Various DOE laboratories throughout the country, other federal agencies, and public and private institutions are involved in SERI's international programs.

Academic and University Research Program. The objective of the research activities is to support, through grants, an extensive basic solar research program in universities and to stimulate the creation of new ideas and concepts needed for the evolution and growth of a vigorous solar industry. A more specific objective is to fund approximately 50 new basic research projects in addition to those initially funded in FY 1980. SERI will act as a surrogate to move those projects that prove to be scientifically feasible to DOE programs that can better finance the development of an idea. At the same time, potentially bad projects will be terminated.

Commercialization Activities. The overall objective of the commercialization activities program is to support market-related actions that serve more than one technology and to provide precommercialization assistance to those solar technologies not yet mature. Specific objectives are to provide relevant information to technical and nontechnical audiences. The scope of the program includes providing information about DOE solar R&D results to relevant technical and nontechnical communities and supporting publicity showing the benefits of solar applications both to technical audiences and to the larger interested public.

The commercialization activities program contains those broadly based projects that have an impact on more than a single technology or that address technologies or applications not yet ready for major commercialization actions. This program provides assistance to, and stimulation of, the technical, economic, and social development of solar energy at both supplier and user levels and within the distribution and support system. The activities are aimed at creating an awareness of, and confidence in, solar energy.

## REGIONAL SOLAR ENERGY CENTERS

The Solar Energy Research, Development, and Demonstration Act of 1974, Public Law 93-473, called for the creation of the Solar Energy Research Institute. SERI was to conduct R&D and related work to achieve commercial viability of various applications of solar energy technology. The Energy Research and Development Administration (ERDA), a predecessor agency to the DOE, announced on March 24, 1977, that it had selected the Midwest Research Institute to manage and operate SERI, to be located in Golden, Colorado.

At the same time, recognizing the regionally diverse nature of solar energy applications, ERDA announced its intent to establish four regional solar energy centers (RSEC). Subsequently organizations representing the northeastern, north central, western, and southern regions of the United States were awarded planning grants to promote the widespread use of solar energy.

In March 1978, management responsibility for SERI was assigned to the Assistant Secretary for Energy Technology. Responsibility for managing the four regional centers was given to the Assistant Secretary for Conservation and Solar Applications. Primary missions were also defined. SERI is responsible for RD&D; the regional centers concentrate on commercialization of solar application. Commercialization includes those activities that will bridge the gap between successful solar technology demonstrations and widespread use of these technologies by both the public and private sectors. These activities are designed to provide a secure and permanent market for the solar industry, to reduce buyer uncertainty, and to accelerate the use of solar technologies. As a result of a DOE reorganization on October 1, 1979, the regional centers and SERI report to the Assistant

Secretary for Conservation and Solar Energy and work closely together to support the DOE program.

The centers are located in Cambridge, Massachusetts (Northeast Solar Energy Center); Atlanta, Georgia (Southern Solar Energy Center); Minneapolis, Minnesota (Mid-American Solar Energy Center); and Portland, Oregon (Western Solar Utilization Network). The regional jurisdiction of each center is given in Figure 18-1.

## Scope of Activities

The market development activities of the regional centers are aimed at identifying and removing institutional and other barriers to solar energy use by encouraging state and local governments to establish appropriate standards, codes, regulations, and incentives through education and technician training programs, technical and marketing assistance

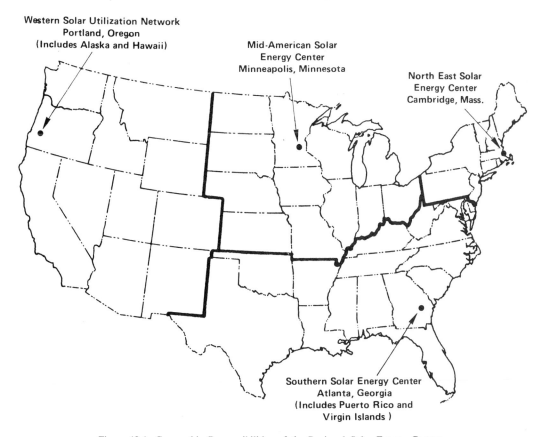

Figure 18-1. Geographic Responsibilities of the Regional Solar Energy Centers.

efforts for business, and consumer-awareness campaigns. Programs are designed and funded to permit the federal government to work together with the key constituencies who can make solar energy happen for the nation: manufacturers, designers and architects, builders, installers, financial officers, state and local government officials, public utility commissioners, and consumers. Programs now structured to emphasize those solar technologies that have reached or are near market-readiness include passive solar design, solar water heating, industrial process heat, small wind, and wood combustion.

Passive Solar. Regional centers promote the development and adoption of available, proven passive heating and cooling designs for residential and commercial buildings. Passive designs vary by climate and by building type. Activities in each program include developing and market-testing passive solar designs in major climate and market areas; providing design assistance to architects and builders of both residential and commercial structures; supporting development of locally adapted, marketable, passive cooling designs; and providing information to the public, the construction industry, and the financial community. Market analysis programs will measure the effectiveness of passive program activities at demonstration sites.

Solar Water Heating. Programs are developed by the regional centers to provide education, training, or demonstration of solar hot-water technology, to reduce or eliminate one or more of the barriers restricting the widespread use of solar hot-water systems, and to actively assist the industry in its marketing efforts. Program objectives include development of a data base on the performance, reliability, durability, and cost of solar water-heating systems and developing information about the commercialization of solar water heating required by institutional groups such as lenders, appraisers, realtors, and local government officials. Programs will also provide information, education, and training services to each group in the product distribution chain: designers, manufacturers, dealers in-

stallers, and consumers. Information programs are designed to "sell" solar water heating to groups of decisionmakers who are essential to obtaining widespread acceptance of solar hot water: homebuilders, homeowners, and business executives in industries that consume large quantities of hot water.

Industrial Process Heat. The primary objective of the regional centers' industrial process heat (IPH) programs is to examine regional solar IPH applications; to identify early markets; to locate special cases for near-term implementation; to provide technical, financial, and barrier reduction assistance in order to stimulate installations as well as to provide feedback to national programs and to the solar manufacturing industry. Programs will be structured to demonstrate the economic feasibility of solar IPH systems in local applications in order to provide experiences that may be easily transferred from one firm to another. Site-specific feasibility studies of the most cost-effective IPH applications will be undertaken, with initial focus on those industries that represent a region's greatest IPH consumption. Analyses of market potential for solar IPH applications will be conducted based upon local economic market and climatic conditions.

Wood. Identification and promotion of wood fuel supply systems, the support of site-specific technical assistance for industries and utilities, and the dissemination of applications data to all market sectors are the primary objectives of the regional centers' wood programs. Generic feasibility studies will be conducted to determine a region's most likely types of wood fuel supply operations for the residential, industrial, and utility markets. These feasibility studies will then be used as base data to develop venture analysis information for use in direct contact with potential suppliers. A residential firewood supply project will be planned and implemented within each region. Assistance will be given to companies in the forest products industry, as well as to other industrial and commercial firms investigating wood-fired heating or cogeneration opportunities. Industrial and commercial use of wood fuels will be promoted through the news media, workshops and conferences, trade and business association contacts, and direct contact with conversion prospects.

Small Wind Energy Conversion Systems. The regional centers support implementation of the Public Utility Regulatory Policy Act of 1978 (PURPA) by monitoring state public utility commissions and local nonregulated utilities as they implement PURPA, sections 201 and 210 (small power producers), and by acting as an information resource to these regulatory bodies to secure fair treatment for small wind-energy conversion systems (SWECS) owners. They will make available existing market-analysis tools for use by state energy offices and local SWECS dealers and distributors; these include wind-resource maps and computation programs to analyze SWECS performance and economics in the region. Outreach efforts will provide information on current economics and performance to potential purchasers, rural electric cooperatives, and at trade/home shows and state/county fairs.

State and Local Programs. Each regional center maintains a state solar-office network to help in achieving program goals. This kind of a network is an important outreach component of the regional center concept. Each state solar office functions independently and reflects the needs of its own state; at the same time, contact with the regional center ensures that the states are included in the center's programs and objectives. These programs allow each center to track state solar legislation, keep current on utility and regulatory activities affecting solar energy, and be aware

of solar programmatic activity at the state and local levels and of the response by state and local governments to federal solar initiatives. Efforts are made to establish working relationship with key individuals and groups at the state and local government levels. Through the regional centers, the states are involved in DOE solar consumer assurance planning, identification of solar consumer problems, and other DOE programs aimed at removing institutional barriers to solar acceptance such as the development of model solar building codes for use by states and localities.

Each fiscal year, DOE will issue a program guide that outlines regional center program direction and emphasis. Following this guidance, the regional centers will submit annual operating plans that will specify the activities and projects they have selected to undertake in adherence to DOE's planning and funding guidance. This process for program definition was initiated for the second half of FY 1979.

Status

The four regional centers are under 5-year (1979–84) operating contracts with DOE. Funding and staff levels and activities are established for each fiscal year. Funding and staff levels for regional centers are summarized in Table 18-2.

Each regional center is governed by a board of directors with an appointment to the board by each of the state governors in the respective regions. In addition, the centers have advisory groups that assist the regional center staff in planning and program develpment. State solar offices work directly with state and local governments to promote solar energy

Table 18-2. Funding Levels for Each Regional Solar Energy Center.

| CENTER | FY 79 ($ IN THOUSANDS) | FY 80 (PLANNED) ($ IN THOUSANDS) | AUTHORIZED PERSONNEL |
|---|---|---|---|
| Northeast | 4,400 | 5,700 | 90 |
| Southern | 2,700 | 5,600 | 55 |
| Mid-American | 3,800 | 5,300 | 60 |
| Western | 2,600 | 5,100 | 30 |
| TOTAL | 13,500 | 21,700 | 235 |

use. The regional centers are able to identify and implement activities that are tailored to the specific needs of each region, as well as to DOE's national Solar Energy Program.

For FY 1980, the regional centers conducted activities in four areas of the solar program. Table 18-3 indicates the expected level of activity for each.

## SMALL AND DISADVANTAGED BUSINESS

The federal government has established a policy of awarding a portion of its total procurements for research, technology development, and demonstration; support services; supplies; construction; and maintenance and repairs to small and socially or economically disadvantaged business. An important legislative initiative requiring procurements in areas of the country that are considered labor surplus areas was pursued on October 12, 1978 (Public Law 95-507). One of the provisions of the amending act requires federal contract proposers to submit subcontracting plans that provide the maximum practical opportunity for small and disadvantaged businesses to participate in federal programs.

The DOE has established a five-point framework to guide its small and disadvantaged business program in the following areas:

- Creation of a Small and Disadvantaged Business Development Task Force
- Establishment of formal procurement goals and objectives for small and disadvantaged business concerns and labor set-aside areas
- Designation of small and disadvantaged

business facilities in each of the major program areas
- Development of a comprehensive source list of small and disadvantaged business concerns and labor surplus areas within each program area
- Development of a public information program to increase small and disadvantaged business participation in procurements

## Solar Research and Development Activities

The DOE Solar Program is firmly committed to increasing the participation of small and disadvantaged business. To meet these objectives, the Solar Program has participated in or initiated the following:

- Solar and other DOE programs participated in a "Conference for Small R&D Companies on Opportunities in the Government," held in Washington, D.C., on October 24–26, 1979.
- Small business set-asides have been awarded for support services and solar thermal innovations. In FY 1980, set-asides were expanded to the solar international area.
- A small business program has been initiated at the Solar Energy Research Institute (SERI) as well as at national laboratories involved in the Solar Program.

Plans for the future include:

- Implementation of several recommended strategies proposed as a result

## Table 18-3. Estimated Funding for Solar Energy Centers Activities by Function.

| ACTIVITY | FY (PLANNED) FUNDS IN THOUSANDS |
|---|---|
| Market Analysis | 3,300 |
| Systems Development Support | 3,200 |
| Market Tests and Applications | 4,200 |
| Market Development and Training | 11,000 |
| TOTAL | 21,700 |

of a study commissioned by the solar program in FY 1978. These would include identification of small business R&D firms to be part of the Solar Energy Information Data Bank (SEIDB); development of orientation programs for DOE and national laboratory procurement staffs; establishment of a national regional advertising strategy to improve outreach to small business; promotion of small business subcontracting, and development of a process to review annual program requirements for small business opportunities.

- Development of a program with SERI and the DOE Small Business Group to include a screening procedure for small business set-aside procurements; identification of small R&D firms to be incorporated into the solar data bank; orientation programs for BEA personal on solar programs and small business loans; training to improve proposal-writing skills, and a workable plan for improving outreach through the Commerce Business Daily, technical publications, and expansion of small business mailing lists for early program announcements.

# 19. Markets and Applications for Solar Technologies

Energy resources are utilized in large quantities in modern societies for many different applications. Conventional energy forms such as fossil fuels, hydropower, and nuclear fission are routinely applied to meet end-use needs directly or to generate intermediate energy forms such as electricity. The supporting infrastructure and end-use equipment for these conventional resources are highly developed and represent extensive investments. Thus, markets for solar technologies will develop as potential users decide between conventional sources and solar systems, either through purchases of original equipment for new installations or for retrofit or early retirement of conventional technologies in existing capital stock. Figure 19-1 illustrates the diversity of the solar resource and identifies technologies that can be applied to satisfy demand in various end-use market sectors.

Rapid commercialization of a variety of solar technologies is necessary if the nation is to achieve the national solar energy goals set forth by the president. Such commercialization must take place in several distinct market areas, including buildings, industry, utilities, and transportation. Each of these market areas faces unique constraints in adopting solar technologies to supplement or replace existing conventional energy systems.

The following topics are discussed for each market area:

- Market components
- Structure of end-use demand
- Equipment selection options
- Market potential
- Constraints

Market components include the institutions, individuals, and corporate entities active in energy equipment selection and operation for that sector. The structure of end-use demand considers the demands for energy in each sector by form, quantity, and quality of energy required. The equipment selection options section considers existing and perceived future energy utilization conditions for both conventional and solar technologies. Market potential is assessed, based on recent DOE analyses of solar technology penetrations. Constraints to market development are noted as they may delay or prevent rapid commercialization of solar technologies.

## Buildings

**Market Components.** Buildings serve two major uses in society: the residential housing of people (living space); and the sheltering of groups of people for commerce, information exchange, or recreation (industrial structures are considered later). Further classification identifies six unique building types: single-family detached; multifamily low-rise; multifamily high-rise; commercial low-rise; commercial high-rise; and specialty buildings such as hospitals, gymnasiums, auditoriums, and meeting halls.

Building users may be private or commercial and may either own or lease their property. Prominent institutions in the building sector include codemaking and enforcement groups, a diversified construction industry, and construction trade unions. Energy is presently supplied to the building sector by natural gas or electrical utilities and fuel-oil firms.

**End-Use Structure.** Major end-use demands (see Table 19-1) in the building sector include thermal energy for space heating and cooking; mechanical energy for fans and refrigerators;

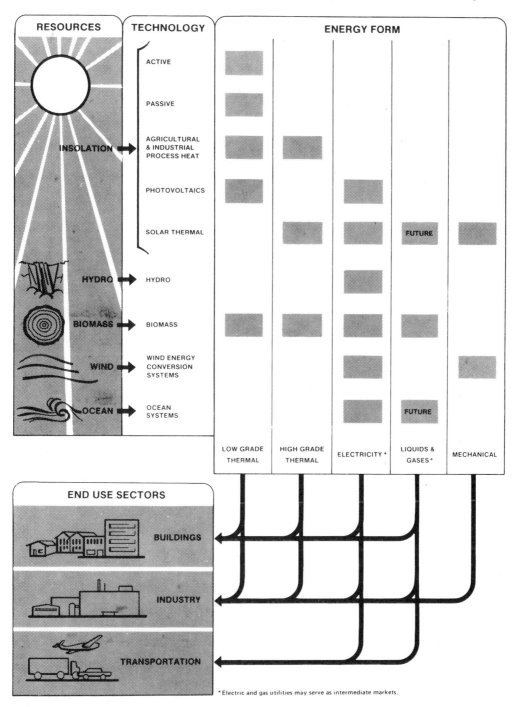

Figure 19-1. Diversity of solar resources and applications.

and electricity for lights, electronics, and other equipment. Common forms of energy used in buildings include liquid and gaseous fossil fuels (thermal applications) and electricity.

Electricity use in buildings has grown rapidly in recent times because of its inherent flexibility, unique applications (such as electronics or lighting), and comparative supply security. However, electricity is relatively ex-

## Table 19-1. Primary Energy Consumption for the Building Sector, 1973.

| ENERGY SERVICE | DIRECT FUEL | ELECTRICITY | TOTAL |
|---|---|---|---|
| Space Heat | 10.44 | 0.32 | 10.76 |
| Water Heat | 1.87 | — | 1.87 |
| Space Air Conditioning | 0.35 | 0.73 | 1.08 |
| Refrigeration | — | 0.66 | 0.66 |
| Cooking | 0.53 | 0.08 | 0.61 |
| Lighting | — | 1.31 | 1.31 |
| Clothes Drying | 0.09 | 0.08 | 0.17 |
| Other Electrical | — | 0.25 | 0.25 |
| TOTAL QUADS | 13.28 | 3.43 | 16.71 |

Source: Ross and Williams. "Assessing the Potential for Fuel Conservation." Institute for Public Policy Alternatives. State University of New York. Buffalo. 1975.

pensive for many space- or water-heating applications. In current market situations, shifts away from electricity or high-energy-quality fossil-fuel forms may be possible for a significant fraction of building energy requirements (predominantly low-temperature thermal energy).

Equipment Selection Options. Solar technologies suitable for building applications span a wide range of resource inputs, conversion technologies, and end-use energy forms, ranging from wind furnaces to wood stoves to solar cooling. A discussion of this diversity, by solar resource, follows. Two major technologies are highlighted in this section: solar technologies for space conditioning and water heating (including active, passive, and hybrid systems) and photovoltaic energy systems. These two technologies are likely to represent the bulk of market opportunities in the buildings sector because of their general applicability throughout the United States. Other technologies discussed here can be important in particular applications.

*Direct solar radiation.* Direct solar radiation can be used in several ways:

- Active solar thermal energy collection
- Passive solar thermal energy collection
- Hybrid solar thermal energy collection
- Photovoltaic power systems
- Combined photovoltaic/thermal power systems
- Solar thermal electric power systems

Active, passive, and hydrid systems directly address the predominant building sector energy demand for low-temperature (§ 100° C) thermal energy for space and water heating. Photovoltaic and solar thermal power systems supply electricity to the building and may incorporate onsite storage or have energy exchange with the grid. Hybrid versions of these technologies can supply thermal energy as well.

Active solar technologies use mechanical energy to pump a working fluid (usually water or air) between the solar collector (usually a flat-plate design) and the load or storage. Depending on the design requirements, a wide variety of operating temperatures is possible. System control is achieved with thermostats, and precise temperatures can be maintained. A backup system using conventional energy is usually incorporated. Collectors can be placed in remote locations, if necessary. Uses include space heating, water heating, and space cooling.

Passive solar technologies use only natural heat transfer mechanisms (conduction, convection, radiation) or direct thermal gain to supply thermal energy to loads. In a direct-gain space-heating design, for example, the sun is permitted to shine directly into the living space to arm the dwelling and its furnishings. A thermosyphon water heater makes use of the fact that warm water is slightly less dense than cold water; thus, the solar-heated water is displaced by cooler storage water into the top of the storage tank. Although currently characterized by a limited range of operating temperatures and a lack of system control, passive systems can be designed to satisfy building thermal loads in a very cost-effective manner. Passive cooling techniques, using night-sky radiation or induced drafts, can also be cost-effective in certain climates.

Hybrid solar technologies rely on passive thermal energy collection features, but employ mechanical transfer of working fluids to improve system control, operating flexibility, and efficiency. A direct-gain structure with underfloor stone/air-heat storage (fed during daylight with a fan) is an example of a successful hybrid system design.

Photovoltaic power systems are not always associated with buildings (they reach many other markets), but they are often described as using rooftop collectors for residences or commercial structures. These systems employ a solid-state semiconductor wafer, which, when exposed to light, generates an electrical current that can be conditioned to substitute for conventional utility power or used as direct current. These systems may supply all or part of the building load or they may be linked directly with the grid and operate on a power exchange basis. Although initial systems costs are presently high, building applications for photovoltaics may earn a favorable rate of return during their lifetimes (which are expected to be about 15 to 20 years) if DOE price reduction goals are met.

Photovoltaic/thermal power systems perform a hybrid function in that they produce both electrical and thermal energy. A typical configuration employs a concentrating collector focused on a photovoltaic cell. The cell will convert about 20 to 40 percent of the incident radiation into electricity; most of the rest of the incident energy must then be removed as heat. This heat, if collected at sufficiently high temperatures, can be applied to building thermal loads. Overall utilization of incident energy may be as high as 80 percent with this method.

Solar thermal electric power systems employ concentrators to drive a heat engine to generate electricity. For onsite generators, a parabolic dish or a linear fresnel concentrator driving an organic fluid Rankine turbine or

Stirling engine would be typical configurations. In many cases, waste thermal energy from the vapor condenser could be used to assist in meeting building thermal energy requirements.

*Wind power.* Two major wind turbine classes are being evaluated for onsite building use: vertical and horizontal axis machines, ranging from 500 W to 100 kW in size. End uses include electricity generation, thermal energy generation, and mechanical energy. Electricity and thermal energy systems would usually incorporate a storage system. Most-likely applications are in homes—urban locations where local wind conditions are favorable.

*Wood power.* Wood is a versatile energy resource with a wide range of potential applications in all market sectors of the economy. Wood energy can provide residential heat, industrial process heat, and electricity from either industrial cogeneration or utility power plants. Wood combustion currently provides more than 1.5 quads of energy—1.1 quads for industrial steam production and approximately 0.4 quad for residential units.

Wood energy has been a common building fuel in the United States for centuries. In regions where wood is inexpensive or available for cutting by users, it is an attractive energy source. Advances in fireplace and stove design have greatly improved the thermal efficiency of wood burning; many rural residences and some commercial buildings are using wood to meet a high proportion of building thermal energy requirements. For example, the town of Soldiers Grove, Wisconsin, is building a new central business district that will use a woodburning central (district) heating system to supplement the passive solar design aspects of the new buildings.

*Solar technology selection.* Equipment selection decisions in the building sector will be influenced by the following factors:

- Partially due to oil price decontrol, fuel oil prices have risen rapidly in the past

2 years, more than doubling the average residential user's space-heating bill.

- Natural-gas hookups for new buildings, although available in recent years, are likely to be constrained in the future. Price deregulation for gas will tend to eliminate the historic price advantage of this popular fuel form.

- Electricity continues to be an expensive energy source for providing space heat, although heat pumps can mitigate this factor. Overall energy expenditures tend to remain high with heat pumps, however, due to whole-house (instead of room-by-room) summer cooling and frequent compressor replacement in some applications.

- Extensive design, testing, and commercialization efforts have improved the performance of active and passive solar heating technologies. Capital costs have remained high for active systems, however, and passive technologies are not considered to be universally applicable in all building types, especially in retrofit programs for commercial buildings.

- If photovoltaic systems costs achieve DOE price goals, building such systems may produce competitively priced electricity for onsite applications in the late 1980s.

**Potential for Solar Technology in the Building Market.** To meet President Carter's 20 percent solar goal for year 2000, active solar domestic hot-water systems must provide 0.08 to 0.10 quad (equivalent fossil fuel) in 1985 or about 0.1 percent of all of the energy required by the United States in that year.* Estimates for the year 2000 range from 0.8 to 0.9 quad, or about 0.9 percent of all the

---

*This chapter uses the Domestic Policy Review (DPR) of Solar Energy as a source for potential solar energy contributions. The DPR used the informed judgment of experts, aided by computer models, to estimate solar contributions for the year 2000. Although the DPR did not publish estimates for the near term (1985), the same computer models (as well as other sources) were used to provide data on potential impacts for the year 1985.

energy required by the United States in that year. This means that one-half of the buildings in this country must have a solar hot-water system supplying 20 to 80 percent of the hot-water energy requirements.

Active solar heating systems may supply 0.15 to 0.25 quad (equivalent fossil fuel) in 1985, or about 0.2 percent of all energy required by the United States. This 1985 goal means that approximately 1 building in 20 must be fitted with a solar heating system supplying 20 to 40 percent of that building's heating energy. Estimates for the year 2000 range from 0.9 to 1.0 quad, or about 1.1 percent of all of the energy required by the United States in that year, displacing about half a million barrels of oil per day. About 1 building in 5 must be fitted with a solar heating system supplying 20 to 40 percent of that building's heating energy.

Estimates of the potential for supplying United States energy requirements by passive heating systems are from 0.3 to 1.0 quad (equivalent fossil fuel) in the year 2000, or about 1.0 percent of all the energy required by the United States in 2000. To reach this goal, 1 new building in 10 must be designed with a passive heating system that supplies 25 to 60 percent of that building's heating energy. Market penetration by 1985 will be small because passive retrofit is difficult and market penetration into the new home industry is expected to be slow.

Estimates of the potential for supplying space-cooling demands by active or passive systems range from 0.15 to 0.20 quad (equivalent fossil fuel) in 1985, or about 0.2 percent of all of the energy required by the United States in that year. This 1985 projection means that about 1 building in 80 must be fitted with a new or retrofit solar cooling system supplying 20 to 40 percent of that building's cooling energy. Estimates for the year 2000 range from 0.7 to 0.8 quad, or about 0.8 percent of all of the energy required by the United States in 2000. This means that about 1 building in 20 must have some form of solar cooling system supplying 20 to 40 percent of that building's cooling energy requirements.

The Domestic Policy Review (DPR) of Solar Energy estimated that the potential for

## Table 19-2. Residential Wood Combustion Goals.

| USE | QUADS/ YEAR | NUMBER OF UNITS | INCREMENTAL UNITS TO MEET THE GOALS |
|---|---|---|---|
| Residential—Heat | | | |
| Current | 0.4 | 7 million | — |
| 1985 Goal | 0.6 | 10 million | 3 million |
| 2000 Goal | 1.0 | 17 million | 7 million |

Source: "National Plan for the Accelerated Commercialization of Solar Energy," MITRE—79W00416. November 1979.

photovoltaic systems to displace fossil fuels is very small in 1985, and 1.0 quad of primary fuel displaced in 2000. This corresponds to 43,000 MW of photovoltaic collectors in place by 2000. Mounting arrays of several kilowatts capacity on 2 million homes during the 1990s would, for example, save about 0.1 quad of fuel by the year 2000.

The market for small wind machines is estimated to be about 4000 units. By 1985, a potential market for 50,000 to 100,000 installed wind machines has been identified for the residential and commercial sectors.

The use of wood for energy is projected to increase by 1985 to 2.45 quads, not quite 3 percent of the projected national energy consumption and, by 2000, to 4.4 quads. Meeting the 1985 goal, an increase of almost 1 quad over current use of wood for energy, will require the installation of significant numbers of wood combustion units. As shown in Table 19-2, this could include an additional 3 million residential units.

The DPR of Solar Energy and other recent government reports have detailed the expected building market impacts for solar technologies. Table 19-3 is a compilation of such data on a national aggregate basis. Analysis suggests, however, that market penetration is likely to be sensitive to regional variations in renewable resource availability and conventional energy system costs. Chapter 20 discusses regional variations in resource availability in detail.

Constraints. Even after solar technologies are cost competitive with conventional energy systems (as some already are, particularly on a life-cycle basis), a series of regulatory, institutional, and marketplace constraints will serve to limit the rate and magnitude of building solar market development.

Table 19-4 lists examples of constraints that are being addressed by a variety of DOE programs to accelerate market penetration of solar technologies in the building market.

## Table 19-3. Building Sector Solar Contribution.

| HEATING AND COOLING | LOADS AND SOLAR CONTRIBUTIONS | | |
|---|---|---|---|
| | FY 1985 | FY 1990 | FY 2000 |
| Residential Buildings | | | |
| National Residential Building Load | 8.08 Quads | 8.76 Quads | 10.35 Quads |
| Solar Contribution | 0.14 Quad | 0.43 Quad | 1.89 Quads |
| % Solar | 1.7% | 4.9% | 18.3% |
| Commercial Building | | | |
| National Commercial Building Load | 7.02 Quads | 7.17 Quads | 7.81 Quads |
| Solar Contribution | 0.15 Quad | 0.32 Quad | 1.14 Quads |
| % Solar | 2.1% | 4.5% | 14.6% |
| Total Solar Building Contribution | 0.29 Quad | 0.75 Quad | 3.03 Quads |
| Total Building Energy Demand | 15.10 Quads | 15.93 Quads | 18.16 Quads |

Source: "National Plan for the Accelerated Commercialization of Solar Energy," MITRE—79W00416, November 1979.

## Table 19-4. Constraints to Market Penetration in the Building Sector.

| MAJOR CLASS OF CONSTRAINT | SPECIFIC CONCERNS |
|---|---|
| Lack of performance and cost data for technology selection | • Potential users<br>• Potential manufacturers |
| Lack of relevant decisionmaking tools | • Potential users<br>• Regulatory officials |
| Lack of relevant technology design tools | • Manufacturers<br>• Architects and builders |
| Limited incentives for owners of leased or rented space | • Capital cost versus operating cost investment decisions favor low capital, high operating cost technology. Retrofit opportunities are also limited. |
| User/utility interface | • Back-up power or gas rates<br>• Buy-back rates for surplus on-site electricity<br>• Private power sources-regulation |
| Consumer protection | • Solar fraud<br>• Equipment warranties |
| State and Local Government roles | • Codes<br>• Subsidies<br>• Regulations |
| Infrastructure requirements | • Skills for installation, operation, and maintenance<br>• Commercial expertise for industry growth |
| Financing of non-conventional technologies | • Familiarity with performance and lifetime characteristics of solar technologies |

## Industry

**Market Components.** Industry is defined here as firms involved in manufacturing, mining, construction, agriculture, forestry, fishing, and public utilities. The number of types of industries is very large, as reflected in the highly detailed Standard Industrial Classifications developed by the Department of Commerce, which include up to four digits (with 10,000 possible unique combinations) to classify such firms. This vast number of firms functions within a basic industrial pattern of resource extraction or harvest, processing, manufacturing, wholesaling, retailing, servicing, and disposal.

Complex trade pattern and relationships have evolved during commercial growth in the United States, and government regulation of industry has served to increase this complexity since the 1930s. This complexity is mirrored in the extremely varied patterns of energy use within industry, as reflected in widely varying energy intensities of production among or even within product lines.

The very largest industries, such as automobile manufacturing and building construction, create the products that directly consume the bulk of our energy resources. The energy supply sector, whch provides automobiles and buildings with energy products, is itself one of the largest industries in the United States. This section discusses the potential for fuel substitution within existing industry, where solar thermal or wood energy would replace

Table 19-5. Primary Energy Consumption for the Industrial Sector (Manufacturing).

| ENERGY SERVICE DEMAND | 1973 PRIMARY ENERGY CONSUMPTION* (QUADS/YR.) | | |
|---|---|---|---|
| | DIRECT FUEL | ELECTRICITY (AS ELECTRICITY) | TOTAL (PRIMARY ENERGY) |
| Process Steam | 10.54 | — | 10.54 |
| Electricity Generation | 0.33 | — | 0.33 |
| Direct Heat | 6.58 | 0.18 | 7.09 |
| Electric Drive | — | 2.34 | 6.48 |
| Electrolysis | — | 0.34 | 0.94 |
| Other Electrical | — | 0.10 | 0.28 |
| Feedstock | 3.99 | — | 3.99 |
| TOTAL | 21.44 | 2.96 | 29.65 |

*Manufacturing only

Source: Ross and Williams, "Assessing Potential for Fuel Conservation," (Buffalo: Institute for Public Policy Alternatives, State University of New York, 1975).

fossil or electrical fuels, and where biomass industries would produce clean and renewable solid, liquid or gaseous fuels.

End-use Structure. At the risk of over-simplification, a large array of industrial manufacturing patterns in end-use and energy consumption are summarized in Table 19-5. Table 19-6 presents estimates of these fuel uses distributed among end-use qualities. These tables suggest that there is a significant expenditure of fossil fuels for process steps requiring temperature differences of less than 100° C, which could be supplied or supplemented by solar energy using today's technology.

Nonmanufacturing industries have specialized energy requirements. Agriculture has two major classes of energy use: on-farm and off-farm. The bulk of on-farm energy is used for motive power and irrigation; off-farm energy is consumed in fertilizer manufacture, commodity transport, and food processing (included in Table 19-5); the latter presently uses large quantities of low-temperature thermal energy.

Mining uses electricity and liquid fuels for most applications, as does the construction trade. Forestry uses liquid fuels for harvesting, but many wood-processing plants use steam generated from wood wastes for thermal and power needs. In addition, sawmills may use either electricity or liquid fuels.

Fishing involves liquid fuel consumption for harvesting and a variety of fuels for processing. Public utilities, such as water supply

Table 19-6. Energy Service Demand for Manufacturing.

| ENERGY SERVICE DEMAND | ANNUAL ENERGY DEMAND | | | | | |
|---|---|---|---|---|---|---|
| | HEATING & COOLING | | END-USE ELECTRICITY | LOSS AT POWER STATIONS | FEEDSTOCK | TOTALS |
| | ΔT < 100°C | ΔT ≥ 100°C | | | | |
| Process Heat | 4.84 | 12.46 | | 0.33 | | 17.63 |
| Electricity, All Uses | | | 2.78 | 5.25 | | 8.03 |
| Electric Drive | | | (2.34) | (4.92) | | ( 7.26) |
| Electrolysis & Other Elec. | | | (0.44) | (0.33) | | ( 0.77) |
| Feedstocks | | | | | 3.99 | 3.99 |
| TOTAL | 4.84 | 12.46 | 2.78 | 5.58 | 3.99 | 29.65 |

Source: Ross and Williams, "Assessing the Potential for Fuel Conservation". (Buffalo: Institute for Public Policy Alternatives, State University of New York, 1975).

or water treatment, use electricity for most needs. Some waste treatment plants use the methane generated in certain processes to provide electrical or mechanical energy for plant requirements.

**Equipment Selection Options.** Solar equipment options in industry span a wide range of types, including:

- Active and passive systems for agricultural and industrial process heat
- Wind power systems
- Photovoltaic power systems
- Photovoltaic/thermal power systems
- Solar thermal electric power systems
- Wood
- Other biomass (ethanol, methanol, methane, exotic plants)

Many of these technologies are discussed in the building market section; additional technologies follow.

*Agricultural and industrial process heat.* A wide variety of solar agricultural and industrial process heat projects are being sponsored by DOE to meet process heat energy demands across a wide temperature spectrum. As these technologies achieve full commercial readiness, industrial firms will evaluate the effectiveness of agricultural and industrial process heat relative to their own specific process requirements.

Equipment selection options range between applications for basic water preheating (usually flat-plate collector technology), low-pressure steam, and high-temperature process heat (provided by concentrator-selective receiver technology). In favorable locations, solar systems using 100,000 to 200,000 ft² of collectors are comparable to a 100 MBtu/hr fossil-fuel boiler.

*Wood.* Woodburning technology is also receiving increased attention. Industrial firms or electrical utilities in certain regions may be able to use woodchips or pellets as a clean-burning and competitively priced replacement for coal, fuel oil, or electricity. Fluidized bed boilers are being commercialized; these will provide increased multifuel capability in such applications.

Industrial use of wood for energy is currently limited to the pulp-and-paper industry—almost 45 percent of the industry's internal energy requirements in 1976 were satisfied by wood combustion.

*Biomass industries.* The potential for biomass energy production and utilization is also being assessed at DOE; technology development and market studies are being pursued to explore the potential for large-scale silviculture, kelp, ethanol, methanol, other liquid hydrocarbons, or biogas production. Economic and environmental analyses are being carried out to support the technology studies. The forest products industry already uses over 1 quad of residues, principally for steam generation.

**Potential for Solar Technologies.** The potential contribution that solar technologies can make toward meeting IPH demand in the future is considered to be large. The Domestic Policy Review (DPR) of Solar Energy estimated that the potential of solar IPH systems to displace fossil fuels is 0.18 quad in 1985 and 2.6 quads in 2000. The estimate of 0.18 quad in 1985 corresponds to a usage of solar by fewer than 1 in 150 factories (approximately 4000 systems). However, by the year 2000, 1 in 7 factories would have to use solar power if the goal is to be met. The year 2000 goal translates into 1.3 million barrels of oil per day saved by solar IPH installations. Advanced IPH systems could generate steam to drive a turbine and produce electrical power, with the waste heat used in industrial processes. The solar technology for this application is under development. The market potential for these solar cogeneration systems adds at least 20 percent to the 2.6 quad goal in the year 2000. Wood energy goals for the industrial sector are presented in Table 19-7.

In his July 16, 1979, message on a U.S. Import Reduction Program, President Carter indicated a 1990 goal of displacing 0.1 million barrels per day of oil by producing fuels from biomass. Assuming a gallon-for-gallon replacement, this is equivalent to 1.5 billion

## Table 19-7. Industrial Wood Energy Utilization Goals.

| USE | QUADS | NUMBER OF UNITS | INCREMENTAL UNITS TO MEET GOALS |
|---|---|---|---|
| Steam-Industrial | | | |
| Current | 1.1 | 350 | — |
| 1985 Goal | 1.6 | 500 | 150 |
| 2000 Goal | 2.3 | 730 | 230 |
| Cogeneration-Industrial | | | |
| Current | — | — | — |
| 1985 Goal | 0.1 | 30 | 30 |
| 2000 Goal | 0.4 | 110 | 80 |
| Low-Btu Gas-Industrial | | • | |
| Current | — | — | — |
| 1985 Goal | 0.1 | 60 | 60 |
| 2000 Goal | 0.5 | 320 | 260 |

Source: "National Plan for the Accelerated Commercialization of Solar Energy," MITRE—79W00416. November 1979.

gallons of ethanol or methanol per year. Such a program would require the installation of at least 26 fermentation plants using corn, two plants for hydrolysis and fermentation of agricultural residues, two plants for conversion of municipal solid waste, and two methanol plants using wood, as shown in Table 19-8. Alternatively, this goal can be met by the installation of 75,000 small, on-farm systems, each producing roughly 20,000 gallons of ethanol per year.

The Solar DPR and other recent government reports have estimated the expected agricultural and industrial process heat and biomass industry technology penetrations. Table 19-9 summarizes these data.

Constraints. The industrial sector is sophisticated in its ability to evaluate the cost-effectiveness of solar agricultural and industrial process heat or biomass industries. One critical constraint to market penetration will be the need for industry to assess the reliability, maintenance costs, and performance variability of solar energy, and the effect of these concerns on overall costs. These and other constraints to market penetration are listed in Tables 19-10 and 19-11. The second table indicates constraints that may apply in biomass production and utilization in industry.

## Utilities

Market Components. Electrical and gas utilities face an uncertain future for their primary fuel resources. Each major fuel source faces supply constraints: natural gas supply is constrained in the midterm; petroleum fuels are rising steeply in price; coal is a fuel with

## Table 19-8. Goals for Synthetic Fuels from Biomass.

| SYNTHETIC FUEL | PLANT CAPACITY (MILLION GAL/YR) | NO. OF UNITS TO MEET 1990 GOAL | TOTAL CAPACITY (MILLION GAL/YR) |
|---|---|---|---|
| Ethanol From Grains | 50 | 26 | 1,300 |
| Ethanol From Crop Residues | 25 | 2 | 50 |
| Methanol From Municipal Solid Waste | 32 | 2 | 60 |
| Methanol From Wood | 57 | 2 | 110 |
| TOTAL | | 32 | 1,520 |

Source: "National Plan for the Accelerated Commercialization of Solar Energy," MITRE—70W00416, November 1979.

Table 19-9. Industrial Sector Solar Energy Projections.

| | 1985 | | 2000 | |
| TECHNOLOGY | QUADS/ YEAR | NUMBER OF SYSTEMS (THOUSANDS) | QUADS/ YEAR | NUMBER OF SYSTEMS (THOUSANDS) |
|---|---|---|---|---|
| Agricultural and Industrial Process Heat | 0.18 | 7.9 | 3.12 | 53 |
| Wood and Other Biomass | 1.65 | 265.0 | 2.8 | 400 |
| Biomass Fuels | — | — | 0.7–1.0 | 1,300 |
| TOTAL | 1.83 | | 6.62–6.92 | |

Source: "National Plan for the Accelerated Commercialization of Solar Energy." MITRE—79W00416. November 1979.

environmental penalties and potential supply disruption; and nuclear fission is involved in a continuing controversy. Thus, utilities are considering new primary energy sources, including direct solar energy, wind power, ocean thermal energy conversion, biomass for electricity, and biomass sources for methane supply. However, utilities have stringent technology requirements that demand high reliability, reasonable investment and operating costs, environmental feasibility, and good operating flexibility.

Typically, investor-owned utility companies are large corporations with a high level of technological expertise. They are sophisticated buyers of technologies and, together with their suppliers and the Electric Power Research Institute and the Gas Research Institute (their research organizations), they represent tightly integrated market components. They supply buildings, industry, municipalities, and electricity-powered services such as rail transportation.

Except for environmental factors and cost, electricity and gas are largely indifferent to supply technologies, assuming that a reasonable measure of supply security can be ensured. The utilities themselves, however, will closely scrutinize potential performance of any solar option.

End-Use Structure. Electrical utilities presently use several types of primary energy resources to generate electricity: geothermal, fossil, fission, and hydropower resources. Table 19-12 described generating capacity and energy delivered by primary resource type for the United States in 1978. Table 19-13 details the planned capacity additions of electricity generation by utilities during the period 1978–88.

The gas utility industry is made up of diverse production, pipeline transport, storage, and distribution firms, most of which are subject to some form of government regulation. Gas utilities presently serve 25.3 percent (down from 30.2 percent in 1973) of end-use consumption in the United States, primarily in the building and industry sectors. In 1978, building and industry each consumed about 8.4 quads of natural gas; electrical utilities used 3.2 quads of gas to produce about 1.1 quads of electricity.

Equipment Selection Options. Thirty-two percent of the energy consumed in the United States in 1978 was used for the generation of electric power. Coal-fired power plants generated 46 percent of this electricity: 16 percent came from petroleum; 13 percent from natural gas; 13 percent from hydroelectric facilities;

Table 19-10. Constraints to Agricultural and Industrial Process Heat Market Development.

| MAJOR CONSTRAINTS | SPECIFIC CONCERNS |
|---|---|
| Lack of Performance and Cost Data for Technology Selection Need for Reliable Service | • Uncertainty for investors • Uncertainty for regulatory officials • Need for long-term storage or backup power |

Markets and Applications for Solar Technologies 357

## Table 19-11. Constraints to Biomass Market Penetration.

| MAJOR CONSTRAINTS | SPECIFIC CONCERNS |
|---|---|
| Land Availability | • Cost and competition for such uses |
| Supply Uncertainty | • Competing uses for feedstocks |
| Environmental Impacts | • Uncertain long-term impacts for mineral depletion, erosion, and machinery |
| Market Logistics | • Problems in transporting low-density feedstock or unique fuels which may not be compatible with existing fuels |

and the rest from nuclear power. The petroleum consumed up to 1.7 million barrels per day, or the equivalent of 20 percent of oil imports.

Electric utility companies can use several different types of solar technologies for electricity generation, including solar thermal electric power, photovoltaics, hydropower, wind power, wood and other biomass sources, and ocean thermal energy conversion.

Solar thermal electric power systems operate on principles similar to those described previously, but on much larger scales. Central receivers (power towers) with distributed reflecting mirrors to concentrate the sun's energy can also be used. The Barstow Pilot Plant (Solar One) will test this concept on a 10-MWe scale. In addition to applications for new, stand-alone plants, solar thermal power systems can be used to repower conventional steam-fired power plants in which solar components are used to supply steam to the existing turbine-generator, and the conventional fuel system serves as a backup. The best locations for generating electricity with solar thermal collection are believed to be in the Southwest, where sunshine is strongest and cloudy days are infrequent.

A number of studies have indicated that the wind resource in the United States is very large. To harness 5.0 quads of wind energy (fuel displacement) would require the construction of about 130,000 large, 1.5-MWe wind machines (or a large number of smaller machines). The logistical problems of manufacturing and installing this many wind machines are by no means overwhelming. During the last 15 years, United States utilities have installed about 300,000 transmission towers, similar in design and construction to the tower for a wind machine.

Wind systems are expected to enter the utility market, first in a fuel-saver mode, where energy from the wind system is used to replace expensive fossil-fired energy from conventional units. Although the wind resource is subject to short-term fluctuations, the wind follows daily and seasonal patterns, and the annual energy from a wind turbine at a specific site is predictable and unlikely to change significantly from 1 year to the next. Studies of wind resources suggest that the energy out-

## Table 19-12. 1978 National Electric Generation Capacity and Primary Energy Consumption.

| PRIMARY | 100 MW | % OF TOTAL CAPACITY | KWH SUPPLIED × 10⁹ | % OF KWH SUPPLIED |
|---|---|---|---|---|
| Coal | | | 980.2 | 43.7 |
| Natural Gas | 451.2 | 78.6 | 308.8 | 13.8 |
| Oil | | | 362.1 | 16.2 |
| Hydro | 70.6 | 12.3 | 310.9 | 13.8 |
| Nuclear | 52.6 | 9.2 | 280.0 | 12.5 |
| Geothermal, Wood & Wastes | — | — | 3.0 | 0.1 |
| TOTAL | 574.4 | 100.0% | 2245.1 × 10⁹ | 100.0% |

Source: *Electrical World,* March 15, 1979.

## Table 19-13. Additions to Projected United States Electric Generation Capacity.

| SOURCE | PROJECTED 4/1/79 FOR 1979–88 CAPACITY TO BE ADDED, MEGAWATTS | PROJECTED 4/1/78 FOR 1978–87, CAPACITY TO BE ADDED, MEGAWATTS |
|---|---|---|
| Coal | 136,243 | 146,206 |
| Oil | 17,380 | 21,072 |
| Natural Gas | 366 | 502 |
| Hydro | 14,503 | 16,945 |
| Refuse | 192 | 807 |
| Nuclear | 110,156 | 116,177 |
| Geothermal | 1,820 | 1,536 |
| Wind | 55 | 10 |
| TOTAL | 285,363 | 308,017 |

Source: "Additions to Generating Capacity 1979–1988 for the Contiguous United States." (DOE/ERA-0020/1 Rev. 1). October 1979.

put of large wind farms, particularly where components of the farm are geographically separated, will have a smoother output because, when some machines are "down" due to a wind fluctuation, others may be "up." Of course, when storage is available, as with hydro facilities, this may be used to firm up the wind capacity. The Pacific Northwest appears to be a good market for wind because of its hydro facilities and wind resources. Operational data from a three-machine cluster of Mod-2 wind turbines integrated with the Bonneville Power Authority system at Goldendale, Washington, will be available in late 1981.

Photovoltaic systems could be used in very large arrays to form large blocks of capacity similar to solar thermal systems. The opportunity for residential and commercial systems is expected to occur prior to the electric utility markets because the cost target for the former sector is higher due to the impact of non-generation costs on consumer electricity bills.

System control, surface cleaning, and grid integration problems are being addressed.

Hydropower technology is well developed and readily available; the economical development of new or retired dam sites with high-efficiency turbines is a major thrust of DOE's program in the Office of the Assistant Secretary of Resource Applications.

Ocean thermal plants can range in size from 50 kWe to 500 MWe, although utilities are likely to be interested in only the larger sizes. Used offshore or at near-shore sites, the plants will require transmission of power along cables to a shore-based substation. Ocean thermal plants have the capability to operate in a baseload mode, although daily or seasonal water temperature variations can result in capacity fluctuations.

*Biomass.* The potential for wood energy use in utilities is summarized in Table 19-14. If electrical utilities were to use photovoltaics and solar thermal power, their likely end-use

## Table 19-14. Projection of Electricity Utility Wood Utilization.

| USE | QUADS | NUMBER OF UNITS | INCREMENTAL UNITS TO MEET THE GOALS |
|---|---|---|---|
| Electricity (Utilities) | | | |
| Current | Negligible (50 MW) | * | Negligible |
| 1985 Goal | 0.05 (700 MW) | 30 | 25 |
| 2000 Goal | 0.2 (3000 MW) | 110 | 85 |

*Less than five individual plants identifiable.

would be for intermediate- or peakpower fuel displacement. Ocean thermal energy conversion could serve as a baseload power source with a significant capacity-displacement potential.

The vast bulk of utility gas supply comes from natural gas wells spread throughout the country, supplemented by gas imported from Canada and liquefied natural gas from other nations.

Although domestic production, (lower 48 states) has fallen in recent years, the process of oil price decontrol promises to call forth substantial new production. Some of this production is likely to come from biomass-derived sources, such as from anaerobic digestion of biomass. This biological process can, with appropriate filters, produce pipeline-quality methane in a cost-effective manner. In Colorado, for example, the city of Lamar has authorized the construction of a large cattle feedlot biogas system (with funds borrowed from USDA), designed to provide supplemental gaseous fuel for a nearby electric power plant. In another application in Iowa, a biogas system is presently injecting gas into the pipeline system feeding the People's Gas Company of Chicago.

In addition to the utility-scale operations, biomass-derived methane can be utilized at much smaller scales on many farms where animal confinement is practical.

## Potential for Solar Technologies.

In the Solar DPR base case, which assumed a continuation of current policies, about 20 to 50 solar thermal power plants (100-MWe) were projected to operate as intermediate-load plants. They would displace about 0.1 to 0.2 quad of oil or gas normally used for this purpose. In the Domestic Policy Review Maximum Practical Case, 100 such plants would operate, plus some number of variously sized dispersed total energy systems. Together, these systems would displace about 0.4 quad of fuels.

MITRE estimates that solar thermal electric power systems could supply 0.3 to 0.5 quad of electric power in 1990 and 1.7 to 2.0 quads of power in the year 2000—roughly 800,000

barrels per day, equivalent to 10 percent of current imports.

The pre-1985 energy savings provided by utility-scale wind machines will be minimal. Successful tests and development of advanced machines could lead to rapid growth in the 1985–90 period. By the year 2000, utility-scale wind systems may save 1.7 to 2.5 quads of equivalent fossil fuel. For this growth to occur between the years 1983 and 2000, at least fifty 2500-kW machines must be in use in the private sector by 1985. An additional 50 machines could be in use by government power agencies, principally the DOE Power Marketing Agencies and the Bureau of Reclamation.

Based upon President Carter's goal of attaining a 20 percent contribution from solar energy by the year 2000, the ocean thermal energy conversion technology must at that time have sufficient installed capacity to displace about 0.3 quad of oil and gas each year.

## Constraints.

Utilities, as regulated industries, are required to satisfy energy demands with high levels of performance reliability and minimum cost. Under such constraints, utilities are conservative in their investments, demanding proven performance, high reliability, and cost-effectiveness. Overall systems-performance characteristics and ease of integration are also considered.

Constraints to market penetration depend, therefore, on technology performance and perceived economics. If solar technologies have the proven ability to perform cost-effectively, utilities will have the technical skills and investment capability to follow through with significant investments.

## Transportation

## Market Components.

The transportation industry represents the largest single sector of economic activity in the United States, directly and indirectly representing some 10 to 15 percent of the GNP and 22 to 25 percent of employment. Ranging from automobile and tire manufacturing to airplanes to oil companies (and some may wish to include the highway construction and repair business as well), the transportation sector is an extremely large diverse group of firms and government agencies meeting the need for transportation services. The energy demands for the sector include the use of automobiles, buses, light and heavy trucks, trains, water carriers, and airplanes. Gas, water, and coal pipelines may also be considered in the transportation sector.

**End-Use Demand.** Except for fixed-rail and pipeline systems, essentially all transportation systems currently employ petroleum-derived liquid fuels as an energy source. This may be in the form of gasoline, diesel fuel, kerosene, or bunker oils. In 1978, the transportation sector accounted for 9.5 million barrels per day of primary energy demand, representing 24 percent of the 77.1 quads of total consumption in that year. The electrical energy consumed by electric locomotives and cars was less than 0.05 quads.

**Equipment Selection Options.** Solar technologies can produce several fuel forms useful for the transportation sector, primarily alcohol fuels (from biomass) and electricity. Advanced solar technologies will produce hydro-gen, which could be used in transportation. Each of the solar generating options mentioned previously could be considered as contributing to electric-vehicle fuel demands, although the number of electric vehicles on the road in the future may not be large. Ethanol (which, when blended with gasoline at a 1:10 unit ratio, is called *gasohol*) and methanol (derived from biomass) can directly displace the use of liquid fossil fuels, with some trade-offs with respect to energy storage density and convenience. The problems involved in moving from an all-petroleum fuel economy to a mixed-fuel economy must be addressed. Gasohol, however, is seen as a means for immediately reducing the nation's dependence on imported petroleum energy.

**Market Potential.** The DOE Alcohol Fuels Policy Review and other recent studies have highlighted the potential for fuels from biomass. As noted in previous sections, President Carter has called for a program to produce some 1.5 billion gallons of ethanol or methanol per year in 1990.

**Constraints.** The constraints to using solar energy forms in the transportation sector have been explored earlier in this chapter.

# 20. Solar Resources

This chapter describes the nature and magnitude of solar energy resources available to the United States. Besides the use of direct sunlight, other renewable forms of solar energy are discussed, including wind power, ocean energy, and biomass. The description of each energy form offers quantitative estimates of the resources and gives geographic distributions in the United States.

An important point, explained further in the resource sections, is that the various solar resources are distributed in a very uneven geographic fashion. Most areas of the country, deficient in one solar resource such as insolation, have an abundance of other solar resources, such as wind or biomass. This explains, in part, why all solar resources are being pursued concurrently by DOE. The regional variability and overlap required to make solar power feasible on a national basis make the concurrent R&D strategy the only reasonable alternative if solar energy is to be a true national resource.

## SOLAR INSOLATION

The insolation resource is diffuse and variable with respect to both geography and time. The temporal variation occurs seasonally, daily, and during the course of the day as the sun traverses the sky and clouds move overhead. Solar radiation is commonly measured as either direct normal insolation, as the amount of energy striking a surface perpendicular to the sun's rays; or as total horizontal (or global) insolation, which includes both direct and diffuse striking a horizontal surface.

Figures 20-1 and 20-2 show the values for the annual average daily insolation—direct normal and total horizontal—in the form of contour maps for the United States. The units are kilowatt-hours per square meter per day (kWh/m²/d).

Solar radiation in the United States is greatest in the Southwest—southern California, Arizona, New Mexico, and western Texas—where direct normal insolation is 7 kWh/m²/d and global insolation is about 6 kWh/m²/d. The areas of lowest direct normal and global insolation are in New England, the Great Lakes region, and the Puget Sound area, where average daily values are 3 to 4 kWh/M²/d.

The maps show that, in most areas, the values for direct normal radiation are higher than those for total horizontal. This is because the direct component is a high percentage of total solar radiation in these areas, and a surface that points at the sun all day receives more direct radiation than a stationary horizontal surface. In the northeast and northwest sections of the United States, the two insolation values are much closer together. The higher number of cloudy days in these regions results in a larger diffuse radiation component in total horizontal insolation. This makes up for the penalty a horizontal surface pays in receiving direct radiation, as compared to a sun-tracking surface. Especially cloudy areas can be noted along the Gulf Coast and in the Piedmont area, where direct insolation is less than total horizontal.

Putting this information in the context of the solar technologies helps to interpret the numbers on the maps. Solar thermal systems use only direct normal insolation. In the best insolation areas, a 10-MW solar thermal system that is 50 percent efficient would need about 30,000 m² of collector area; a typical trough-collector spacing, for example, would cover about 14 acres. Obviously, the amount of insolation has a significant effect on the economics of solar thermal systems; as a result, the initial domestic markets for high-temperature systems are expected to be predominantly in the Southwest.

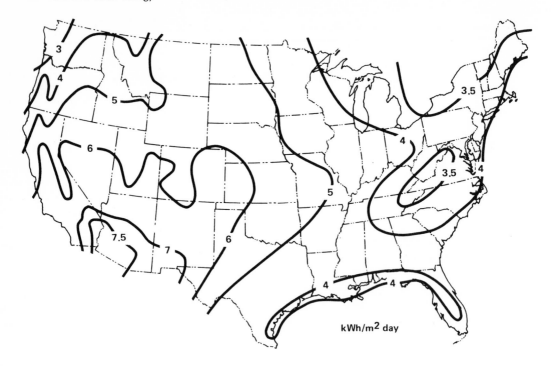

Figure 20-1. Average Annual Solar Radiation: Direct Normal Component.

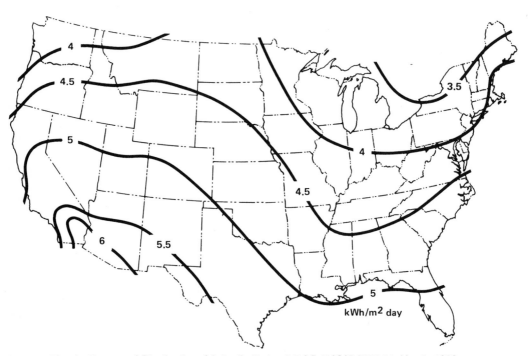

Source:  "On the Nature and Distribution of Solar Radiation," DOE, HCP/T 2552-01, March, 1978.

Figure 20-2. Average Annual Solar Radiation: Total Horizontal Component.

Photovoltaic cells use both the direct and diffuse components of solar insolation. A residential photovoltaic system, with an average output of 2.5 kW (about 10 peak kilowatts) and a 10 percent overall efficiency, would require about 45 m² of photovoltaic cells in the southwestern United States. The array size would not vary directly with insolation, however, because size is a function of several other factors, such as the energy load size and configuration, cost of competing fuel, and utility rate structure.

## BIOMASS RESOURCE POTENTIAL

Potential biomass energy feedstocks include (1) residues from agricultural, livestock, and forestry operations, and (2) woody, agricultural, and aquatic crops grown specifically for conversion to energy.

Agricultural residues include leaves, stalks, roots of vegetables, grains, and other agricultural products. The primary residues are corn stover (cured stalks) and wheat and rice straw, which comprise about 60 percent of the dry tonnage of residues generated each year. Residues are often returned to the soil to supply nutrients and to maintain its structure. Some residues, particularly from corn and sorghum, are used for livestock feed.

Animal manures are a source of biomass with a high moisture content and a high fraction of biodegradable material. Little, if any, pretreatment is required for anaerobic digestion of this feedstock. Primary sources of manure are beef feedlots and dairy, hog, and poultry farms.

Forest residues are defined as logging wastes and mill residues. Logging wastes are generally left in the forest, and residues resulting from mill activities are either burned at the millsite to provide process heat and electricity, converted to chips and wood pulp, manufactured into particle board, or discarded.

Crops such as corn, sugar cane, and sweet sorghum can be grown specifically for energy production. Similarly, silvicultural tree farms could be developed, using advanced cultivation and harvesting techniques to increase yields and reduce costs. Aquatic farms are

also potential sources of energy feedstocks, including kelp and marine and freshwater algae.

Biomass is a highly dispersed and site-specific resource. It is bulky and contains 50 to 90 percent water by weight. Transport over long distances is not economical, and conversion into usable energy must, therefore, take place close to the source of the biomass. However, after its energy density has been increased by conversion to liquid or gaseous fuels and chemicals, transportation over long distances becomes feasible.

The dispersed character of the biomass resource tends to limit its use to particular regions and restricts the scale of economically sensible processing facilities. Further restrictions are highly competitive uses of biomass as food, fiber, and soil amendment; storage requirements; cost of inventory; and spoilage. As a consequence, biomass conversion systems are not likely to exceed input capacities of 1000 to 2000 dry tons per day.

Table 20-1 summarizes the estimated energy potential of biomass feedstocks. The quantities of residues available for energy production depend upon many factors, including alternative uses, prices, collectability, and transportation costs. The development of agricultural or silvicultural energy crops are influenced strongly by competing land uses, water availability, capability of marginal lands to produce energy crops, climatic constraints, production, harvesting, and transportation costs.

## WIND POWER

Unlike solar radiation, wind power can vary widely over a relatively small area. There are many factors that need to be considered in obtaining representative estimates of wind power and its geographic distribution. These factors include atmospheric density variations with elevation and temperature (the air density in Denver is about 15 percent less than at sea level), height of wind measurements above the ground, and local terrain influences. Because wind power is proportional to the cube of the wind speed, small variations in wind speed result in significant differences in wind power. An increase in the average wind speed by a factor of 2 yields roughly 8 times the

## Table 20-1. Estimated Biomass Energy Potential (in Quads).

| TECHNOLOGIES / RESOURCE BASES | AGRICUL-TURAL CROPS | AGRICUL-TURAL RESIDUES | ANIMAL RESIDUES | FOREST CROPS | FOREST RESIDUES | TOTALS |
|---|---|---|---|---|---|---|
| CURRENT TECHNOLOGY AND PRACTICE | | | | | | |
| Total Energy Content | 7.0 | 7.0 | 3.0 | 5.4 | 3.6 | 26.0 |
| Present Usage | 0 | 0.1 | 0.1 | 0 | 1.2 | 1.4 |
| Additional Recoverable | 0 | 1.5 | 0.3 | 0 | 0.8 | 2.6 |
| EXTENSIONS (TO RECOVERY) | | | | | | <2.1 |
| Idle Lands | <1.0 | | | | | |
| Forest Wastes | | | | | <1.0 | |
| Fresh Water Plants | 0.1 | | | | | |
| ADDITIONS (TO BASE) | | | | | | >12.0 |
| Forest Management | | | | 6.0 | | |
| Agricultural | 3.5 | | | | | |
| Silviculture Farms | | | | 3.7 | | |
| Aquaculture Farms | * | | | | | |

Source:   Charles C. Carson and Carolyne Hart "Considerations for Biomass Energy Systems, Draft Report" Sandia Laboratories, Albuquerque, N. Mex. 1979.

*Unknown, but potentially very large

wind power. Therefore, selection of sites for wind data collection and methods for taking the above factors into account can be very important.

The average annual wind power at 50 meters above exposed areas is shown over the contiguous United States in Figure 20-3. Exposed areas refer to locations which are unobstructed to the wind, such as hilltop locations over regions of gently rolling and hilly terrain, and capes and open shoreline sites along coastal regions. The estimates are considered to be lower limits for exposed sites. A few isolated areas may have 50 to 100 percent

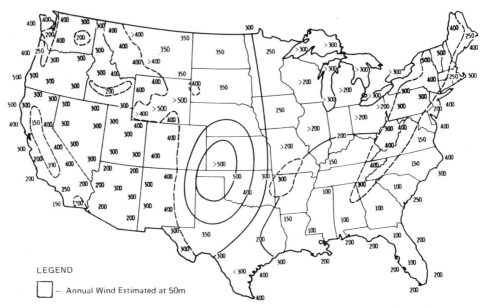

LEGEND

☐ – Annual Wind Estimated at 50m

Source:   Elliot, D.L., "Synthesis of National Wind Energy Analysis", Battelle Pacific Northwest Laboratories, PNL–2220–WIND–5, 1977.

Figure 20-3. Mean annual wind power in the United States.

greater wind power than the map shows. Thus, the estimates must not be construed as representative of all sites within an area. Specifically, the estimates are not representative when topographical features block or channel airflow.

In mountainous regions, the estimates are the lower limits expected for typical, well-exposed sites along mountain summits, ridges, and gaps. Over some isolated mountains and ridges, the wind speeds are enhanced considerably, and the wind power may be a factor of 2 or 3 times greater than the values indicated on the map. The slope and orientation of a ridge to the prevailing winds greatly affect the wind power. Thus, relatively low isolated ridges, gaps, and saddles may have greater wind power than higher mountain summits. All mountainous regions, with the exception of some areas in the Southwest, are estimated to have considerable wind power potential (at least 300 W/m²). However, even the Southwest may have some exposed sites where the winds are enhanced with high wind-power potential.

In nonmountainous regions, the mean annual wind power is high (400 W/m²) over the central and southern Great Plains; offshore and exposed coastal sites in the Northeast and Northwest; and in parts of Wyoming, Montana, and the south Texas coast. High wind power also exists along the exposed coastal and offshore areas of Alaska, with the greatest amounts appearing along the Aleutian Peninsula and Islands. Moderately high wind power (300 to 400 W/m²) can be expected over the northern Great Plains and exposed sites along the Great Lakes and in Hawaii.

Seasonal patterns of wind power are similar to the annual patterns. The seasons of maximum wind power are given in Figure 20-4. Over the eastern one-third of the nation, maximum wind power occurs during the winter and early spring. A spring maximum occurs over the Great Plains, the north-central states, the Texas coast, most nonmountainous areas in the West (e.g., the basins and broad valleys), and offshore areas of central and southern California. Winter maxima occur over all the mountainous regions, except for some areas in the lower Southwest where the spring wind power is almost as large.

Source:   Elliot, D.L., "Synthesis of National Wind Energy Analysis,"
          Battelle Pacific Northwest Laboratories, PNL—2220—WIND—5, 1977

Figure 20-4. Seasons of maximum wind power in the United States.

## OCEAN ENERGY

The available solar energy in the ocean is in the form of both dynamic (waves and currents) and static (thermal and salinity gradient) systems that are continuously renewed by direct and indirect action of the sun. An enormous amount of solar energy is supplied daily to the open ocean, and technology is now being developed to convert a small portion of this energy for use in generating electricity or in energy-intensive processes. Obviously, there are preferred locations where these resources are most abundant and can be most easily and economically tapped. As these locations are exploited and as prices of competing fuels rise, less favorable parts of the resource will be tapped. Any ocean energy resource must be defined in terms of people's ability to use it; thus, the geographic boundaries of the known resources reflect only a preliminary assessment of the resource base.

## Ocean Thermal Resource

The global ocean thermal resource can be mapped, as shown in Figures 20-5 and 20-6, in terms of annual average temperature differences between the surface waters and the water at a depth of 1000 meters (3300 feet). The most valuable ocean thermal resources should probably offer average temperature differences of 20° C (36° F) or greater. The contours of greatest interest in the figures, therefore, are those for 20°, 21°, 22°, 23°, and the nominal 24° C contour that corresponds to temperature differences of 24° C or larger. A considerable region of the globe thus appears advantageous for ocean thermal exploitation, including many locations accessible to land via submarine cable and extensive ocean areas, where ocean thermal plant-ships could produce energy-intensive products at sea. This region happens to coincide geographically with the locations of numerous developing nations, and, if commercially feasible, they could provide a substantial addition to world energy resources. For applications where mooring of the plant is required, there are practical upper limits of about 2000 meters

for tolerable mooring depths. In such cases sites of interest would be limited to regions where ocean depths range from about 1000 to 2000 meters.

Although the ocean thermal resource at a given location is quite stable from day to day, there is a seasonal variation of the resource. The amplitude of variation increases with distance of departure north and south of the equator. There are substantial ocean thermal resources in the Gulf of Mexico available to the continental United States via submarine electrical cable. Also, islands such as Puerto Rico, the Virgin Islands, Guam, and Hawaii have excellent ocean thermal resources very close to their shores.

The seasonal variation of the ocean thermal resource, and, therefore, of the output power from OTEC plants, would tend to match the seasonally varying electrical load in the southern United States. The combination of appropriate mixes of seasonally varying ocean thermal power with fixed baseload power (such as from coal and nuclear power plants) would match the seasonal load variation well.

## Current Velocity

While the oceans are in constant motion, the mean surface current speed seldom exceeds 1 knot, with the overall average probably close to half that amount. The only major oceanic region where the surface current exceeds 2 knots is in the Florida current, where the gulf stream turns northward and runs between the Bahama Islands and the coast of Florida. The annual maximum steady flow occurs during the early summer months (3.8 knots), with a minimum in the late fall (2.6 knots). However, the high-speed core of this surface current is relatively narrow and, if one integrates the velocity over a depth and distance commensurate with multiple turbine generators, these mean speeds drop to 2.7 and 2.2 knots, respectively. Significant temporal and spatial variations occur in regions not highly bounded by the land masses. Tidal oscillations may generate strong currents in shallow coastal areas. However, these are usually not steady currents, but cycle through maximum and minimum speeds, making the economics of

## ΔT(°C) BETWEEN SURFACE AND 1000 METER DEPTH

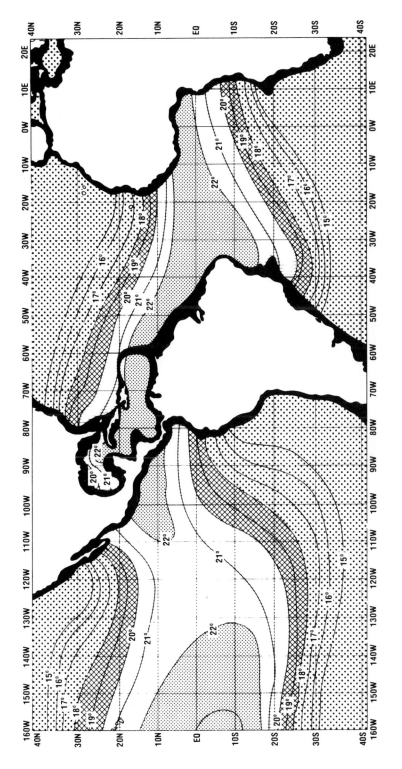

31 AUGUST 1978

Average of Monthly △T's Less Than 18°C
Average of Monthly △T's More Than 18°C, Less Than 20°C
Average of Monthly △T's More Than 20°C, Less Than 22°C
Average of Monthly △T's More Than 22°C, Less Than 24°C
Average of Monthly △T's Greater Than 24°C
Water Depth Less Than 1000 Meters
Source: Ocean Systems Branch, Department of Energy

Figure 20-5. Global ocean temperature change from 160° west to 25° east latitude.

367

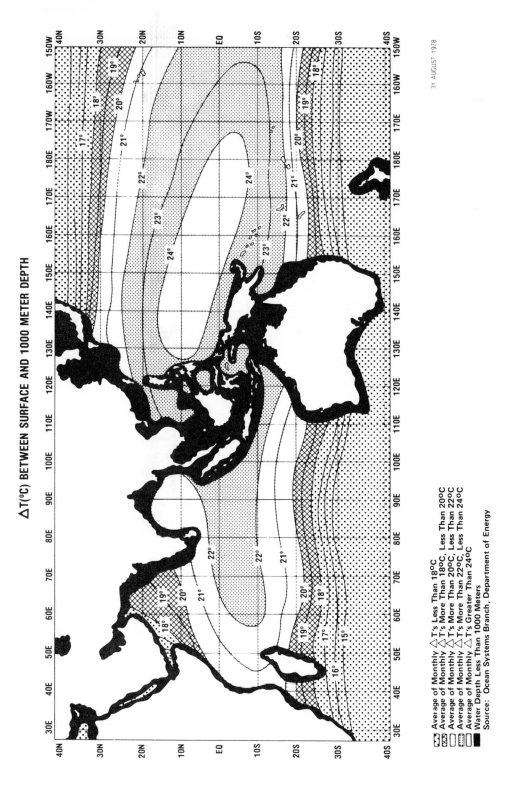

ΔT(°C) BETWEEN SURFACE AND 1000 METER DEPTH

Average of Monthly △T's Less Than 18°C
Average of Monthly △T's More Than 18°C, Less Than 20°C
Average of Monthly △T's More Than 20°C, Less Than 22°C
Average of Monthly △T's More Than 22°C, Less Than 24°C
Average of Monthly △T's Greater Than 24°C
Water Depth Less Than 1000 Meters
Source:  Ocean Systems Branch, Department of Energy

31 AUGUST 1978

Figure 20-6.  Global ocean temperature change from 30° east to 150° west latitude.

using them less attractive than for using steady currents. the potential of this ocean resource is not thought to be large.

## Ocean Wave

Ocean wave energy is distributed in nearly direct proportion to ocean wind velocity. There are some differences in distribution resulting from the fact that waves will propagate long distances without significant decay. Consequently, wind direction plays an important role in determining wave direction; in the middle latitudes, where significant storms occur, the prevailing westerly winds drive waves toward the eastern part of the oceans. This results in areas such as the Pacific Northwest coast of the United States and the northwest coast of Great Britain and Europe having the greatest wave energy resource. In the tropical regions where easterly trade winds dominate, significant waves are generated; however, their magnitude is not as great as in the northern latitudes.

## Salinity Gradients

In theory, power from salinity gradients can be extracted in several ways, using the energy potential that exists across a selective membrane between two solutions of different salinity. One method, osmosis, is physical and makes direct use of the flow that occurs across the membrane that permits water but not sodium, chloride, or other ions to pass through. Two water masses of two different salinities are required. Possible sites originally considered for this concept are the mouths of rivers discharging their waters into the ocean, lakes located in warm regions where continuous evaporation produces high salinity, salt domes, or solar ponds. Energy conversion is achieved by allowing the difference in hydrostatic pressure to build (because of the membrane). This difference in pressure is used to drive a hydroelectric turbine.

Because of the osmotic pressure difference, a 240-meter waterfall theoretically exists at the mouth of every river and stream in the world. Few dams are this high. At present, river water irreversibly mixes with ocean water with no social gain. However, if half of the flow of the Columbia River could be converted into electricity at only 30 percent efficiency, 2300 MW would be produced. Where the Jordan River empties into the Dead Sea, the energy density is even more spectacular. The nearly saturated brines of the Dead Sea have an osmotic pressure of about 500 atmospheres, corresponding to a dam more than 5000 meters high. The mouths of the Mississippi, Amazon, and Congo Rivers and the Great Salt Lake are other examples of locations with high potential for salinity gradient energy conversion.

Salinity gradient technology is at a relative early stage of development. The magnitude of the resource available to the United States, the feasibility of the technology, the environmental consequences, and the economics as a whole remain to be defined.

# 21. Solar Energy Policy and Related Legislation

The United States is at a pivotal point in its energy/economic history. In the past, economic growth has been fueled by abundant supplies of inexpensive energy. Economic planners could assume an uninterrupted supply of energy from sources as varied as oil, natural gas, coal, and nuclear power. However, recent events have drastically altered the energy situation in the United States.

The rapid escalation of prices by the OPEC nations has caused significant economic problems for the United States, and there is little

Table 21-1. Status of Solar-Related Legislation as of December 17, 1979.

| SUBJECT | HOUSE | SENATE | STATUS |
|---|---|---|---|
| BUDGET | | | |
| DOE Civilian Programs Authorization. FY 80 and 81 (H.R. 3000) (S. 688) | Passed 10/24/79 | | H.R. 3000 awaiting conference S. 688 awaiting action on Senate floor |
| DOE Appropriations for FY 80 (H.R. 4388) | Passed 6/18/79 | Passed 7/18/79 | Enacted 9/25/79 P.L. 96-69 |
| Supplemental Appropriations FY 79 for National Alcohol Fuels Commission (H.R. 4289). | Passed 6/6/79 | Passed 6/26/79 | Enacted 7/25/79 P.L. 96-38 |
| USDA Appropriations — $500 million for loan guarantees for the production of alcohol fuels (H.R. 4387) | Passed 6/19/79 | Passed 7/19/79 | Enacted 11/9/79 P.L. 96-108 |
| SBA appropriations for energy loans including solar (H.R. 4392) | Passed 7/12/79 | Passed 7/19/79 | Enacted 9/24/79 P.L. 96-68 |
| Department of Interior Appropriations — $1.5 billion for synfuels including alcohol fuels. $1 billion for solar commercialization (H.R. 4930). | Passed 7/30/79 | Passed 10/10/79 | Enacted 11/27/79 P.L. 96-126 |
| INCENTIVES | | | |
| Solar Energy Development Bank making long-term low-interest loans. Authorization of $450 million through FY 82 for such loans (H.R. 605) (S.932) | | S. 932 Passed 11/16/79 | S. 932 in Conference H.R. 605 Reported from House Committee (H. Rpt. 96-625) |
| SBA FY 80 appropriations: energy loan fund (H.R. 4392) | Passed 7/12/79 | Passed 7/19/79 | Enacted 9/24/79, P.L. 96-68 |
| Solar Tax Credits of 50% for residences and 30% for business investments (H.R. 3919) | Passed 6/28/79 | Passed 12/17/79 | Reported from House Ways & Means Committee 6/22/79 (H. Rpt. 96-304) Reported from Senate Finance Committee 11/1/79 (S. Rpt. 96-394) Awaiting conference |
| RESEARCH AND DEVELOPMENT | | | |
| SPS Research, development, and evaluation program (H.R. 2335) | Passed 11/16/79 | | Reported from House Committee (H. Rpt. 96-151) |
| Wind Energy RD&D Program (H.R. 5892)(S. 932) | Passed 12/4/79 H.R.5892 | Passed 11/8/79 S 932 | Conference Underway Reported from Senate Committee 12/14/79 (S Rpt. 96-501) |
| DEMONSTRATIONS | | | |
| Military construction authorizations: solar installations in military facilities (S. 1319) | Passed 10/24/79 | Passed 7/30/79 | Enacted 11/26/79 (P.L. 96-125) |
| DOD construction appropriations: solar installations (H.R. 4391) | Passed 6/18/79 | Passed 11/13/79 | Enacted 11/30/79 (P.L. 96-130) |

Table 21-2. Status of Gasohol-Related Legislation.

| SUBJECT | HOUSE | SENATE | STATUS |
|---|---|---|---|
| GASOHOL | | | |
| Guaranteed loans for fuel alcohol-producing plants (H.R. 3905) | | | Reported from House Committee (H. Rpt. 96-515, Pt. 1) |
| Program for federal purchase of synthetic fuels (H.R. 3930) (S. 932) | Passed H.R. 3930 6/26/79 | Passed S. 932 11/16/79 | S. 932 now in conference H. Rpt. 96-165 S. Rpt. 96-394 |
| USDA guaranteed loans for alternative fuels pilot plants (H.R. 3580) | | | Reported from House Committee 6/8/79 (H. Rpt. 96-259) |
| International Sugar Stabilization Act: Incentives for production of energy products from sugar crops (H.R. 2172) | | | Reported from House Agriculture Committee (H. Rpt. 96-125, Pt. 1) Reported from House, Ways and Means Committee (Pt. 11) |
| Acceleration of the development and production of biomass (S. 932) | | Passed 11/16/79 | In Conference |
| Post Office directed to use gasohol to maximum extent in their vehicles (H.R. 79) | Passed 9/7/79 | | Reported from House Committee (H. Rpt. 96-   ) |
| EDA directed to give grants for gasohol plants (H.R. 2063) (S. 914 passed in lieu) | Passed 11/14/79 | Passed 8/1/79 | H.R. 2063 reported from House Committee (H. Rpt. 96-180) |
| | | | S.914 reported from Senate Committee (S. Rpt. 96-270) |
| DOD directed to by and use gasohol to maximum extent in its vehicles (H.R. 4040) (S. 428 passed in lieu) | Passed 9/14/79 | Passed 5/31/79 | Enacted 11/9/79 P.L. 96-107 |
| Funding and deadline extended for National Alcohol Fuels Commission (H.R. 4249) | Passed 6/30/79 | Passed 10/24/79 | Enacted 11/9/79 P.L. 96-106 |
| $500 million authorized for pilot energy projects particularly alcohol fuels (S. 892) | | Passed 6/14/79 | Reported from Senate Committee (S. Rpt. 96-188) |

likelihood of future price relief. The problem is heightened by increasing hostility toward the United States in Iran and other Middle East nations, which has cast uncertainty on the availability of fuel from these nations, even at their escalated prices. In short, the cost and availability of once-inexpensive and plentiful energy have become major obstacles to the orderly growth of the United States economy. As a result, it has become increasingly apparent that the United States must look to alternative sources to satisfy its energy needs.

## ENERGY POLICY

Attention has been devoted in recent years to the development of an energy policy that will reduce current dependence on imported oil and ensure enough supply to support continued economic growth. An important first step in the development of a national energy strategy was the consolidation of the federal energy-related activities under a single structure, the Department of Energy (DOE).

The legislation that created DOE requires the Administration to submit national energy plans to Congress every two years. These plans outline national energy policies and strategies for resolving near- and long-term energy problems facing the country.

## National Energy Act

President Carter submitted his first National Energy Plan to Congress on April 20, 1977, and followed it with a proposal for a National Energy Act nine days later. The purposes of the National Energy Act are to reduce the growth in energy demand; reduce the petroleum imports; extend supplies of natural gas; encourage the use of coal and renewable energy resources; and promote energy conservation. The National Energy Act's goals for 1985 include a reduction in energy demand growth to less than 2 percent per year, improved insulation of 90 percent of American homes and all new buildings, and installation of solar systems in more than 2.5 million homes.

## EXISTING LEGISLATION

Although solar energy bills were introduced in Congress as early as 1951, no legislation was enacted until the Ninety-Third Congress (1973–74). Two important laws that were enacted in 1974 are the Solar Heating and Cooling Demonstration Act (Public Law 93-409) and the Solar Energy RD&D Act (Public Law 93-473). The first provides for commercial demonstration of solar heating and cooling; the second one establishes an Office of Solar Energy Research within the federal government.

Congress passed the National Energy Act in late 1978, in five acts: the Public Utility Regulatory Policies Act of 1978, the Energy Tax Act of 1978, the National Energy Conservation Policy Act, the Powerplant and Industrial Fuel Use Act of 1978, and the Natural Gas Policy Act of 1978. There is no single, unified portion of the energy package dealing with solar energy. Instead, references are scattered through the legislation. The most important are found in the National Energy Tax Act and the National Energy Conservation Policy Act.

# BIBLIOGRAPHY

## Energy—General

Clark, Wilson. *Energy for Survival: The Alternative to Extinction*. New York: Doubleday, 1975.

*Energy in America's Future*. Baltimore, Maryland: Johns Hopkins University Press, 1979.

*Energy: The Next Twenty Years*. Cambridge, Massachusetts: Ballinger Publishing Co., 1979.

Ford Foundation. *A Time to Choose: America's Energy Future*. Energy Policy Project of the Ford Foundation. Cambridge, Massachusetts: Ballinger Publishing Co., 1974.

Hammond, Allen L., Metz, William D., and Maugh II, Thomas H. *Energy and the Future*. Washington, D.C.: American Association for the Advancement of Science, 1973.

Kahn, Herman, et. al. *The Next Two Hundred Years*. New York: William Morrow & Company, 1976.

Maddox, John. *Beyond the Energy Crisis: A Global Perspective*. New York: McGraw-Hill, 1975.

National Research Council. *Energy in Transition 1985–2010*. Final Report of the Committee on Nuclear and Alternative Energy Systems. San Francisco: W. H. Freeman & Company, 1979.

Science and Public Policy Program, University of Oklahoma, Norman, Oklahoma. *Energy Alternatives, A Comparative Analysis*. Washington: Council on Environmental Quality, 1975.

Thirring, Hans. *Energy for Man: From Windmills to Nuclear Power*. 2d. ed. New York: Harper & Row, 1979.

Workshop on Alternative Energy Strategies. *Energy: Global Prospects 1895–2000*. New York: McGraw-Hill, 1977.

## Energy Conservation

Clagg, Peter. *New Low-Cost Sources of Energy for the Home*. Charlotte, Vermont: Garden Way, 1975.

Clark, Wilson. *Energy for Survival: The Alternative to Extinction*. New York: Doubleday, 1975.

Dubin, Fred S. Energy for architects. *Architecture Plus 1* (6): 38–49, 74–75 (July 1973).

Dubin, Fred S., and Long, Chalmers, G., Jr. *Energy Conservation Standards*. New York: McGraw-Hill, 1978.

Fond, K. W., et al. (eds.). *Efficient Use of Energy*. New York: American Institute of Physics, 1975.

Griffin, Charles William. *Energy Conservation in Buildings: Techniques for Economical Design*. Washington: Construction Specifications Institute, 1974.

Hirst, Eric. *Energy Consumption for Transportation in the U.S., (ORNL-NSF-EP-15)*. Oak Ridge, Tennessee, Oak Ridge National Laboratory.

Hittman Associates. Residential Energy Consumption.

*Single Family Housing: Final Report*. Washington: U.S. Department of Housing and Urban Development, 1973.

Large, David B., ed. *Hidden Waste*. Washington: Conservation Foundation, 1973.

Large, David B., ed. 1973. *Minimum Energy Dwelling*. Washington: Energy Research and Development Administration Office of Conservation Division of Buildings and Community Systems.

Smith, Thomas W. *Household Energy Game*. Madison, Wisconsin: University of Wisconsin, Johns Hopkins Marine Studies Center, 1974.

Stobaugh, Roger, and Yergin, Daniel, eds. *Energy Future: The Report of the Harvard Business School Energy Project*. New York: Random House, 1979.

## Solar Energy

Adelson, E. H. *Solar Air Conditioning and Refrigeration*. Isotech Research Labs., 1975.

American Institute of Aeronautics and Astronautics. *Solar Energy for Earth*. New York: American Institute of Aeronautics and Astronautics, 1975.

Anderson, Bruce. *Solar Energy: Fundamentals in Building Design*. New York: McGraw-Hill, 1977.

Beckman, William, et al. *Solar Heating Design: By the F-Chart Method*. New York: John Wiley & Sons, 1977.

Brinkworth, Brian Joseph. *Solar Energy for Man*. New York: John Wiley & Sons, 1973.

Chalmers, Bruce. The Photovoltaic Generation of Electricity. *Scientific American 235(4):* 34–44 (October 1976).

Clark, Wilson. *Energy for Survival: The Alternative to Extinction*. New York: Doubleday, 1975.

Corliss, William R. *Direct Conversion of Energy*. Oak Ridge, Tennessee: U.S. Atomic Energy Commission, Office of Information Services, 1964.

Daniels, Farrington. *Direct Use of the Sun's Energy*. New York: Ballantine Books, 1974.

Davis, Albert J., and Schubert, Robert P. *Alternative Natural Energy Sources in Building Design*. Blacksburg, Virginia: Passive Energy Systems, 1974.

Duffie, John A., and Beckman, William A. *Solar Energy Thermal Processes*. New York: John Wiley & Sons, 1974.

Duffie, John A., and Beckman, William A. "Solar Heating and Cooling." *Science 191:* 143–149 (January 16, 1976).

Federal Energy Administration. *Buying Solar*. Washington: U.S. Government Printing Office, June 1976.

Gay, Larry. *The Complete Book of Heating with Wood*. Charlotte, Vermont: Garden Way, 1974.

Glaser, Peter E. *Beyond nuclear power: the large-scale*

*use of solar energy*. New York Academy of Sciences. *Transactions*, ser. 2, 31 (No. 8): 951–967 (December 1969).

Grey, J. Solar heating and cooling. *Astronautics and Aeronautics* 33–37 (November 1975).

Halacy, D. S. *Earth, Water, Wind & Sun*. New York: Harper & Row, 1977.

Hoke, John. *Solar Energy*. New York: Frankin Watts, 1978.

Keyes, John. *Harvesting the Sun to Heat Your House*. New York: Morgan & Morgan, 1975.

Kreith, Frank, and Kreider, Jan F. *Principles of Solar Engineering*. Washington: Hemisphere Publishing Corporation, 1978.

Lucas, Ted. *How to Build a Solar Heater*. New York: Crown, 1980.

Mazria, Edward. *The Passive Solar Energy Book*. Emmaus, Pennsylvania: Rodale Press, 1979.

Meinel, Aden B., and Meinel, Marjorie P. Physics looks at solar energy. *Physics Today* 44–50 (February 1972).

Michels, Timothy I. *Solar Energy Utilization*. New York: Van Nostrand Reinhold, 1979.

Rau, Hans, *Solar Energy*. New York: Macmillan, 1964.

Russell, Charles R. Solar energy. In *Elements of Energy Conversion*. New York: Pergamon Press, 1967.

Skurka, Norma, and Naar, John. *Living with Natural Energy: Design for a Limited Planet*. New York: Ballantine Books, 1976.

Solar Energy Institute of America. 1979. *Sun Language*. Washington: Solar Energy Institute of America.

Stobaugh, Roger, and Yergin, Daniel, eds. *Energy Future: The Report of the Harvard Business School Energy Project*. New York: Random House, 1979.

Stoner, Carol, ed. *Producing Your Own Power: How to Make Nature's Energy Sources Work for You*. New York: Random House, 1975.

Williams, James Richard. *Solar Energy: Technology and Applications*. Ann Arbor, Michigan: Ann Arbor Science Publishers, 1977.

Wilson, David A., and Rankins, William H. *Practical Sun Power*. Lorien House, 1974.

Yanda, Bill, and Fisher, Rick. *The Food and Heat Producing Solar Greenhouse*. Sante Fe, New Mexico: John Muir Publications, 1980.

## Wind Energy

Blackwell, B. F., and Feltz, L. V. *Wind Energy–A Revitalized Pursuit* (SAND-75-0166). Livermore, California: Sandia Laboratories, March 1975.

Carter, Joe. Wind power for the people. *Environment Action Bulletin 6* (12); 4–5 (June 14, 1975).

Clark, Wilson. "Energy from the Winds." In *Energy for Survival: The Alternative to Extinction*. New York: Doubleday, 1975.

Eldridge, Frank R. *Wind Machines*. New York: Van Nostrand Reinhold, 1980.

Golding, Edward W. *The Generation of Electricity by Wind Power*. London: E. & F. N. Spon. 1976.

Hackleman, Michael A. *Wind and Windspinners: A Nuts*

*and Bolts Approach to Wind-Electric Systems*. Sangus, California: Peace Press, 1975.

Hamilton, R. *Can We Harness the Wind? National Geographic* 148: 812–829 (December 1975).

Hunt, V. Daniel. *Wind Power: A Handbook on Wind Energy Conversion Systems*. New York: Van Nostrand Reinhold, 1981.

Inglis, David R. Wind power now. *Bulletin of the Atomic Scientists* 31(8): 20–26 (October 1975).

Johnson, C. C., et al. *Wind power development and applications. Power Engineering* 78(10): 50–53 (October 1974).

*Mother Earth News Staff. Handbook of Homemade Power*. New York: Bantam Books, 1974.

Putnam, Palmer Cosslett. *Power from the Wind*. New York: Van Nostrand Reinhold, 1974.

Putnam, Palmer Cosslett. *Energy in the Future*. New York: Van Nostrand Reinhold, 1953.

Reynolds, John. *Windmills & Watermills*. New York: Praeger, 1970.

Simmons, Daniel M. *Wind Power*. Park Ridge, New Jersey: Noyes Data Corp., 1975.

Sorensen, Bent. Energy and resources. *Science* 189(4199): 255–260 (July 25, 1975).

Steadman, Philip. *Energy, Environment and Building*. London: Cambridge University Press, 1975.

Stokhuyzen, Frederick. *The Dutch Windmill*. New York: Universe Books, 1963.

Stoner, Carol H. *Producing Your Own Power: How to Make Nature's Energy Source Work for You*. Emmaus, Pennsylvania: Rodale Press, 1974.

Torrey, Volta. *Wind Catchers: American Windmills of Yesterday and Tomorrow*. Brattleboro, Vermont: Stephen Greene Press, 1976.

Wolff, Alfred R. *The Windmill as a Prime Mover*. New York: John Wiley & Sons, 1885.

## Water Energy

Clegg, Peter. *New Low-Cost Sources of Energy for the Home, with Complete Illustrated Catalog*. Charlotte, Vermont: Garden Way, 1975.

Creager, William Pitcher, Justin, Joel D., and Hinds, Julian. *Engineering for Dams*. New York: J. Wiley & Sons, 1945.

Davey, Norman. *Studies in Tidal Power*. London: Constable & Co., Ltd., 1923.

Handbook of Homemade Power. *Mother Earth News Staff*. New York: Bantam Books, 1974.

Macmillian, Donald Henry, *Tides*. New York: American Elsevier, 1966.

Morton, M. Granger (ed.). Ocean Thermal Energy Conversion. *Energy and Man: Technical and Social Aspects of Energy*. New York: IEEE Press, 1975.

Paton, Thomas Angus Lyall, and Brown, J. Guthrie. *Power from Water*. London: L. Hill, 1961.

Rash, Don E., et al. *Energy Under the Ocean: A Technology Assessment*. Norman, Oklahoma: University of Oklahoma Press, 1973.

Reynolds, John. *Windmills and Watermills*. London: H. Evelyn, 1970.

Ross, David. *Energy from the Waves*. Oxford, England: Pergamon Press, 1979.

*Tidal Power*. Proceedings of the International Conference on the Utilization of Tidal Power, Nova Scotia Technical College, 1970. Gray, T. K., and Gashus, O. K. (eds.) New York: Plenum Press, 1972.

Wilson, Paul N. *Water Turbines*. Palo Alto City, California: Pendragon House, 1974.

## Organic (Bioconversion) Fuels

Anderson, Russell E. *Biological Paths to Self-Reliance: A Guide to Biological Solar Energy Conversion*. New York: Van Nostrand Reinhold, 1979.

Bureau of Mines Circular No. 8549. *Energy Potential from Organic Wastes: A Review of the Quantities and Sources*. Washington: U.S. Bureau of Mines, 1972.

Center for Metropolitan Studies. *Capturing the Sun Through Bioconversion*. Proceedings of a Conference, March 10–12, 1976, Washington, D.C. Washington: Center for Metropolitan Studies, 1976.

Clark, Wilson. "Solar Bioconversion." In *Energy for Survival: The Alternative to Extinction*. New York: Doubleday, 1975.

Golueke, Clarence G., and McGauhey, P.H. "Waste Materials." *In Annual Review of Energy*, Jack M. Hollander (ed.). Palo Alto, California: Annual Reviews, Inc.

Hunt, V. Daniel. *The Gasohol Handbook*. New York: Industrial Press, 1981.

Jackson, Frederick R. *Energy from Solid Waste*. Park Ridge, New Jersey: Noyes Data Corp., 1974.

Lowe, Robert A. *Energy Recovery from Waste; Solid Waste as Supplementary Fuel in Power Plant Boilers*. Washington: U.S. Environmental Protection Agency, 1973.

National Research Council, Canada. Research Plans and Publications Section. *Wood and Charcoal as Fuel for Vehicles*. 3rd ed. Ottawa, 1944.

University of Oklahoma. *Energy Alternatives: A Comparative Analysis*. The Science and Public Policy Program. Organic Farms. Washington: U.S. Government Printing Office, May 1975.

# Index

A *Guide to Commercial Scale Ethanol Production and Financing*, 208
absorption refrigeration cycle, 71
Academy of Contemporary Problems, 55
acid rain, 7
activated carbon for sludge digestion, 66
Active Solar Heating and Cooling Program, 193, 197, 204, 275-283, 336, 338
  budget, 277-278
  impacts, 307
  programs, 277
  strategy, 275-277
  structure, 277-283
  systems development, 317, 319
active solar heating systems, 272-283
advanced collector technologies, 227
advanced feedstock production, 214-215
advanced materials/cells research (solar), 224
Advanced Mechanical Technology, Inc., 75
advanced open-cycle OTEC concepts, 263
Advanced Research and Development (AR&D) Subprogram, 223-226
advanced solar energy research, 336, 339
Aero Vironment Corp., 264
Aerospace Research Corporation, 208, 220
Africa, 330
agricultural and industrial process heat (AIPH), 193, 197, 203-204, 264-272, 354
  activities, 269
  budget, 268, 270
  constraints, 40, 356
  impacts, 306-307
  management, 267
  market analysis, 270-271
  milestones, 270
  strategy, 266-267
  structure, 267-268
  systems development, 317
agricultural residues, 363
agriculture, 115-121
AGRIMOD, 118
air conditioners, 9, 71
air-fuel ratio control project, 101
air quality, 302, 308
Ai Research Manufacturing Co., 99
Alaska, 4
alcohol fuels, 19-20, 60, 147
alcohol-gasoline blends, 140
alcohol production systems, 209, 315
alternative farm power systems, 117
Alternative Fuels Utilization Program (AFUP), 140, 146-147
alternative materials, 103

alternative power for handtools project, 115
aluminum, 106-108
  industry, 106
  reduction of, 107
aluminum-silicon alloy, 107
Amana Refrigeration, Inc., 74
Amazon River, 369
American Papago Indian reservation, 217
American Society for Testing Materials, 95
American Waterway, 144
Ammonia, 257
ammonia manufacturing plants, 117
amorphous silicon, 224, 297
AMTROL, Inc., 75
anaerobic digestion, 59, 64, 208, 212, 314, 359, 363
ANFLOW, 61, 65, 85
animal manures, 363
Annual Cycle Energy System (ACES), 76, 83-84
Annual Survey of Manufacturers, 118
apartments, conservation in, 51
appliances, efficiency of, 33, 71
Appropriate Energy Technology Small Grants Program (AETSGP), 177, 185-186
aquatic biomass production, 296, 363
aquatic weeds, 215
Argonne National Laboratory (ANL), 36, 80, 303
Arthur D. Little, 74, 103
ASHRAE Standard, 45, 75-90
Asia, 217
"Assimilate Partitioning in Green Plants," 300
Association of American Railroads, 144
Atlantic City, New Jersey, 67
automobile dealers, conservation in, 51
automobile service and repair shops, conservation in, 51
automobiles and use of gasoline, 9, 19-20
Automotive Propulsion Research and Development Act of 1978 (Title III, P.L. 95-238), 139, 143
Automotive Technology RD&D program, 139, 143
  programs and milestones, 144-147

B&K LTD, 129
Bahama islands, 366
bakeries, conservation in, 51
balance of payments, 137
balance of system (BOS), 218, 225
balance of trade, 13
barriers, legal and regulatory, 13
Basic Energy Sciences (BES), 299-300
Bayer process, 107
beet-sugar processing, 119
bibliography, 373-375
biochemical conversion, 212-214
biocides, 306

bioconversion, bibliography, 375
bioconversion processes, 59, 66
biofouling, 260
biofuel, 214
biogas, 359
biological sciences, 300
biomass converter, 211
biomass energy systems, 193, 197, 203, 204-217,
    295-296, 336-337
  budget, 206-207, 211, 314
  impacts, 302-303
  international programs, 331
  milestones, 211
  organization, 206-207
  program strategy, 205-206
  program thrust, 295-296
  technology development, 314-315
biomass feedstocks, 296
biomass industries, 354
biomass-to-ethanol conversion, 209
biomass transport, 363
biophotolysers, 215, 296
black chrome, 298
black cobalt selective absorbers, 298
black liquor, 113
blast gasifier, 106
boilers, 101
  coal-fired, 191
  oil-fired, 101
boomtown phenomenon, 8, 46
bottoming cycles, 103, 154
Brayton heat pump program, 100
Brayton/Rankine heat pump, 71, 85
Brazil, 331
Brookhaven National Laboratory (BNL), 36, 304
budget, energy conservation, 25
building construction energy use in, 44
Building Energy Performance Standards (BEPS), 42, 84
building envelopes, 44
buildings, energy use projection in, 82
Buildings and Community Systems Program, 25, 33-85
  milestones, 42
buildings sector, 17-18, 82
  conservation programs, 17-18
  energy consumption, 33, 348
  solar contribution, 18, 351

cadmium, 225
cadmium sulfide, 224, 297
California, University of, 296
capital stock, 28
  industrial, 86
  turnover of, 9-10
carpools, 161, 167
Carter, President Jimmy, vii, 6, 13, 52, 79, 157, 175,
    196, 200, 206, 264, 326, 330, 354-355, 359, 371
cell efficiency, 297
cells, 315
  cadmium sulfide photovoltaic, 315
  photovoltaic, 315
  silicon, 315

cellular glass, 298
cellulose, 66, 213
cellulosic feedstocks, 213
cement industry, 108
  high-temperature processes, 109
Central Integrated Systems, 61
central receivers, 247-248, 252, 265
chemical industry, 109
Chemical Production from Waste CO program, 97
chemical science, 300
circleline lamp, 74
Clark University, 85
Clarksburg, West Virginia, 54
Claude cycle, 262
Clean Air Act, 216
Clean Water Act, 61, 67
clothes dryers, energy efficiency in, 71
coal, 7
  cofiring, 215
  production of, 6
coefficient of performance (COP), 75
cogeneration, 21, 101-102, 213, 253
  demonstration program, 104
cogeneration district heat, 61, 62
coil coating, 124
cold corrugation program, 113
collector overheating, 306
collectors, 227, 247
Columbia River, 369
combined solar/heat-pump systems, 274
combustion, 66
  efficiency of, 29-30
commercial buildings, energy use in, 28
commercial printers, conservation in, 51
commercialization, 92, 97
Community Services Administration (CSA), 162
Community Systems Program, 46-85
comprehensive community energy management program
    (CCEMP), 68
computer-controlled papermill project, 113
concrete-block curing, 115
Congo River, 369
conservation energy, vii, 5, 6, 8, 27-32, 67, 83
  barriers to, 11, 27
  bibliography, 373
  economic considerations, 28
  fuel oil, 74
  government role, 12-16, 29-32
  investments, 28
  potential, 27-29
  strategy, 31-32
conservation and solar energy, program summary, 17-24
conservation technology deployment and monitoring,
    121-127
Consolidated Edison, 44
consumer education, 41
consumer products, 68-81
consumption, energy, 3, 27-30
contamination, 306
  chemical, 306
  microbial, 306

Conversion of Plastics to Fuel Oil program, 97
cooling, 118
copper, 108
Corporate Average Fuel Economy (CAFE), 142
corrosion, 260
cost-definition R&D, 201
cost-reduction R&D, 201
Cothane Process for Producing Methane, 96-97
crop-drying activity, 116-117, 269
crystalline silicon technology, 297
current velocity, 366, 369

dairies, conservation in, 51, 117-118
Darrieus vertical axis, 235
Davis, California, 32
daylighting/sunlighting program, 45
Dead Sea, 369
decentralization, 10
Definition Planning activity, 259-260
demolition of cars, 150
Denmark, 293, 332
Department of Commerce (DOC), 32, 59, 163, 179,
    329
Department of Energy (DOE), 4, 13, 16, 33
Department of Energy Act of 1978—Civilian Applica-
    tions (P.L. 95-238), 39
Department of Energy Reorganization Act (P.L. 95-91),
    38, 42, 43, 143
Department of Health and Human Services, (HHS), 17
Department of Health, Education and Welfare (HEW),
    162, 179
Department of Housing and Urban Development
    (HUD), 17, 32, 38, 40, 43, 53, 163, 179
Department of Labor (DOL), 179
Department of Transportation (DOT), 20, 32, 58
dessicant heating and cooling systems, 275, 278-279,
    281
direct combustion systems, 207-208, 215-216
direct heating, 284
Direct Reduction of Aluminum Program, 134
direct solar space- and water-heating systems, 274,
    277-278, 280, 281
dishwashers, energy efficiency in, 71
Distributed Integrated Systems, 61-62
Distributed Receiver subprogram, 251, 253-254
district heating and cooling, 48, 64
DOE regional offices, 180-183
Domestic Policy Review (DPR), vii, 10, 196, 235, 264,
    283-284, 326
driver awareness, 152
Driver Energy Conservation Awareness Training Pro-
    gram (DECAT), 153
drying, 118
Dutchess County, New York, 65
dynamometer testing, 145

ecological effects, 305
Economic Development Administration (EDA), 55, 57
economic growth, 3
educational television, 330
efficiency, energy, 27-28, 35, 41

Electric and Hybrid Vehicle RD&D Act of 1976 (P.L.
    94-113), 140, 144
Electric and Hybrid Vehicles (EHV) Program, 143-144
    milestones, 147-151
    RD&D, 140-141
electric car, 20
electric generation capacity, 358-359
electricity, 19, 82, 140-141
    generation of, 247
electricity utility wood utilization, 358
electrochemical cell efficiency, 225
electrofluid dynamic driven generators, 297
electromagnetic interference, 304-305
Emergency Building Temperature Restrictions (EBTR),
    78-79
emergency conservation, 11
Emergency Energy Conservation Act (EECA) (P.L.
    96-102), 159, 163
emissions, 139, 303
end-use sectors, energy strategy for, 27-28
energy, 19-28
    benefits of, 14
    cost of, 3, 14, 27, 29
    curtailment of, 27
    demand for, 6, 9, 17, 19, 27-28, 33, 35
    renewable, 13, 33
    sources of, 196
    supply of, 6-7
    use of farms, 115-116
energy and petroleum, domestic consumption of, 137
energy audits, 167
energy availability, 121
energy cascading, 67
energy consumption, 105-106
    in 1985, 114
    of major industries, 105
Energy Conversion Technology (ECT), Subprogram,
    184-185
energy-efficiency, 8, 115, 126-127, 138-139
Energy Conservation and Production Act of 1976 (Title
    III, P.L. 94-385), 38, 43, 121, 144, 159, 160, 164
Energy Conservation Polymers program, 97
Energy Extension Service (EES), 51, 159, 163, 164
    goals/objectives, 161-162
    milestones, 168
Energy Future, 8
Energy Impact Assistance, 26, 189
Energy Impact Scoreboard System (EISS), 125
Energy Information Campaign, 26, 190
energy-integrated farm systems program, 116
Energy Management Partnership Act (EMPA), 26, 163,
    173-174
energy management programs, 55
Energy Partnership for American Cities Program, 51
Energy Policy and Conservation Act (EPCA), (P.L.
    94-163), 39, 42, 83, 88, 93, 94, 121, 126-127,
    144, 159, 160, 164
energy problem, 3-11
Energy-Related Inventors Program (ERIP), 175-178,
    183-184
Energy Reorganization Act of 1974 (P.L. 93-438), 38, 143

energy savings, 88, 131, 171-172
  goals for, 34-35, 81-84, 132-133
Energy Tax Act of 1978 (P.L. 95-618), 94, 122, 372
Energy Utilization Systems (EUS), 75
Environmental Assessments (EA), 302-303
Environmental Development Plan (EDP), 134-135, 157
environmental hazards, 10, 27
Environmental Impact Statement (EIS), 301-302
Environmental Protection Agency (EPA), 20, 58, 59,
    108, 124, 134, 143, 151, 304
Environmental Readiness Document (ERD), 301-302
enzymatic hydrolysis, 59, 65, 213
equipment, energy-efficient, 14
ethanol, 208, 213, 355
eutectic salts, 305
evacuated tube collectors, 267
Executive Order 12003, 39, 78, 83
Experimental Process System Development Unit, 227

Farmers Home Administration (FmHA), 67
federal buildings plan, 77
Federal Clean Air Act, 67
Federal Energy Administration Act of 1974 (P.L.
    93-275), 143
Federal Energy Management Plan (FEMP), 39, 78, 80
Federal Highway Administration (FHWA), 163
Federal Nonnuclear Energy Research and Development
    Act of 1974 (P.L. 94-577), 38, 94, 143
Federal Photovoltaic Utilization Program (FPUP), 221,
    233, 322
Federal Railroad Administration, 144
Federal Register, 121
federal sector, buildings in, 33
fertilizer, 117
financial assistance, 19
flat-plate collectors, 227
Florida, 366
  citrus industry, 119
florists, conservation in, 51
fluidized-bed boilers, 100-101
fluidized-bed principle, 100
food-processing industry, 118
food sterilization, 118-119
foreign oil, U.S. dependence on, 33
forest residues, 363
fossil-fuel supplies of, 27
fossil fuels, 30
Fourdrinier machine, 114
France, 293, 329
free enterprise, 14
free market, 12
freeze crystallization-acid project, 110
freezers, energy efficiency in, 71, 74-75
freight transportation, 154
fuel cell, 66
fuel "extenders," 140
Fuel from Farms: A Guide to Small-Scale Ethanol
    Production, 208
fuel shortages, 54
fuel switching, 20
furnaces, efficiency of, 30
furniture manufacturing, conservation in, 51

gallium arsenide, 224, 297
gas, 82
Gas Mileage Guide, 144, 151
Gas Research Institute, 67, 75
gas turbine, 139, 145, 157
gasohol, 20, 206, 208
  legislation for, 371
gasoline, curtailment of, 11
gasoline-like synthetic fuels, 138
gasoline service stations, conservation in, 51
GASPAK, 119
Gedser windmill, 332
General Electric, 75
general retailing, conservation in, 51
General Services Administration (GSA), 43
Georgia, 254
Georgia Institute of Technology, 125
geosciences, 300
Gill, Dr. Gurmukh, ix
Global Marine Development, Inc., 262
goals, energy program, 13-16, 87
Great Britain, 293
Great Salt Lake, 369
Greece, 293
greenhouse effect, 7
gross national product (GNP), 3, 4, 8, 28, 175, 359
Guam, 366
Gulf of Mexico, 260, 366

Haber-Bosch process, 117
Hall cells, 108
Hall-Heoult process, 107
Hawaii, 366
headbox-papermaking project, 113
heat engine program, 138, 139-140
  programs and milestones, 144-146
heat-exchanger concepts, 263
heat loss, 76
heat pumps, 30, 63, 70, 75-76, 84, 85, 100
  gas-fired, 75-76
  solar assisted, 280-282
heating and cooling, energy consumption in, 69
heating, home, 9
heliostats, 248, 333
hemispherical bowls, 247, 253
highlift devices, 297
highway system, transportation energy consumption of,
    137
home heating, energy efficiency in, 71
home oil, 30
hospitals, conservation energy in, 34
hot-water heaters, energy use in, 69
hydrocarbon-bearing plants, 296
hydrocarbon oils, 305
hydrocarbons, 140
hyperfiltration-textiles project, 112

Illinois, University of, 64
Import Reduction Program, 354-355
incentives
  economic, 313
  financial, 15, 17, 21, 31, 129, 313

tax, 14
indirect heating, 284
Industrial and Powerplant Fuel Use Act, 20
industrial boilers, 9, 28
industrial cogeneration optimization program (ICOP), 103
Industrial Energy Conservation Program, 25, 87-136
    accomplishments, 131-132
    budget history, 127
    environmental impacts, 133-135
    federal role, 89
    issues, 135-136
    policies, 129
    program-effectiveness, 130
    program impacts, 128-133
industrial energy consumption, 86
industrial process heat, 338
industrial processes, 30, 86
    activities, 105
    efficiency of, 104-121
industrial programs, 86-136
    activities, 94-104
    legislative framework, 94
Industrial Reporting Program, 126
industrial sector, 267-268, 352-359
    end-use structure, 353-354
    energy consumption in, 18-19, 353
    equipment, 354
    market components, 353-354
    policy, 31, 86, 267-268
    R&D expenditures, 90
Industrial Sector Technology Use Model (ISTUM), 122, 130
industrial wood energy, goals of, 355
information, public, 15, 17
information systems, 339-340
initial systems experiments (photovoltaic), 230-231
insulation, 84
Integrated Community Energy Systems (ICES), 63-64
    coal-using, 61, 63
    grid-connected, 61, 64, 85
    heat-pump, 48, 61, 64
inter-governmental coordination, 167
International Energy Agency (IEA), 67, 252, 254, 283, 328, 332
International Photovoltaic Plan, 329
International Solar Applications Program, 322-323, 325
    budget, 331
    goals, 327-330
    activities, 331-333
investment cost, 28-29
investments, conservation, 12, 84, 88
Iran, 5
Iraq, 5
irrigation projects, 116
isolated heating, 284
Israel, 283, 293, 331
issues, solar, 313
Italy, 283, 329, 332

Japan, 329
Jordan River, 369

kelp, 210
Kraft pulping, 113

La Que Center for Corrosion Technology, 261
land-based aquatic biomass production, 214
land use, 67, 302-303, 310
landfills, 59
Latin America, 217
laundries and dry cleaners, conservation in, 51
Lawrence Berkeley Laboratory, 36, 296, 306
life-cycle cost, 28, 81
lighting, energy consumption in, 69
lighting efficiency standards, 161, 167
lignocellulosic material, 215
line-focusing concentrators, 265
liquefaction, 212
liquefied petroleum gas, 264
liquid desiccant systems, 280-281
liquid metal, 253
liquid propane gas, 116
liquid sodium, 305
load factors, 141
Lockheed Corp., 260
loom modification research, 112
Louisiana, 88
low-Btu fuel, 64
low-Btu gasification system, 207-208, 216
"low cost/no cost" approach, 72
Low-Cost Solar Array (LSA) Project, 225
Low Energy Cement project, 95-96
low-temperature and end-product processes, 10

Madras rotor, 297
mandatory fuel switching, 129
manufacturing, energy service demand for, 353
Maritime Administration, 144
market analysis, 232, 310-313
market development (solar), 323-325
    activities, 324-325
    budget, 324
    goals, 323-324
Maryland, 76
Massachusetts Institute of Technology (MIT), 107, 215
materials science, 300
meatpacking industry, 119-120
mechanical dewatering program, 113
Mercer County, North Dakota, 68
methane, 59, 65, 359
methanol, 355
Mexico, 283, 293, 329
Michigan, University of, 304-305
Minimum Energy Dwelling (MED), 44, 83
Minnesota, University of, 85
Mississippi River, 369
mist and foam, 299
MITRE, 312
MOD-OA WECS, 236, 244, 246
MOD-1 WECS, 244
MOD-2 WECS, 244, 320
modeling, solar economic, 312
molten salt, 253
mooring depths, 366

Mor-Flo Industries, Inc, 75
Motor Vehicle Information Cost Savings Act (P.L. 92-213), 144
Mt. Laguna, 221
multijunction concentrator cells, 225
Multisector Programs, 175-188
  accomplishments, 188
  funding, 186-187
  impacts, 187-188
  objectives, 175-177
  strategy, 177-183
municipal wastes, 58

NASA Jet Propulsion Laboratory, 142
NASA Lewis Research Center, 142, 220, 304
National Bureau of Standards, 43, 75, 121, 163
National Center for Appropriate Technology, 179
National Coil Coaters Association (NCCA), 124
National Energy Act (NEA), 39, 71, 83, 92, 121, 371, 372
National Energy Conservation Policy Act (NECPA) (P.L. 95-619), 43, 85, 94, 121, 159, 166, 202, 233, 304, 372
National Energy Extension Service Act (NEESA), (P.L. 95-39), 159
National Energy Plan (NEP), 14, 33, 34, 82
National Energy Tax Act, 372
National Environmental Policy Act (NEPA), 173
National Park Service, 179
National Science Foundation (NSF), 179
National Solar Heating and Cooling Information Center, 324
National Solar Message, 196
national solar strategy, 198-200
natural gas, 106, 117, 119, 124
Natural Gas Policy Act of 1978, 372
New Car Fuel Economy Program, 151
New Concepts Testing and Evaluation Program, 151
Newark, New Jersey, 45
nitrogen-based carburization project, 106
Nonhighway Research and Development (NHR&D) project, 153-155
Norris Cotton Federal Building (Manchester, New Hampshire), 43
North Atlantic Treaty Organization (NATO), 179
nuclear science, 300

Oak Ridge National Laboratory (ORNL), 36, 74, 82, 303
ocean currents, 256, 263-264, 299
ocean depths, 366
ocean power systems, 263
ocean systems, 21, 193, 197, 203, 256-264, 299, 338
  budget, 259
  impacts, 306
  international activities, 333
  management, 259
  strategy, 257
  structure, 258-259
  technology, 316-317
ocean temperature, 367-368

ocean thermal and total energy, bibliography, 374-375
ocean thermal electric conversion (OTEC), 256, 257-264, 299, 316-317
  engineering test and evaluation, 261-262
  technology, 260-261
  Utility Users Council, 260
ocean thermal plant-ships, 366
ocean thermal resource, 366
ocean wave energy, 369
ocean wind velocity, 369
Office of Buildings and Community Systems (DOE), 17-18
Office of Conservation and Renewable Energy, 13, 16
Office of Inventions and Small Scale Technology (OISST), 175-188
Office of Solar Applications, funding, 205
oil, 5-6
  imported, 5, 6, 27, 33
  residential use of, 82
  U.S. supply of, 5
oil embargo, 175
oil recovery, unconventional, 6
oil shale, 8
on-farm systems, 208-209
organic bottoming cycle, 145
Organization of Petroleum Exporting Countries (OPEC), 3, 370-371
OSMOSTS, 369
Others Local Government Buildings Grant Program, 159
  funding, 173
  goals/objectives, 160

petroleum, 137-138
  consumption of, 106, 137, 138
  savings of, 151
  U.S. dependence of foreign sources, vii, 19, 137
paper-drying, 113
parabolic dishes, 247-248, 253, 255
parabolic troughs, 247-248, 251, 253
passenger transportation, 154
passive and hybrid solar heating and cooling, 193, 197, 204, 283-293
  activities, 289-293
  budget, 289
  impacts, 307-308
  strategy, 286-289
  structure, 289
  systems development, 318-320
passive cooling, 285
passive solar, 17
peak watt (WP), 217
Persian Gulf, 5
petroleum-refining industry, 111
Pfeffer, Dr. John, 64
photochemistry, 297
photophysics, 297
photovoltaic cells, 363
photovoltaic concentrator, 297
photovoltaic energy systems, 193, 197, 203, 217-234, 297, 336-337
  budget, 223

impacts, 303-304
international activities, 331-332
program strategy, 219-221
program structure, 221-223
Photovoltaic Material/Device Evaluation Laboratory, 225
photovoltaic modules, 217
photovoltaic plants, 21
photovoltaic principle, 218
photovoltaic technology, 18
Pittsburgh, University of, 125
Planning Analysis and Social Science, 339
polycrystalline silicon, 224, 297
"Polymer Materials Basic Research Needs for Energy
    Applications," 300
Pompano Beach, Florida, 60, 64
Portland, Oregon, 32
Power, Dr. J. Michael, ix
Powerplant and Industrial Fuel Use Act, 216, 372
prices, energy, 12, 28
    decontrol and deregulation, 15
Procedyne, Inc., 107
Production Systems subprogram, 209-210
Project Review Board, 133
public information, 34
Public Utilities Regulatory and Policy Act of 1978, 235
Puerto Rico, 366
pulp and paper industry, 112-113
PUROX, 65
pyrolysis, 65, 66

R&D, vii-viii, 15-16, 17
    constraints on, 13
    ethanol, 24
    private sector, 91
Rankine cycle, 144, 254
Rankine cycle-steam recompressor heat pump project,
    100
Rankine/Rankine system, 71
RD&D, 19
    cost-shared, 91
real-estate training programs, 73
recuperators, 98-99
refrigerators, energy efficiency in, 71, 74-75
refuse conversion to methane (REFCOM), 64
Refuse-Derived Fuel (RDF) Project, 96
regional solar energy centers, 340-344
    funding, 343
    industrial process heat, 342
    passive solar, 342
    small wind energy conversion systems, 343
    solar water heating, 342
    state and local programs, 341
    wood, 342-343
regulations, energy, 15, 31, 35
regulatory issues, 74
residential/commercial buildings, energy use in, 33-34,
    284-286
residential/commercial sector energy policy of, 31
Residential Conservation Service (RCS), 43, 77
residential wood combustion, goals of, 351
Resource Conservation and Recovery Act, 304

retrofit, 34, 70, 74
    powerplant, 63
return on investment (ROI), 133
right-turn-on-red, 161, 167
Rocky Flats Test Center, 315-316
roof-felt application project, 113-114
roof pond, 291
rural development, 67
Rural Electrification Administration, 234

Safety Analysis Report (SAR), 301-302
sailwing-type systems, 297
salinity gradients, 256, 260, 263, 299, 306, 369
salt domes, 369
Sandia Laboratories, 251, 254
sanitization, 118
Saudi Arabia, 283, 328
schools, conservation energy in, 34
Schools and Hospitals Grant Program, 159
    funding, 173
    goals/objectives, 160
    milestones, 168
Schuchuli, Arizona, 331
Seacoast Test Facility, 261
Seattle, Washington, 32
seawater, 257
seaweed, 214
sectors, most energy-intensive, 126
semiconductor material, 218
semiconductors, 300
sewage sludge, 64
shell-tube heat exchanger, 260
silicon, 218
silver/glass mirror system, 298
silviculture techniques, 209-210
Small and Disadvantaged Business, 344-345
small business, 49
Small Business Administration (SBA), 51, 179
Small Business Energy Cost Reduction Program, 50
Small-Scale Appropriate Energy Technology Grants Pro-
    gram, 175
    federal strategy, 177
Smith-Putnam wind turbine, 234
sodium hydroxide, 305
soil erosion, 302
solar cells, 218-224
Solar Cities Project, 290
solar domestic hot water
solar-driven absorption heating and cooling systems,
    274, 278, 280-281
solar-driven Rankine heating and cooling system, 274,
    278, 280-281
solar energy, 193-372
    bibliography, 373-374
    legislation for, 370-372
    national goal, 196-198
    state and local programs, 21
Solar Energy Information Data Bank (SEIDB), 339-340,
    345
solar energy program, vii, 5, 193-375
    barriers to, 12

solar energy program (*continued*)
  budget, 193-195
  Government role in, 12-16, 200-201
  organization of, 201-202
  technologies, 203-293
Solar Energy RD&D Act of 1974 (P.L. 93-473), 340
Solar Energy Research Institute (SERI), 208, 209, 252, 297, 323, 330, 336-340
solar enhanced oil recovery, 265
Solar Federal Buildings Program, 322
Solar Industrial Process Heat Program, 268, 269, 354-355
Solar Photovoltaic Energy Research, Development, and Demonstration Act of 1978 (P.L. 95-590), 233, 372
solar ponds, 369
Solar Program, 294-333
  implementation, 334-345
  international, 233, 243
  participants in, 334-336
solar radiation, 305, 348, 361, 363
solar research, high-risk, 224-225
solar resources, 347, 361-369
  biomass, 363
  insolation, 361, 363
  ocean energy, 366-369
  wind power, 363-365
solar technologies, 346-360
  buildings, 346-360
  constraints, 346
  equipment, 346, 348-350
  markets, 346-360
  potential form 354-355
solar technology
  contributions of, 197
solar thermal electric plants, 21, 359
solar thermal energy systems, 18-19, 193, 197, 203, 247-256, 298-299, 337
  budget, 250-251
  impacts, 305-306
  international activities, 332-333
  strategy, 249-250
  structure, 250-251
  technology, 254-256, 316
solar thermal systems, 361
solid desiccant systems, 280
solid waste disposal, 304
Southern California Gas Co., 44
space conditioning, 70
space heating and cooling, 79, 84, 289-290
  demonstration program, 321-322
space-heating systems, 276-277
Spain, 253, 283, 293, 328, 329, 332
Springfield, Massachusetts, 85
Standard Industry Code (SIC), 126
standards, energy, 15-16, 17, 31, 34, 71
state and local governments, role of, 32
State and Local Programs, 25-26, 159-174
  federal role, 162
  federal strategy, 162-164
  funding, 170

  management responsibility, 170
State Department, 329
State Energy Conservation Program (SECP), 159, 164
  goals/objectives, 160-161
  milestones, 166-168
Steam gun, 66
steamlift cycle, 262
sterilization process, 117-118
Stirling-cycle heat pump, 99
Stirling engine, 139, 144-145, 157, 255
Stirling/Rankine system, 71, 85
storm windows, 84
Strategic Petroleum Reserve (SPR) Program, 11, 14
submarine cable, 366
sugar-processing industry, 119
Supplemental State Energy Conservation Program (SSECP), 163
surface waters, 366
Switzerland, 293
synthetic fuels, 7-8, 20, 140, 147, 355
Synthetic Fuels Corporation, 6
Systems Development (Solar), 317-320
  activities, 318-320
  budget, 318
Systems Engineering and Standards subprogram (solar), 227-229

"targets of opportunity", 29
tax credits, 87
taxes, 87, 129
technologies, energy, 29
Technology Base (Solar Program), 294-300
  budget, 295
  expenditures, 296
  goals, 295
technology, clearinghouse, 52
Technology Development (Solar) 313-317
  activities, 314-317
  budget, 314
  goals, 313-314
television sets, energy efficiency in, 71
Tennessee, University of, 125
Tennessee Valley Authority (TVA), 43
Tennessee Valley Authority National Fertilizer Development Center (TVA/NFDC), 117
Tests and Applications subprogram (solar) 229-231
Texas, 254, 274
textile industry, 111-112
thermal efficiency standards, 161, 167
thermal storage, 44
thermochemical conversion, 211-212
thermostats, 79, 84
thin-film polycrystalline solution, 224
"Thin-Film Problems and Research in Energy Systems", 300
thin glass, 298
tidal oscillations, 366
titanium, 259
topping cycles, 103
tornado wind system, 297
total energy (hybrid) systems, 227-228, 315

toxic gases, 303
Toxic Substances Control Act, 304
training and education, energy programs in, 41
*Transportation Energy Conservation Data Book,* 155
transportation fuels, 137
transportation networks, 7-8
Transportation Program, 25, 137-158
   benefit/cost analysis, 156-157
   federal role in, 141-142
   funding, 155
   impacts, 155-158
   issues, 157
   legislative framework, 143-144
   major program thrusts, 143-155
transportation sector, 359-360
   constraints, 360
   end-use demand, 360
   energy policy, 31
   energy use in, 19-20, 137
   equipment, 360
   market potential, 360
Transportation Systems Utilization (TSU) Program, 141, 144
   milestones, 151-153
TRW, 103
turbocompound diesels, 145

United Nations, 324
urban transportation, 67
urban waste-processing systems, 66
U.S. Agency for International Development (DOE), 329
U.S. Coast Guard, 144
USDA Forest Service, 210, 216
U.S. Department of Agriculture (USDA) 32, 40, 151, 208, 210, 238, 267, 269, 315, 359
U.S. Postal Service, 153
U.S.S.R., 329
utilities sector, 355-359
   conservation in, 20-21
   end-use structure, 356
   equipment, 356
   market components, 355
   potential for solar technologies, 359
   solar energy and, 21
utility impacts, 310

vanpools, 161, 167
vapor-liquid entrainment, 263
vehicle performance, 141-142
Vehicle Systems Program, 139-140
vent dampers, 72
ventilation, 44
Vermont, 234
Virgin Islands, 366
vortex extraction devices, 297

waste, elimination of, 27
waste-derived fuels, 65
waste energy recovery, 95-102
   recovery, 65, 98, 102
   reduction activities, 95
waste heat, 29, 30, 98
Waste Lube Oil Recovery, 96
Waste Tire Conversion project, 96
wastes, urban, 34
water absorption, 112
water heaters, 75
   energy efficiency in, 71
water pumping, 116, 330
water quality, 302
water use, 302-303
wave energy, 256, 257, 263, 299
Weatherization Assistance Program (WAP), 159, 164
   funding, 172
   goals/objectives, 162, 172
   milestones, 170
   regulations, 169
wholesale distributing, conservation in, 51
wind energy bibliography, 374
wind energy conversion systems, 193, 197, 203, 234-246, 297-298, 337
   activities, 240
   budget, 239, 241
   impacts, 304-305
   international activities, 332
   objectives, 240
   research and analysis, 239-240
   strategy, 236-239
   structure, 239
   technology, 242-244, 315-316
wind power, 363
   buildings use of, 349
wind resources, 21
wind speed, 363
wind systems, 234-246
   horizontal axis, 235
   innovative, 240
   intermediate-scale, 246
   large-scale, 235, 238, 243, 246
   small-scale, 235, 243, 245-246
   vertical axis, 235
window analysis, 45
winds, 236, 241
wood-burning plants, 21
wood commercialization, 215-217
wood energy, 354
wood power, buildings use of, 349
Wood Residue Industrial program, 98
Wood Hole Oceanographic Institute, 214, 296
worker safety, 305
World Bank, 328